POWDER, PEOPLE *and* PLACE

Badger Ordnance Works and the Sauk Prairie

by Michael J. Goc

New Past Press, Inc. Friendship, WI 53934

POWDER, PEOPLE *and* PLACE

Badger Ordnance Works and the Sauk Prairie

by Michael J. Goc

Publishing Services
New Past Press, Inc., Friendship, WI 53934
Publishing Assistant, Indexing
Carol Huber Podoll

Page Design, Copy Editing
Diane Egbert, Sandhill Studio, Green Lake, WI 54941

Cover Design
Jay Jocham, Big Flats/Hancock, WI

Cover Photos
Prairie, Paula Bunch; plant aerial, Badger Army Ammunition Plant

ISBN 0-938627-56-2

Library of Congress Cataloging-in-Publication Data

Goc, Michael J.
Powder, people and place : Badger Ordnance Works and the Sauk Prairie / by Michael J. Goc.
p. cm.
Includes bibliographical references and index.
ISBN 0-938627-56-2 (paperback)
1. Badger Ordnance Works (Wisc.)–History. 2. Sauk County (Wis.)–Environmental conditions. I. Title.

UP524.W6 G63 2002
623.4'55'09–dc21

Printed in Canada

Dedication

~ To the Powdermakers ~

Table of Contents

I Prairie

The Millennial Prairie 15
Kindled by Human Hands 23
"Brilliant Chains of Liquid Fire" 30
The Great Prairie Engine 33

II Settlement

Prairie Breakers, Rail Makers 39
York Staters, Germans and Swiss 46
"Every Neighbor Was My Co-worker" 53
Modern Times 59

III Powder Plant

The Fixer 73
Rearming for War 77
Not in Massachusetts 80
Patriotic, All of Them 93
Condemned 98
Only the Dead were Undisturbed 104
The Table of Explosive Distances 112
A Construction Chronology 122
Badger M-2 132
Powdermakers — Polly and Pete 136
Getting to Work 147
Inside the Gate 153
Making Smokeless 160
Making Rocket 168
Boom, Bridge, Bypass and Bar 173
No Room in the Chicken Coop 186
Photos 195

Badger Village 243
The Condemnation Settlement 253
Decontamination by Destruction 262
What to do with Badger Ordnance? 271
"A Little Shangri-La" 274
The Company Town 282
Korea vs. World War II 286
Reactivation And Operation 288
New Powder, New Plant 298
On Hold Between Wars 302
At War Again 308
Working for the Boys at War 318
From Badger Village to Bluffview Courts 324
Focusing on the Obvious 330
Target BAAP 334
A New Powder Plant 343

IV Legacy

The Evil and the Good 353

Notes 362
Appendices
    Family Names/Properties Purchased for BAAP 373
    Periods of Operation of Badger's Production Areas 377
    Production 378
    Commanding Officers 379
    Resident Managers 380
    Casualties 381
Bibliography 384
Index 393
About the Author 404

Acknowledgements

The Badger History Group was conceived in the spring of 1998, only a few breaths after the United States Army announced that it was decommissioning the Badger Army Ammunition Plant. The opportunity was too rare, the necessity too great, for local people aware of the historical significance of the Sauk Prairie and the powder plant to let it slip away. Here was history that was too important to be left to chance. As with any volunteer group, the membership of the BHG has changed over the years, with Michael Mossman, Lisa Hartmann, Carol Sorg, Dave Fordham, Wayne Hatz, Lola Huber, Verlyn Mueller, Orie Eilertson, Peter Shrake, Tammy Kochaver, Bob France, Michael Bunch, Paula Bunch, Phil Hasheider, Jill Hunger, Al Gallagher, Donald Kindschi, Curt Meine, Keir Sterling and Tom Every sharing the work of research, preservation and education to a greater or lesser degree.

This book is only one of the many projects they have undertaken on their own and in cooperation with other local and state organizations and with the plant itself: organization of a reunion of plant employees; creation of an information-rich website; production of a nationally distributed video history of the plant; co-sponsorship of a comprehensive lecture series on the Prairie and the plant; the erection of a memorial to those who gave up their land and homes for Badger; exhibits of historical and contemporary photos; production of a video history of Badger Village/Bluffview; fund-raising for all of the above and more. BHG founder Michael Mossman served on the Sauk County BAAP Re-Use Committee and other members have provided input to "the Re-Use" and other official committees. The group has also conducted research, written and reviewed preservation proposals and consulted with local, state and federal agencies seeking historical information on the plant. The long-term goal of the group is to create a museum/learning center which will capture and preserve the story of the powder plant, the farm community that preceded it and the native human and non-human communities that preceded them. For their participation in this book project the author owes them a warm expression of gratitude.

Thanks is also owed to the parent organizations of the Badger History Group, the Sauk Prairie Historical Society and Sauk County Historical Society and to many individuals; first to Erhart Mueller, in memoriam; to those who shared their memories in interviews with the author and with members of the Badger History Group; to Wayne Hatz, for his careful reading and critique of the manuscript; Paula Bunch, for generously sharing her photographs; Jack Holzhueter, State Historical Society of Wisconsin emeritus, for suggesting that the link between Leo T. Crowley and Badger Ordnance be investigated; Dave Hendrickson, Wisconsin Department of Natural Resources, for reading the opening chapter and amending its excesses; Carol Sorg, the one-woman research department, who answered every call to search out, look up and track down. It is to Carol that we owe the completion of the "chicken coop," Marachowsky and land acquisition stories, to list just a few. Carol also tracked records of casualties at Badger, discovering the name and final resting place of Herman Beach, the young Jamaican powdermaker who died in a cold place many miles away from his tropical home. A special word of gratitude is also reserved for Dave Fordham, Commanding Officer's Representative at the plant, and to his predecessors who preserved the historical records without which this story could not be told. Thanks also to his staff on the "Army side" of the administration building, and the contractor's personnel on the "Olin side." Like Dave, they were unfailingly helpful and hospitable to a curious stranger visiting during an already tense time made worse by heightened concerns for security.

Despite the best efforts of this worthy fellowship, errors may appear in the text. If so, they are my own.

Michael J. Goc, February, 2002

Funding

Funding for this book was provided by a grant from the revolving fund for publications of the Sauk County Historical Society, by a grant from the Sauk County Committee on the Arts, Humanities and Historic Preservation, by an anonymous donor and by Carol Sorg in memory of Dayl Sorg, United States Marine Corps, 1943-1946.

Introduction

Every place has its history, but some have more history than others. The Sauk Prairie is one. In terms geologic and geographic, agricultural and industrial, social and political, the 14,000 acres identified as prairie in the southeast corner of Sauk County, Wisconsin, bear an extraordinary historical lode.

Had the United States Army not built the Badger Ordnance Works, the Sauk Prairie would stand out for its natural history. The terminal moraine of the Wisconsin glacier traverses the ancient grassland on its way from the Wisconsin river to the Baraboo Hills. In the 12,000 years since the last glacier receded, the Prairie evolved into a transition zone where the flatlands of a grass-covered outwash plain meld into the million-year-old quartzite of the wooded Baraboo Hills.

Had the glacier, the quartzite and the hardwood Hills not been present, the Sauk Prairie would be significant for its human history. Using fire as their shaping tool, native people worked the raw materials of soil, water and green plants to create the grasslands that gave the Prairie its name. Archeologists pursuing evidence of the people who inhabited the Sauk Prairie in pre-settlement times need not excavate in search of village sites, stone circles, potsherds or other conventional Old World artifacts. They have only to examine the New World grasslands, which were the monumental handiwork of the Ho Chunk, the Sauk, the Mound Builders and people unnamed who preceded them.

The farm community that supplanted the natives and erased their work as surely as the powder plant rubbed out the farms has its own evocative history. It is an American success story with a dramatic and tragic ending. It tells of immigrants who establish themselves on the land and create the institutions of farm, church, school, shared labor and neighborliness that define pre-World War II rural America. In memory it is an idyllic family farm community, abruptly snuffed out in its prime, leaving a story of mythic power and a bard to record it in Erhart Mueller. Like the athlete dying young, the farm community of the Sauk Prairie was cut down before signs appeared of its impending decline. It remains forever young and vibrant, with brow unfurrowed by the passing of time. In the decades after the war, family farm communities like the Sumpter of Erhart Mueller matured, withered and died in every nook and corner of rural America. On the Prairie the demise came fast and early in 1942, instead of gradually over the next sixty years. Those who mourn the loss of the traditional family farm community of the Sauk Prairie also mourn its passing from the entire United States.

The Prairie farmers did not go gently into exile from their homes and the story of their resistance is also part mythic and part historic. Unlike American citizens of Japanese heritage who also lost their homes and property, these Americans, predominantly of German ancestry, were able and willing to exercise their right to challenge their government even in a time of war. By sacrificing their homes they helped defend the United States. By resisting their government they helped defend the freedoms in whose name World War II was fought.

The powder plant itself is the great protagonist of the Prairie epic. It is a sprawling subject, with a database covering fire ecology, fireworks and the firing mechanism of the M-16 rifle. And politics, of the Democratic and Progressive parties in the 1930s, rearmament in the 1940s, anti-war protest in the 1960s, environmental activism in the 1970s and post-Cold War disarmament in the 1990s. Its story is one of work, of duty performed, responsibility accepted, by the men and women who came in the gate by day and night to manufacture nitrocellulose-based propellants — not gun powder, napalm or TNT, as occasionally and erroneously perceived.

Their work was risky, and the powdermakers took casualties, although not as many as they might have if awareness of safety had not been designed into the plant's layout, drummed into the workers during initial training and repeated while they were on the job. Space was the first element of safe design and explains why Badger required the confiscation of 10,565 acres of land. Not only was the plant isolated in the country, manufacturing areas were further insulated by buffers of unoccupied ground on the perimeter. Individual steps in the production process took place in single structures separated from each other by a calculated measure of empty space.

The requirements for safe, accident-free operation to protect the lives and limbs of the employees themselves and to preserve a manufacturing plant vital to national defense created an uncommon workplace culture. It combined the relatively simple and straightforward mixing and refining of raw materials — cotton, wood pulp, acids, nitroglycerin — into a finished product with the constant knowledge that some of these materials were hazardous at the very start of the process and all of them became increasingly dangerous as they moved down the line. Badger had a fire department whose first task was not to fight fires but to instruct employees in fire prevention and monitor their safe performance on the job. The guard force was trained to secure the perimeter of the plant from trespassers, but spent much more time checking employees for unsafe activities — like striking matches and smoking cigarettes. Strategic security was also a consideration. Potential employees were screened before hiring and many were investigated. Those suspected of sympathy with enemies foreign and domestic — Nazi and Communist — were weeded out, along with those who showed signs of personal instability or poor work habits.

The Badger Ordnance story is local history writ large and world history compressed to a few thousand acres of Wisconsin ground. Every war—hot or cold—that the United States fought from 1941 to 1991, along with the consequences of those wars, came home to the Sauk Prairie by way of the powder plant. It brought death and destruction, economic progress and prosperity. Social phenomena that characterized the second half of the 20th century: the dominance of car culture, the suburbanization of the countryside, tension and progress in racial relations, baby boomer generational conflict, the movement for equal opportunity for women, the modern redefinition of the family, experimental educational experiences for children, the challenges faced by an aging population, environmental pollution, rural economic development in a post-agricultural age and on and on, came to Sauk County via the powder plant. Herein lies the nugget of the story. Badger reveals to us that no place is so isolated, no spot on the planet possessed of such solitude that it is not linked to every other place and shares responsibility for its fate. The cow farmer in Sumpter, the chemist at Olin Corporation, the powdermaker from Jamaica, the 'copter pilot launching 2.75-inch rockets in combat over Vietnam, his mother from Merrimac at work on the E-Line, come together at the powder plant on the Prairie.

Badger Ordnance has left a broad and deep footprint on the Sauk Prairie and on the history of the United States in the 20th century. Its story is as broad and deep as the prairie soil out of which it has grown. This book is only a first telling.

Michael Goc
Spring 2002

I

Prairie

To the north, east and south, the broad Wisconsin trails its lazy, tortuous way throughout the land, basking like a silvery serpent, beneath the sun's glorious beams, while to the west extends Sauk Prairie, the richest portion of Sauk County...Any lover of nature will acknowledge the perfection and beauty of the whole picture, and perchance, may indulge a sigh that all the world, and every place in particular, is not so happily conditioned.

Consul W. Butterfield, 1880[1]

The Millennial Prairie

The happily conditioned Sauk Prairie flows over one of a series of terraces that step up from the sandy bed of the Wisconsin River to the quartzite face of the Baraboo Hills. For Increase Lapham, Wisconsin's pioneer man of all science, it was the "sack" prairie, not the Sauk, because it was shaped like a miller's grain sack, half standing with its top open to the Baraboo Hills, half lying with its tail trailing west to the golden sandstone towers of the bluffs above Honey Creek. From the bluffs, the prairie border bends northeast-north to the foot of Baxter's Hollow, then flanks the south face of the Baraboo Hills east across the glacial moraine into the Town of Merrimac. Once over the moraine, it zigzags south almost to the riverbank above Prairie du Sac, then curves west back to the bluffs along the edge of the oak savanna on the terrace above Honey Creek. This is the millennial prairie as it was, untroubled by Yankee plowman, Town of Sumpter Roadmaster, Hercules Powder Company engineer or Century 21 country lot subdivider.[2]

A *Sauk Prairie Gazetteer* reads as follows:

> 14,000 acres of airy loam on top of sand, gravel, silt and clay washed out of the last glacier topped by wind-borne loess from the ancient Mississippi Valley.[3]
>
> 600 plants—"a beautiful green carpet, dotted with innumerable flowers of great variety," including draba, whose humble, early, spring blossom Aldo Leopold identified as no more than a "postscript to a hope," and silphium, whose proud stalk and bright yellow blossoms led prairie travelers to call it the compass plant because it "always" faced south—except when it didn't.[4]
>
> 1,000 invertebrates—black ants, emerald beetles, digger bees, dragonflies, thumblebugs, spittlebugs, grasshoppers; about one million insects and arachnids per acre in August; among them a regal fritillary butterfly, with orange and deep ocean blue checkered wings; an orb weaver spider spinning a new web nightly to glisten with drops of morning dew; a goldenrod spider camouflaged yellow waiting on a yellow flower for its prey.[5]

100 birds—from the tiny dickcissel, named for its dick, dick, cissel call, heard less in Wisconsin as its grassland nesting habitat continues to disappear, to the turkey vulture, silent except for the sound of the wind rippling its feathers as it soars on the updrafts flowing off the Bluffs and over the grass.[6]

30 reptiles and amphibians—bull snakes, up to eight feet long, combative when threatened and feared by prairie plowmen; American toads, whose skin is not a bright red, white and blue, but a dull reddish brown camouflage; blue racer snakes, pale blue back, white belly, gliding through the grass with head cocked high, alert for prey; legless slender glass lizards with tails that shatter when pulled; tiger salamanders, speckled instead of striped, that breed while still in their larval stage.[7]

30 non-human mammals—from the prairie vole, a low-profile, extant prey species; to the prairie wolf, a high-profile extirpated predator, on whose head Sauk County paid a bounty of $2.00 until 1895, when the bountiful supply diminished.[8]

1 human species—present in the vicinity for about 12,000 years; an actor on the prairie ever since.

Microorganisms, on and in the soil, responsible for plant decomposition and soil building, equal in weight to all the vegetation among them.

Water, absorbed by roots, respirated by leaves, measured by the millions of gallons.

Solar energy, converted to green growing matter, in therms by the billions.

The infinite spectrum of color detectable by human and non-human eyes.

Sounds innumerable, of locust's rasp, meadowlark's song, prairie chicken's drum, bluestem swirling in a summer's wind.

Impressive statistics, but the Sauk Prairie is only an appendage of the grassy regime that once dominated 1.5 million square miles of central North America. From the banks of the Wabash in Indiana, to the foot of the Rockies in Colorado, from Alberta in Canada, to Coahuila in Mexico, prairie in all its tallgrass, short-grass, and sparse-grass varieties, quilted over the landscape. When they first encountered it in the 1600s, English empire builders and their Yankee cousins, accustomed to oak groves, sheep pastures and piney woods, had no word to describe the transcontinental vastness of grass, so they adopted the inadequate but exotic-sounding French-Canadian term for "meadow." An alien landscape required an alien language. *La prairie* entered English parlance and American consciousness.[9]

But the prairie existed long before its name was sounded in any European tongue. The result of cyclical climatic meanderings, oscillations in precipitation and temperature, over a continent of space and centuries of time, then steered by human intervention, the immense grassland ebbed and flowed over the North American heartland. When the climate warmed and rain and snowfall diminished so fires sparked by lightning and kindled by men and women could carry over the landscape, the prairie grew. East of the Mississippi River, a wedge-shaped peninsula of grass covered more than three million acres in southern Wisconsin, nearly all of Illinois and most of northern Indiana. As recently as 500 years ago, aided by climate and people, the prairie and its kindred savanna landscape advanced through the flatlands of northern Ohio, to maximize the realm of the grass until it stretched from the back of the Appalachians to the face of the Rockies.

The Sauk Prairie is a cape, a point, on the north face of the prairie peninsula and almost as far north as prairie grows east of the Mississippi. An outpost on the tallgrass frontier, it is separated from the northern forest of yellow birch, black spruce and white pine by the hardwood forest that crowns the Baraboo Hills and a varying boundary of diverse grassland terrain that includes grass marshes, sedge meadows, pine barrens and the oak savanna that the white pioneers called openings.

The Sauk Prairie was ringed by oak savanna that spread down the slope of the Baraboo Hills to the north and occupied a lower terrace in the Town of Prairie du Sac to the south. Savanna also met the prairie in front of the bluffs to the west and from the east where it stretched across the Wisconsin River, and down the Fox River Valley to Lake Winnebago. Natural scientists define an oak savanna as a prairie growing on relatively fertile soil whose grassy carpet is "opened" by widely spaced groves of mature white or burr oak trees. A literary artist, British novelist Frederick Marryat, on tour in the Fox River Valley in 1837, saw the savanna as "beautiful beyond conception. Its features alone can be described; but its effects can only be felt by being seen....To form an idea of these oak openings, imagine an inland country covered with splendid trees about as thickly planted as in our English park; in fact it is English park scenery..." Looking west from his campsite on the lofty escarpment east of Fond du Lac, Marryat could also see the transition from prairie to savanna to forest. "I never saw a more beautiful view....From the high ground upon which our tents were pitched, we looked down to the left, upon a prairie flat and level as a billiard table, extending, as far as the eye could scan...to the right the prairie gradually changed to oak openings and then to a thick forest."[10]

Marryat was in eastern Wisconsin, where the transition zone from prairie to forest was thin and readily defined. Had he camped at Point Sauk, the highest spot in the Baraboo Hills, and climbed a tree for a squirrel's eye view, he would have seen a more complex panorama. He could look down on the Sauk Prairie to the south, observe the "park scenery" of the savanna on the slope, but the northern forest was well beyond the range of human vision or squirrel's eye. In between were seventy miles of diversity: the southern hardwood forest of the Baraboo Hills; more forest, plus swamp and savanna in the Baraboo River Valley; dry prairie in the Town of Delton; oak and pine barrens on the arid sand "islands" of Adams and Juneau counties mixed with grass marshes and sedge meadows that stretched north into Wood, Portage and Marathon counties. Only then did the north woods of tall trees and taller tales, of cathedral pines and Paul Bunyan, begin its sway.

Aldo Leopold recognized the jumble of landscapes and the diversity of plant life in the forest-prairie transition zone as consequences of "the prairie war" that ensued when the glaciers began their advance south from the Arctic. In the cooler and wetter periods of glacial advance, the forest marched south from Lake Superior to plant black spruce in central Illinois and, along the way, seed a tamarack swamp on Honey Creek just west of the Sauk Prairie. In drier and warmer periods, the prairie advanced north through Wisconsin, seeding bluestem in the rotten granite of Rib Mountain and carrying "Missouri" goldenrod to the Sauk Prairie. About 6,500 years ago the dynamic paused at a stage that survived until the American settlement of the 1830s. North of the transition zone, big woods stretched to Hudson's Bay. South of the zone, big grass ran far into Illinois. About midway in between the Bay and the foot of the Illinois grassland, ringed by oak savanna, down the bluff from the Baraboo hardwoods, evolved the Sauk Prairie.[11]

Long before it evolved as part of the prairie/forest mosaic, the Sauk Prairie lay atop a confluence of geological frontiers. The Prairie scratches at the foot of the ancient quartzite highland of the Baraboo Range on the north, brushes the edge of the younger bluff and coulee country on the west, and straddles the line of demarcation separating glaciated Wisconsin from the ground where the glacier did not tread.

The quartzite of the Baraboo Hills was formed more than 1.5 billion years ago. It is one of the oldest and hardest, most dense and durable of rocks on earth. The quartzite deposits are so deep that, if they had not been buried by sediments laid down by ancient oceans during succeeding millions of years, the Baraboo Hills would stand about twice as high above the horizon as they do today. Despite their height, the Hills were, in the course of millions of years, buried entirely by ocean sediment that fused into sandstone. As the ocean receded and the sandstone eroded, the peaks of the quartzite resurfaced as hard rock islands in an ancient salt sea. During one phase of the quartzite islands-in-the-sea period, the famous "pudding" rocks of the Baraboo Hills were formed. Tsunami-sized waves repeatedly crashed into the quartzite shore, breaking off chunks of rock. Over many millions of years continuous wave action smoothed the chunks into rounded boulders, some as large as wagons, some as small as pebbles that fit neatly into a child's hand. In time, the quartzite boulders were imbedded in softer, sandier ocean sediments like fieldstone embedded in mortar to make a stone wall. To the geologists who excavated them in the 1800s, these conglomerate mixes of quartzite and sandstone resembled a European-style pudding in which chunks of meat, vegetables or fruit were embedded in a thick sauce, so they called it pudding stone.[12]

The oceans that buried the quartzite mountains and shaped the pudding stone also deposited the sand, iron and lime that form the bluff and coulee terrain west of the Sauk Prairie. To paraphrase the children's song, inch by inch, layer by layer, sand of varied color, size and texture, alternating with metallic deposits and the mineral remains of marine animals, grew Castle Rock, Leykauf's Bluff, Ferry Bluff and thousands of other rocks, bluffs, mounds and "mountains" unnamed in southwestern Wisconsin. As they grew, they eroded. Rain pelted, runoff gouged, waves crashed, wind rasped. With the softer stone eroding faster than the harder, the sandstone acquired its zigzag, top-heavy, layered profile.

The processes of deposition and erosion continued for several hundred million years. Oceans flooded and drained, the wind blew or not, climates fluctuated as the continents drifted around the planet. The rock formed by these processes varied from place to place and time to time. By the beginning of the Pleistocene Ice Age period, two-three million years ago, the 14,000 acres where the Sauk Prairie developed were underlain by a thick deposit of sandstone similar to that of the bluffs farther west, but not as high, hard and resistant to erosion. Rainstorms and snowmelt washed it away, windstorms carried it off. What might have been 14,000 additional acres of Sauk bluffs weathered down to form the sub-floor of the Sauk Prairie.

Between two and three million years ago, a lengthy period of global cooling began. Temperatures fell over the northern hemisphere and, in the sub-polar regions of northern Canada, more snow and ice accumulated than melted during the brief, chilly summers. Fed by warm, moist air blowing up the center of the continent that turned to snow and ice when it met frigid air from the north, the ice cap expanded until it extended from the North Pole to the Ohio River Valley. The ice sheet descended on Wisconsin from the northeast and north, with three individual lobes spreading over the state like cookie dough rolled out on a baking sheet. The Lake Michigan lobe covered southeastern Wisconsin and much of Illinois. The Lake Superior lobe gouged out the bed of the great lake and moved south until high ground in west central Wisconsin halted its progress on a line running roughly west from Stevens Point to the Mississippi River. When it paused, the Lake Superior lobe spared southwestern Wisconsin the scouring, leveling and excavating the glacier forced on other parts of the upper Midwest. Had the Lake Superior lobe not halted, southwestern Wisconsin would look more like southeastern Wisconsin, with crags and bluffs chiseled down to rolling hilltops and narrow river valleys filled into broad marshes and shallow lakes.

In between the Lake Michigan and Lake Superior lobes, the Green Bay lobe bulldozed through east central Wisconsin. About 15,000 years ago, the glacier entered eastern Sauk County and halted its advance. A warming trend began, the ice face melted back and the glacier released the burden of rock, soil, and organic debris it had carried for hundreds of miles. When released by the softening ice, this debris, or till, formed ridge-like moraines that snake over the earth and mark the glacier's progress in thousand-yard folds of earth north of Devils Lake State Park, or in the wider rolling pock-marked moraine that bisects the Sauk Prairie. The end moraine of the Green Bay glacier crosses the Wisconsin River to enter Sauk County at Wisconsin Dells. It runs south to West Baraboo, shifts southeast to Devils Lake, winds through the Hills, then travels down the south face onto the Sauk Prairie, where it continues south until arching east to recross the Wisconsin on the site of the Prairie du Sac dam.[13]

The view east from the tip of any of the bluffs on the west side of the Sauk Prairie in the last glacial period would have been breathtaking. No one has yet verified that a man or woman actually stood on a bluff now in Sauk County and saw an Ice Age glacier, but people were present in North America at the time and might have experienced it. If so, a proto-North American standing on a bluff above Otter Creek could look a few miles in the direction of sunrise and see, running north to south from the Baraboo Hills to Prairie du Sac, the edge of the ice sheet that stretched east to the Atlantic Ocean and north beyond Hudson's Bay. A great, grey front of ice, the glacier tapered upwards to the east, rising a mile or more in height and reducing any other mountain nearby to molehill status. From the foot of the ice sheet, a water-soaked flats fanned out, similar to the delta of a river, but multiplied by any number of rivulets, streams and full-born torrents of meltwater braiding back and forth across it. At times a cascading flood bore out of the glacial mass, cut a channel in the flats and pushed downstream a Mississippi's load of ice, sand, gravel and rock. At other times the flow slowed and shifted to fill channels, build sandbars and deposit gravel in beds.

Rarely static, the face of the ice sheet shifted as the climate chilled or warmed. It carved new river channels and dammed its meltwater into frigid, freshwater lakes. Glacial Lake Wisconsin, the largest and longest lasting of these temporary lakes, came into being when the glacier moved west onto the Baraboo Range and trapped meltwater flowing down the future valley of the Wisconsin. The resulting lake covered most of Adams and Juneau counties, as well as parts of Waushara, Portage and Wood, and extended bays up the valleys of Hulburt, Dell and Narrows Creek, plus the Baraboo River as far as Wonewoc. When the ice sheet retreated a short distance east off the Baraboo Hills, it also flooded what is known as the Lewiston Basin of glacial Lake Wisconsin. This body of water covered the townships where Adams, Columbia and Marquette counties meet and extended north about fifteen miles. Another shift of the ice gave birth to glacial Lake Merrimac, which formed in what became the bed of the Wisconsin River about where the present Lake Wisconsin is today, but with an outlet that flowed west through a hole in the moraine and across the southern edge of Sauk Prairie into a greatly swollen Otter Creek.

All this moving water carried immense amounts of silt, sand, clay and gravel that was deposited in lake beds and outwash plains. In one phase, so great was the outwash from the glacial moraine when it ran across the Sauk Prairie that it dammed the flow of Otter Creek into a lake that filled Stones Pocket. Farther downstream, outwash flowing across the Sauk Prairie pinched the outlet of Honey Creek and backed up its water into glacial Lake Black Hawk, which flooded the valley around Witwen and Black Hawk, and extended upstream almost to Denzer, Leland and Plain. The lakes eventually filled with silt, sand and clay: up to sixty meters deep in parts of Lake Black Hawk, "a few to a few tens of meters" in Stone's Pocket and ten meters in Lake Merrimac. On the outwash plain that became the Sauk Prairie the glacial meltwater deposited "tens of meters" of silt, sand, clay and gravel.[14]

All of this flooding, damming, gouging and filling took place in front of the glacier. Behind its face, the ice sheet also reworked the terrain. In the eastern part of the Town of Sumpter and in Merrimac, the glacier left what a modern soil scientist called a "hummocky topography," formed when ice buried beneath glacial till melted and the sand and rock above collapsed. It looked "like a 'chopped up sea,' " to William Canfield, Sauk County's first historian and natural scientist. It was a sea of chopped stone and silt, whipped into potholes and hillocks, hummocks and kettles, by the working edge of the glacier.[15]

While the glacier that stopped on the Sauk Prairie had the greatest impact on it, ice farther away also made its mark. In the millions of years in which ice sheets slid back and forth across the Upper Midwest, they exposed vast areas of bare soil to the wind. The periods of glacial retreat were also periods of monumental dust storms that carried fine silt from west of the Mississippi River into Wisconsin. In parts of the state the wind-borne silt, or loess (luss), lies yards deep on the ground. On the Sauk Prairie, the loess deposits run up to eight feet deep. Fine-grained and mineral rich, the loess joins the glacial outwash to form the subsoil of the prairie and the foundation of its fertility.

Millions of years of geologic and climatic activity was floridly summarized by county historian Consul W. Butterfield in 1880. "Nature, in shaping this portion of the country and in contributing a variety of qualities or properties to the soil, seems to have been unusually prodigal...in her bequests." By the time the glacier began to recede 15,000 years ago, prodigal nature had bequeathed the broad valley of the Wisconsin River and the narrow channel of Otter Creek. It sculpted the hills and hummocks back of the terminal moraine, smoothed the outwash plain in front. Raw and barren, the outwash flats did not look like much of a bequest 15,000 years ago, but the glacial edge held all the promise of Eden on the day before Creation. It was "one vast, sparsely vegetated, gravelly, sandy or muddy seedbed, an ecologically open habitat just waiting to be invaded by a rain of seeds and propagules."[16]

Which seeds and propagules took root in the new seedbed was determined by climate over time. First to sprout, blossom and survive in the immediate post-glacial environment, which resembled that of Alaska's North Slope today, were the mosses, lichens and low-growing plants and shrubs that constitute tundra. They provided browse enough to support populations of caribou, musk ox and elephantine woolly mammoths. As they bloomed, died and decayed, the tundra species, along with the bodies and manure of animals, birds and insects, provided organic matter that was processed by microbes, mixed with outwash minerals and wind-borne loess, to make soil. By 12,500 years ago, only 2,500 years after the glacier began to recede, the climate dried and warmed enough to support a boreal forest of spruce, fir, black ash and birch inhabited by mastodon, giant beaver, elk, deer and black bear, as well as other forest animals and birds found in Wisconsin woodlands today. The process of soil-making continued, as organic matter, minerals and microbes mixed to add an onion skin layer of fertility each year. Another 2,500 years passed, the glacier receded entirely from Wisconsin, the climate continued to warm but grew wetter and the boreal forest was replaced by a pine, elm, basswood and maple forest which resembled the "north woods" encountered by the American settlers of northern Wisconsin in the 1800s. Several more millennia passed, the climate fluctuated, continued to dry and, on the outwash plains and lower terraces, oak trees and prairie grasses began to replace the north woods species. The rate of change could be surprisingly swift. In Illinois — the Prairie State —it has been estimated that two million acres of woodland converted to two million acres of prairie in 500-800 years. On the colder and wetter edge of the prairie peninsula in Sauk County, the conversion from southern hardwood forest to grassland started later and took longer than in Illinois. Not until about 6,500 years ago did the various forms of grassland — oak savanna, tallgrass prairie, grass/sedge wetlands — start to become the dominant land cover in Sauk County; in about 1,000 years, it succeeded.[17]

In addition to the 14,000 acres of the Sauk Prairie itself, 13,000 acres of dry prairie occupied the river flats of the "Wisconsin Desert" near Spring Green. Cassell Prairie spread over 1,200 acres in the town of Troy, Pleasant Prairie claimed 2,200 acres north of Baraboo and Peck's Prairie accounted for another 1,600 acres east of the city. Grass marshes, sedge meadows and wet prairies made up another 70,000 acres. The largest portion of ground — near a quarter million acres — was in oak savanna, where prairie grasses grew between groves of mature oak trees. It was not the "forest primeval...bearded with moss...indistinct in the twilight," as scripted by Longfellow, but an "English park" where sturdy oak limbs arched leafy branches over breezy, sunlit meadows.[18]

Kindled by Human Hands

Climate and soil conditions had to be suitable, but the grasslands of North America would not have developed as they had without human action. "Rather than a climatic change the general encroachment of grassland into forest may have reflected the penetration of nomadic hunting cultures...enlarging the range of their prey," writes Stephen Pyne, pre-eminent scholar on the use of fire in non-industrialized cultures past and present. "This expansion could have been accomplished only through a change in habitat, partly due to climate, but largely achieved through the application of anthropogenic [human-made] fire." Pyne cites the arrival of the American bison, or buffalo, east of the Mississippi River, which occurred about 500 years ago, and its spread as far east as Pennsylvania, as a result of native people creating grassland habitat for it. If—and it is a big if—buffalo appeared on the Sauk Prairie, they turned up at this time because of fires kindled by human hands.[19]

The earliest evidence of humans in the vicinity of the Sauk Prairie was found a few miles west at the Raddatz Rockshelter near the Natural Bridge stone arch. Charcoal excavated at the rear of the shallow cave is presumed to be left over from fires kindled by humans 11,000-12,000 years ago. At the time, the dominant ground cover of the hills where the shelter is located and of the outwash plain to the east was the boreal forest of spruce and birch inhabited by giant beaver and tusked mastodon. More concrete evidence of human presence dates from about 2,000 years later and comes in the form of a deer antler sharpened for use as a projectile point to be fitted onto the shaft of an arrow or spear and a scraper made from basaltic stone not native to the shelter. By the time these tools were made, 9,500 years ago, the climate outside the shelter was warmer than a millennium earlier but still wet, and the landscape was in its north woods phase, reminiscent of Vilas County in 1850. By the Western calendar based on the presumed date of the birth of Jesus, the bone and stone tools at the Raddatz site were made around 7500 B.C. In other parts of the world, agriculture was already underway and farm animals domesticated, although not yet in North America.[20]

From 7500 B.C. on, the Raddatz site sheltered humans, if not continuously, but regularly for another 9,000 years. Its significance, and that of the Durst Rockshelter a few miles west, the Zech Rockshelter a few miles north, the Hulburt raised-garden beds west of Wisconsin Dells, plus the effigy mounds and other archeological sites on and around Sauk Prairie, lies in the testimony they give to 12,000 years of human presence. For, if people were present, they had an impact on the environment they shared with plants and other animals. As stated by historian William Cronon, "everything we know about environmental history suggests that people have been manipulating the natural world on various scales for as long as we have a record of their passing."[21]

Rich and varied, prairie and savanna vegetation requires relatively fertile soil and a climate with dry seasons in which fires can spread over the landscape and wet seasons to foster the growth of grass, shrubs and trees for fuel. Abundant and consistent precipitation, as in the tropical rain forest or the north woods of Wisconsin, feeds thick woodlands and suppresses the fires that would destroy them. For fire, recurring fire, is essential for grassland development. "Grass, with fire as its ally, can beat timber," wrote Wisconsin geologist Thomas Wight in 1877. "Timber can beat grass when it has no fire to fight."[22]

An infernal scythe, fire mows stems and stalks and trunks at ground level. Plants whose root systems can sustain regrowth after their tops are burned off, and trees that resist fire or even take advantage of it to reseed themselves, survive to dominate the landscape. Prairie types such as needle grass and big bluestem are among the oldest of grasses and have the largest concentration of genes that encourage adaptation to fire. They evolved to succeed in an environment where fires burned regularly. Similarly, the trees of the pine barren produce cones that require intense heat to open, while the white and burr oaks of the savanna have roots that resprout if they burn off when young and thick bark to resist fire when the trees are mature.[23]

The fire necessary to create and maintain grasslands in North America has two sources — lightning and people. Records of lightning strikes on the Sauk Prairie 6,500 years ago are not available, but as surely as rain fell, lightning struck. Whether lightning struck often enough in the right place and at the right time to burn back the forest over a period of several thousand years is problematical. Recent data from nine southeastern Wisconsin counties indicate that lightning can cause up to five fires a year in grasslands and, based on observations in northern Wisconsin, twice as many in forests. For the million-plus acres of grassland in southeast Wisconsin, five to ten fires a year probably would not prevent forest trees from shading out prairie grasses. Yet prairie and other grasslands were the dominant ground cover of the southern two-thirds of Wisconsin. The fires that nurtured the grassland here must have had a source other than random bolts of electricity cascading earthwards from the sky.[24]

Direct evidence of proto-North Americans firing the Sauk Prairie is not available. Fire pits, traces of charcoal or layers of ash 6,500 years old have yet to be discovered. However, as the archeology of the Raddatz Rockshelter shows, humans have been building fires near the Prairie not just for 6,500 years, but for 10,000, probably for 12,000 years. It is not far-fetched to surmise that in one instance, by accident or design, based on reasoning mundane, frivolous or sublime, a woman, man or child ignited ground cover already evolving in that direction and helped create a prairie. With that act, repeated many times over, a not-so-simple, skin-clad denizen of the Stone Age had an impact on the Sauk Prairie as profound, as complex, and more enduring than that of the plow men of the 1840s or the Army engineers of World

War II. The work of the Army has survived about sixty years but no more than a portion of it is likely to see another sixty, and none is likely to be used for its original purpose. The farm community the Army partially displaced is about 160 years old and, like all of the United States, has since changed immensely. The prairie landscape created by prodigal nature and Stone Age humans flourished for six millennia.

For the non-Europeans who inhabited North America in historical times, and probably for the maize farmers and mound builders of 1,000 years ago, as well as the prairie pioneers of 6,500 years ago, fire was a universal cultural presence. More than a tool as defined by Euro-American culture, fire was spiritual and practical, a multi-faceted gift from the creator that, according to one Ho Chunk tradition, was presented to man and woman to help them shape and understand their world.[25]

Fire warmed and illuminated cold, gloomy rockshelters on the post-glacial frontier and bark hogans in historical times. It cooked and preserved food, both deer meat served in a turtle shell bowl at Raddatz in 3000 B.C. and maple sap boiled to sugar in a Ho Chunk camp on the Baraboo River in 1800 A.D. Fire hardened clay for pots, softened metal for tools, hollowed logs for canoes. It ignited sacred tobacco to honor ancestors and welcome strangers to the hearth. Fire cleared ground for the mound builders who, starting 1,200 years ago, raised 20,000 mounds and excavated an uncounted number of intaglios throughout Wisconsin, including nearly 300 mounds on or near the Sauk Prairie. Whether burial sites, spirit effigies, clan totems or art for art's sake, if they were to be seen from only a short distance away, the earthworks had to be kept clear of brush and trees. They had to be burned. Not until American settlers halted the fires in the mid-1800s did stripling maples sprout in the 100-yard wings of earthen eagles and groves of quaking aspen obscure processions of marching bears aligned 1,000 years before.[26]

The uses of fire to procure food and animal hides were both subtle and direct. Fires were lit in grass or woodland to drive large game — deer, elk, bear — into the traps, snares, spears, clubs and arrows of hunters. Fire also drove smaller game — turkey, grouse, rabbit, lizard, squirrel — into hunting range. It uncovered edible eggs in nests and small mammals in their lairs. In forests, fire burned off the underbrush to reduce cover where quarry sheltered. It facilitated tracking and enabled hunters to move through the forest without alerting their prey. Several generations of American children, raised on western novels, comic books and Hollywood movies, aspired to skulk stealthily through forests of their imagining, never cracking a twig or crackling a leaf, "just like the Indians did." Genuine, flesh-and-blood "Indians" wanted to move quietly too, and did it by firing the dried twigs and leaves in front of them.[27]

Indeed, fires were lit to clear brush and to make travel in general easier for people who lived a semi-nomadic life and, with only a few exceptions in historical times, had no beasts of burden other than dogs. Carrying a canoe over a portage, bearing a pack of household goods, an infant child, or both, from camp to camp, even lugging a bundle of sticks for the night's fire, was easier on the fire-cleared ground of a savanna or a prairie than in a brushy forest. It is not surprising then that trails predating American settlement crisscrossed the Sauk Prairie and wound throughout the encircling savanna.

Fire also supported hunting by expanding edge habitat favorable to game animals and birds. Just as some modern shooters wink at the law and spread shelled corn in a clearing to coax whitetail deer out of a nearby woods, ancient hunters fired the edge to improve the browse and bring deer into clear firing range. In some instances, fire was also used to create cleared refuges free of mosquitoes and flies which attracted deer and elk to hunters on stand.

Deer were a necessity of life for the inhabitants of pre-Columbian Wisconsin, going back to the Archaic people of the Raddatz and Durst Rockshelters, who left more remains of whitetails in their caves than of any other animal. Deer were equally important to the Woodland Indians of 500 years ago, whose lifestyle, it has been estimated, required the annual harvest of as many as six whitetails per person.[28]

The population of Wisconsin prior to first contact with Europeans has been estimated at 60,000-70,000 people, with the grasslands in the southern half of the state supporting many more people than the northern forest. If each person needed six deer per year for food, shelter, tools, clothing and footwear, the annual kill of animals matches that of a typical Wisconsin gun deer-hunting season of the 1990s. Even if the population estimate and the six-deer-per-person ration were halved to 35,000 people needing three deer each, the kill would match the annual gun deer harvest of the 1970s. Speculating on the size of the deer herd, on the pre-Columbian population of Wisconsin and on how those people managed big game is a hazardous venture. Yet it is safe to say that a deer herd only a fraction as large as that of the late 20th century required a bountiful store of aspen buds, acorns, new grass and other green feed. Fire was the tool native people used to raise that feed. On the forest edges fire forced shrubs and tree roots to sprout succulent browse favored by deer. On the savanna it maintained mature oaks heavy with acorns. On the prairie it rejuvenated and thickened the grass for additional deer browse. As researchers at Madison's Curtis Prairie have learned, a prairie burned in late fall or early spring produces about twice as much herbage in the next growing season as a prairie left unburned. For people who avidly hunted deer and elk, firing the prairie was the equivalent of a modern cattle raiser adopting a farm practice that doubles the yield on a field of corn. It is possible to view the people who lived in Wisconsin prior to European contact as game managers who used fire to raise more and better feed for their livestock.[29]

Fire was as much a tool of horticulture as of hunting. Burning rejuvenated edible berry bushes and vines in woodland openings, on forest edges and on cranberry marshes. Although earlier people had cultivated food crops, farming increased dramatically in Wisconsin after the introduction of corn about 1,200 years ago. Fire was used to clear woodland for farms in the classic slash-and-burn fashion, whereby large trees were girdled by axing through the bark, brush piled against the trunks to concentrate the flames and the whole ignited. Afterwards the ground was worked into raised beds, where soil, roots, charcoal and ash were mixed together into a fertile plot. The Hulburt Creek ridged fields, which may have been as large as 200 acres, were worked in this manner about 1,000 years ago.[30]

After the harvest, fire was used again as a purgative to dispose of crop residue that harbored harmful insects and diseases and to burn off sprouts from tree roots that survived the initial firing. Clearing land with fire also provided the fuelwood necessary to survive a Wisconsin winter. Standing trees killed by the fires were more easily felled with stone axes or toppled over on their own so they could be reduced to size for the hearth.

Fire was also a weapon of war. Raiding parties scorned their enemies by leaving campfires free to burn enemy ground or deliberately fired enemy territory as an offensive tactic. As a defensive measure, brush and grass obscuring the view on village perimeters were burned off to prevent surprise and to remove fuel for an enemy's fire attack. Village sites were selected with fire defense in mind. One survey of Ho Chunk villages in Wisconsin in historical times found nearly all of them protected by a natural firebreak of lake, marsh or river. Villages were downwind of the Fox River and Lake Butte des Morts in Winnebago County, on the island now occupied by Neenah and Menasha, on a peninsula jutting into Lake Koshkonong or surrounded by wetlands at Horicon.[31]

Evidence of any sizable native American village on the Sauk Prairie or, as would be more likely, on its edge, has yet to be found, perhaps because it is on the wrong side of the Wisconsin River. During the spring and fall fire seasons winds are most likely to blow out of the southwest, west or northwest, so fires tended to burn from the bluffs to the river. A village located on the Sauk County side of the Wisconsin would be more vulnerable to fire than a village across the river. The well-known exception to this premise was the "great town of the Saukies" with "about ninety houses each large enough for several families" located on the site of Sauk City in the mid-1700s, but it is exceptional for more than its location. The Sauk were newcomers to Wisconsin, refugees displaced by fur trade war from their homes across Lake Michigan in the 1670s. They came to Sauk City about a half-century later, once again in flight, this time from the French and Indian armies that invaded east central Wisconsin in pursuit of the Fox Indians. Allies of the Fox, the Sauk were probably more concerned about an enemy coming down the

Wisconsin from Portage than one coming over the prairie to the west and therefore preferred a site overlooking the river. Be that as it may, the Sauk village was still separated from the prairie by a buffer of savanna and "four hundred" acres of ground cleared for farming that could act as a firebreak. After coming to Wisconsin, the Sauk showed more interest in farming than they had in Michigan. Building a large village adjacent to extensive farm fields was more of a traditional Ho Chunk practice than traditional Sauk, but immigrants habitually adopt new modes of behavior. So, like the Ho Chunk, the Sauk built a large farming village, lived in it during the growing season, raised and stored food for the lean times of late winter and early spring. Living in uncertain times, in an unfriendly place, pursued by a powerful, merciless enemy, yet attracted by good farmland, the Sauk accepted the risk of settling on the west bank of the Wisconsin, even though it was downwind of the flammable prairie. They stayed about fifty years. About the time the American War of Independence ended and Great Britain ceded Wisconsin and other parts of the "Old Northwest" to the new United States, the Sauk left the banks of the Wisconsin and moved to an equally fertile farm village site on the Rock River in Illinois.[32]

The Ho Chunk Indians were already present in Wisconsin when the Sauk migrated here in the late 1600s, and remained here after the Sauk left in the late 1700s. At the time of their first contact with Europeans in the mid-1600s, the Ho Chunk were concentrated in the Lower Fox River Valley and along the shores of Lake Winnebago in east central Wisconsin. Early French chroniclers report them as great in number and fierce in defense of their territory. Their villages, which consisted of winter-tight, multi-family dwellings of poles, reeds and bark, were fairly substantial and once described as so large that a person living at one end of town did not know what was going on at the other end. The Ho Chunks had developed settlements of such size because they were farmers who took sustenance from the traditional three sisters of Native American agriculture—maize, beans and squash. Their villages were invariably located on fertile soil and skirted by extensive farm fields. They also hunted, fished, gathered and, in season, small family bands ranged out of the main villages to harvest berries or pursue deer. Following the waterways in dugout canoes, or traveling overland, the Ho Chunk could have readily come to Sauk County, whose food-rich grasslands they might already have been managing by means of fire.[33]

Shortly after their first contact with a European in the person of Jean Nicolet, who they reputedly feted with a lavish outlay of choice beaver tail in 1634, the Ho Chunk suffered a disastrous decline. Through disease that had traveled west either with the European adventurers or Indian fur traders, or through warfare with Illinois Indians, the Ho Chunk population was so reduced that the tribe all but disappeared. One oral tradition holds that as few as fifty male members of the tribe survived, another that only a single boy was spared slaughter at the hands of the Illinois tribes. Oral tradition is not bookkeeping and the numbers recited are not meant to be taken literally, but they do indicate the extent of the catastrophe the Ho Chunk "men" suffered 350 years ago. (The Ho Chunk women suffered as grievously as the men, but neither Indian nor European chroniclers made more than peripheral mention of them). It is possible that the severe population loss experienced by the Ho Chunk and other midwestern Indians in the 16th and 17th centuries enabled the bison to expand its range east of the Mississippi River. If so, then the buffalo that may have grazed on the Sauk Prairie in the 1600s were there for a very sad reason. There were not enough Indians left to hunt them. In the historical chronicles of the 1700s and early 1800s, the Ho Chunk play a minor role compared to the Fox, Sauk, Chippewa and Menominee. By the 1830s, with Americans moving into southern Wisconsin, the Ho Chunks, whose population was still only a fraction of what it had been prior to European contact, were living in small villages and hunting camps scattered throughout the Upper Fox River Valley, the middle portion of the Wisconsin Valley and along the Baraboo, Yellow, Plover and Black Rivers.[34]

In the wake of the Black Hawk conflict of 1832-33, all Indian tribes in Wisconsin signed treaties of dubious legality by whose terms they relinquished title to land here. The Chippewa and Menominee were able to renegotiate their treaties and obtain reservation land in the state, the Ho Chunk were not. They split into two factions. One accepted reservation land in Iowa and Nebraska; the other "non-abiding" faction, led by Yellow Thunder, refused. Five times between 1840 and the early 1870s, the non-abiders were forcibly removed by contractors under federal supervision who raised posses of local men. They loaded Ho Chunk men, women and children into steamboats and onto rail cars, or drove them on foot across Wisconsin and Iowa. Every time they were removed, Yellow Thunder and his band returned. They dispersed throughout the infertile, sparsely populated regions of northeastern Sauk, Adams, Juneau, Wood, Portage and Jackson counties. It is one of the many sad twists of Ho Chunk history that a people who once inhabited fine, prosperous villages and farmed Wisconsin's richest soil, now eked out a livelihood in the barrens and marshes of the state's poorest agricultural region. Nonetheless, their will to remain in Wisconsin prevailed over that of the Americans to remove them. As early as 1849, believing that a property owner and taxpayer could not be evicted from his land, Yellow Thunder himself purchased forty acres in the Sauk County town of Delton. It did not protect him or his people from the removal posses. The passing of time and their own persistence were of better service to Yellow Thunder's band. By the mid-1870s, the United States had given up its attempts to remove the Ho Chunk and had enacted the Indian Homestead Act, which enabled individual Indians to acquire unclaimed government land for little more than a filing fee. Yellow Thunder's forty, which he had purchased at market value, had already become a gathering place for Ho Chunks and remained one throughout the 19th century. On the Sauk Prairie, from the time of American settlement, the occasional Indian visit was more tolerated than welcomed and, by the 1860s, a rare event.

"Brilliant Chains of Liquid Fire"

No record has been found of a fire on the Sauk Prairie from the period of Sauk residence on the Wisconsin, from the Woodland Indian phase that preceded it, nor from the term of the Ho Chunk that followed. Sumpter historian Erhart Mueller reported that in 1838, the bluffs above Honey Creek had few trees, "Due to burning by the Indians, who wished to improve their hunting grounds": but he was talking about the bluffs, not the Prairie.[35]

Records of grassland fires lit by Indians elsewhere in Wisconsin are more plentiful. Near Lake Pepin in 1833, "...the Prairies were on fire...the Indians had taken advantage of the driving wind to fire the country for a great many miles...the country opposite blazed like tinder in the wind...the scene was fearfully grand." In Kenosha County in 1835, "...in the latter part of September. We began to see the advancing fire towards evening on the prairie....The wind was from the southwest and pretty strong, and the fire progressed rapidly....The roaring terror came through the woods with an awful grandeur. Large trees as well as all smaller vegetation, quickly fell before the ruthless invader." Near Fort Winnebago at Portage in 1836, "The leaves have fallen, annual plants have become dry, and the fields, the swamps, the forests and the prairies are set on fire by Indians and hunters. The smoke arising from them is abundantly sufficient to produce all that peculiar redness of the sky so common to Indian summer."[36]

In an often-quoted passage recounting a fire on a western prairie, the painter George Catlin depicted how a blaze on the bluffs around the Sauk Prairie might have appeared: "Over the elevated lands and prairie bluffs, where the grass is thin and short, the fire slowly creeps with a feeble flame, which one can easily step over; where the wild animals often rest in their lairs until the flames almost burn their noses, when they reluctantly rise and leap over it, and trot off amongst the cinders...These scenes at night become indescribably beautiful, when their flames are seen at many miles distance, creeping over the sides and tops of the bluffs, appearing to be sparkling and brilliant chains of liquid fire...hanging suspended in graceful festoons from the skies."

On the prairie itself, thick with dried grass in autumn or early spring, a fire would resemble an 1828 blaze reported by H.A. Gleason: "The line of flame rushed through the long grass with tremendous violence and a noise like thunder; while over the fire there hovered a dense cloud of smoke. The wind, which even previously had been high, was increased by the blaze which it fanned; and with such vehemence did it drive along the flames, that large masses of them appeared actually to leap forward and dart into the grass, several yards in advance of the line. It passed me like a whirlwind, and with a fury I shall never forget."

For the prairie to successfully compete with forest trees, fires had to recur, if not annually, at minimum every decade. When he described the "mineral lands" of southwestern Wisconsin in 1852, geologist David Dale Owen reported "those parts of the western country which were originally prairie, and where the fires have been kept out for twenty years or more, are now covered with thick groves of small trees.... The American aspen in the whole district of the mineral lands, seems to be the pioneer tree which first invaded the prairie.... When once the prairie sward has been broken by this kind of tree, others come in one after another; the prairie changes to the thicket and in a few years it becomes the vast wilderness, the boundless contiguity of shade." In accounts penned by American settlers of the 1840s, no mention is made of trees growing on the Sauk Prairie. If trees were absent in 1840 and, considering how fast and tall an aspen or even a burr oak can grow in twenty years, then the Sauk Prairie must have been fired at least once in the 1830s and every decade previous.[37]

Except to attribute it to their habit of driving game, neither the deliberate nature of the Indians' firing of the countryside, nor its consequences, were recognized by the Americans who displaced them. While David Dale Owen was noting the conversion of prairie to "wilderness, the boundless contiguity of shade," the Scotch immigrant and budding environmentalist, John Muir, was luxuriating in what he called "that glorious Wisconsin wilderness." Young Muir grew up on the family homestead about forty miles northeast of the Sauk Prairie in the fire-suppressed, and already brush-choked, oak savanna of southern Marquette County. Like nearly all of his contemporaries of the American pioneer generation and most of their children for years to come, Muir believed that the landscape he encountered when he arrived in Wisconsin was a product of the Christian god's will expressed through geology, climate, plants and animals, but absent human intervention. Rare is the perception of Edmund Rendtorff, who first saw the Sauk Prairie in 1841: "It was no unbroken wilderness that met my gaze. The idea struck me that Indians had lived here and that still an ancient people before them occupied this lovely prairie." The majority view was voiced by the author of "The Prairie," a verse selection that appeared in *The Wisconsin Farmer* newspaper in 1862:

"Behold the prairie, broad, and wild, and free
Ocean of emerald grass and golden flowers
Tis God's own garden unprofaned by man."

The people who had for centuries inhabited "God's own garden" were perceived as mere spectators in nature's theater, not as actors in it. Like Adam and Eve in the Garden of Eden, they reaped where they did not sow and randomly plucked the fruits of the earth bestowed upon them through no efforts of their own.[38]

In fact, the landscape Muir and other newcomers encountered in the grassland regions of Wisconsin was very much created and managed by the men and women who lived there. Blueberries gathered in a forest opening were improved with the aid of fire that opened the canopy and applied ashen fertilizer to the roots below. Deer bagged for winter sustenance were fat on aspen browse sprouting on a burned woodland edge. The settlers steered plows to raise corn for cows. The Indians set fires to raise acorns for whitetails. The comparison is apt, but the relationship between prehistoric people, their historical descendants and the land they inhabited was much more complex than that of a 19th-century American farmer and his fields. It is the difference between an aboriginal culture, ancient and familiar with its environment, and an immigrant culture centered on market values, uncomfortable in new surroundings, and compelled to refashion it.

A complete picture of the aboriginal relationship with the North American landscape has yet to be painted. The use of fire as an all-purpose land management tool as described here is a relatively recent interpretation of evidence available to scholars in the last century who overlooked it or drew different conclusions from it. Much of Indian tradition was lost in the upheavals of the fur trade and American settlement periods which brought war, disease and displacement as well as direct assaults on traditional culture from politicians, missioners and teachers. The catastrophic population loss suffered by the Ho Chunk and other native tribes in the wake of contact with Europeans is sufficient to explain the loss of much ancient knowledge, especially subtle and complicated relationships involving place, calendar, plants, animals and weather. When the folk die out, so does folk wisdom. In addition, there is a justifiable reluctance on the part of many Indians to share what they have preserved. Nonetheless, the logic is persuasive. Prairie, savanna and other grasslands required regular burning to survive for the thousands of years they covered central North America, including the Sauk Prairie. Lightning did not strike often enough to keep the trees from overtaking the grass. The people living on the grasslands could improve their lives by maintaining them. They did so by means of fire. The "contiguity of shade" may have been a wilderness untouched by human hand. The Sauk Prairie was not.

The Great Prairie Engine

Although possessed of a single name, the Sauk Prairie encompassed many prairie types: dry prairie with less fertile, sandy, fast-draining soil; wet prairie, where heavier soil held water close to the surface; mesic prairie, which might also be called the calendar photo prairie, because here all the prairie components combined to produce the greatest diversity of prairie plants, the tallest grasses and the showiest flowers. To Wisconsin prairie scientist John Curtis the mesic was a prairie "of unequaled splendor."[39]

For Frederick Herman Lueders, who published the benchmark survey of the plants of the Town of Prairie du Sac in 1895, it was the "undulating prairie." The phrase suits the gently rolling ground and also the predominant vegetation growing there, i.e., grass, that rolled in waves and troughs as the wind pushed through it. Although comparatively few in number of species, grasses made up as much as three-quarters of the prairie ground cover. Of these, none was more prolific than big bluestem. Its roots delved twelve feet into the earth to extract subsoil nutrients, weather droughts, resprout after a fire and nurture stems whose turkey foot seedheads rose eight feet into the sky. Bluestem was the fabled grass of the western adventure yarn — high as a horse's ears, where ranchers' livestock wandered and homesteaders' children strayed, never to return. It was the tall grass of the tallgrass prairie, yet, despite its name, bluestem is not really blue. Predominantly green, the stalk is also tinged with yellow, red and purple, which, as part of a prairie stand undulating in the wind, melds into a hazy, shifting, blue wave. One bluestem is not blue, a bluestem prairie is.[40]

Sharing prairie ground with the bluestem was Indian grass, about as tall, and whose Latin name translates as swaying sorghum. Also present was switchgrass, a member of the millet family with an Asian cousin that was one of the first domesticated crops, but in North America seems to have been reserved not for farmers, but for grassland birds to harvest at summer's end. There was prairie dropseed, whose name describes what happens as soon as its seeds ripen, and needle grass, June grass, sideoats and hairy grama, plus little bluestem, whose three-foot-tall stem is little only in comparison with its big blue cousin.

The prairie carpet was woven of grass, but it was a carpet pierced from April to October by more showy plants. Small shrubs and herbaceous plants — forbs — penetrated the grass cover and added stark, arresting color to the subtleties of the green. As Edmund Rendtorff lyricized, "Look at the prairie from the end to the commencement of winter, full of flowers, changing almost every week, one closing to make room for another — more blossoms than leaves or spears of grass." One after another the blossoms came, hundreds of varieties, the greatest number of which belonged to the composite family: asters, sunflowers, silphiums, daisies. So numerous were they that, as John Curtis stated, "On floristic grounds, the prairie should be called a daisyland instead of a grassland."[41]

Not a daisy, but first to bloom on the grassland was the pasque flower, whose pale lavender and yellow blossoms often opened while snow still hung in the hollows of the Hills. Next came the deep blue prairie violet, the blue and white shooting star and the first golden yellow puccoon, pink phlox and blue lupine. Late spring and early summer brought the yellow cream of the false wild indigo, the pink of the rose, the bright orange of butterfly weed, the white of the anemone and flowering spurge, plus the radiant blue of the spiderwort. By midsummer the lavender and green of the lead plant blossomed, along with the white and purple prairie clovers. With them came the provocatively named rattlesnake master, also called snakeroot, which was white, and two specimens of another "snake root," the magenta and the purple blazing star. In August and September, room was made for more composites with infinitely varying shades of yellow and purple. The goldenrods, of which there were at least four on the Sauk Prairie, and six milkweeds were joined by the coneflowers, compass plants, asters and three sunflowers. The season of flowers ends humbly with the gentian and spectacularly with the bright yellow, flat-topped flowers of the stiff goldenrod.

Just as the flower is only one part of the plant, the prairie flowering season was only one feature of a complex natural system where soil, sun and water interacted with plants, animals, insects and microbial life to create a unique micro-environment. Competition for nutrients, water, light and reproductive advantage was furious. It prompted some plants to grow tall faster than others, capture the sun first and shade out their neighbors, or to send roots deeper into the earth to tap stores of water and minerals beyond the reach of their competitors. Plants with thick, deep taproots, like the silphium, anchored themselves solidly in place and fed new growth when their old stems were burned, frozen or lopped off by grazing animals. Several prairie plants produced toxins that prevented other species from spreading into their space, while others colonized and monopolized a patch of ground, which limited flowering to plants on the edges of the colony so it would continue to expand. Prairie grasses and forbs reproduced vegetatively, by means of lateral shoots rising from their roots, or sexually, via flowers pollinated by wind, insects or birds, or left nothing to chance and did it both ways.

The prairie was a plant-eat-plant world, but it was also a place of cooperative relationships between plants and plants, plants and insects, plants and animals. The pollination of plants by insects feeding on nectar is a well-known example of cooperation; so is the grazing of animals that ate prairie plants, then deposited manure to fertilize new growth, and the prairie birds that consumed seeds and passed them undigested to sprout on new ground. There is also evidence that, as they fed, grazing animals released an enzyme that stimulated the roots of the plants they were consuming. On the other hand, overgrazing as in a farm pasture was harmful and eventually fatal to the prairie. Timeliness also mattered, with grasses as well as flowers. Cool season grasses — needlegrass, prairie dropseed — tolerated the chilly spring soil and matured early before they could be shaded out by taller warm season grasses. In turn, the later, warm season grasses shaded, cooled and delayed the drying of prairie earth during searing summer weather.

Soil microbes fertilized the prairie by breaking down roots and helping leguminous plants extract nitrogen from the air so it could be stored in root nodules and made available to stimulate the growth of other plants. Legumes such as lead plant, tick trefoil, bush and prairie clover accounted for as much as seven percent of the plant volume on mesic prairies and helped build fertility that native plants tapped for centuries and decades of farm plowing could not exhaust. Legumes tended to increase along with fires and were among the first to colonize ground that had been disturbed, as the prairie often was even before the arrival of the Yankee plowmen. Never seamless, the prairie sward was churned up by the occasional deer or elk, by rodents constantly burrowing underground lairs, by predators digging after them and by all of the mammals, from badgers to wolves, when they trampled trails. Ants were the most extensive excavators, turning over so many square yards of new ground so consistently that, in time, they plowed up the entire prairie, altering and renewing as they passed.[42]

The prairie was home to specially adapted populations of mammals and birds. The elk, black bear and wolf were the largest mammals and they were all extirpated from the Sauk Prairie shortly before or after the American settlement. If bison ever inhabited the prairie, as they may have when the Indians nearly died out, they were hunted out as soon as Indian populations recovered. Estimates of pre-Settlement bird species on all varieties of grassland run as high as 300, but many birds ranged over more than one kind of habitat, making strict classification difficult. One reasonable list of summer residents of the Sauk County grasslands includes the swallow-tailed kite, red-tailed hawk, harrier, kestrel, prairie chicken, bobwhite quail, sandhill crane, upland plover, mourning dove, short-eared owl, nighthawk, short-billed marsh wren, upland sandpiper, loggerhead shrike, bluebird, bobolink, dickcissel, eastern and western meadowlark, eastern kingbird and the field, song, grasshopper, vesper and Henslow's sparrow.[43]

They all adapted to the steady winds, the absence of trees and the extensive visibility over many square miles of ground that were intrinsic to the grassland environment. Consequently, birds of prey were more numerous over the grass than in the trees. The prairie skies were dotted with soaring, gliding raptors, eyes fixed on the ground to detect any movement of prey before dropping into a lethal dive. Although raptors tended to hunt silently, the compact, powerful kestrel commonly called as it hovered overhead, one of many grassland birds that were vocal flyers. In the absence of trees, prairie birds like the bobolink, meadowlark and dickcissel sang while in flight, unlike nearly all woodland birds that needed a perch to trill. Also vocal, if not melodious, the kingbird lived up to its Latin name of "tyrannus" by shrieking as it drove intruders away from its nesting ground then swooped down on a seed or berry and carried it off while still in flight. The prairie also resounded to the mating calls and dance of the prairie chickens, which were once as common in the upland prairies of Wisconsin as puddle ducks in the wetlands. Not surprisingly, grassland birds were also more likely to nest on the ground than woodland birds, which made them extremely vulnerable to the habitat loss that accompanied the conversion of the prairie to farmland.

The grassland bird habitat disappeared because grassland was also excellent habitat for farmers. The lavish topsoil of the Sauk Prairie began with sand washed out of the glaciers and dust dropped by the wind, then evolved into a local variant of the most fertile soil type on the planet. Prairie soil was so rich in humus that it could have been bagged and sold for potting soil, so teeming with microbes it was as much animate as not, so aerated and uncompacted that a Yankee pioneer jumping out of a democrat wagon parked on the prairie sward could set ripples radiating through the turf as if he had stepped off a pier into a pond. All wealth begins with the soil, as the prairie populists were fond of proclaiming a century ago. They were talking about American farmland, but the statement works for unplowed prairie as well. Prairie soil was a growing bank of natural wealth where for centuries microbes, plants, insects, animals and people deposited more resources than they withdrew. The prairie of 1830 was richer than the prairie excavated by the mound builders of 830 A.D., which was itself richer than the prairie traversed by the rockshelter dwellers of 830 B.C.

A journalist of the 1830s looking for a cutting-edge metaphor better suited to his mechanical age than a bank might compare the prairie to a steam engine, fuelled by sunlight and rain, maintained by fire, spinning out a heavy yield of bacteria, bluestem, box turtles, meadowlarks, elk and, although he probably would not have mentioned them, Ho Chunks. Continuing down that metaphorical track, he would also say that in the 1830s, the great prairie engine, after running for 6,000 years, began to sputter and die.

II

Settlement

What a change time has made here!

Edmund Rendtorff, 1861[1]

Prairie Breakers, Rail Makers

William "Uncle Billy" Johnson was born in Troy, New York, in the inaugural year of George Washington's presidency, 1789. Part of the first generation born to citizenship in the United States, and stirred by the same itch to get up and go that afflicted so many of them, he moved west with the frontier: to Pennsylvania as a baby in the 1790s, to Ohio as a young man in 1810, to Wisconsin at middle age in 1836. He paused at Belmont in Iowa County long enough to help build the meeting house that served as the first capitol of Wisconsin Territory, but kept his eye on the frontier and his ears open to news of it. When word reached Belmont that the United States had signed a treaty with the Ho Chunk whereby the Indians relinquished title to the lands north and west of the Wisconsin River, the frontiersman was ready.

In December 1838, Johnson, his son Roswell and at least one compatriot bucked a team of oxen up the trail from Belmont and attempted to drive the beasts across the frozen Wisconsin at Sauk City. The ice was too soft to bear the weight of the cattle but the two Johnsons, leaving their oxen and companion behind, made their way across. They hiked up the Honey Creek Valley for a few miles, ascended the bluffs, then struck back east. When they reached a point where the bluffs, the prairie and Otter Creek met, Uncle Billy decided he'd found his new home. It has been said that he claimed about half the Sauk Prairie and maybe he did, but his purchase was recorded as 640 acres in Section 15, Town of Sumpter. Some of this land later became the Staff Village of the Badger Ordnance Works, while the original Johnson homestead, which Uncle Billy lived on until he died just shy of his 101st birthday in 1890, became the country home of Governor Emmanuel Phillip and, later, the Valley of Our Lady Monastery.

Uncle Billy knew it was important to be first on the frontier—first to scope out a parcel of land, first to stake out its boundaries and first to protect it from claim jumpers. He could not purchase his claim until after it had been surveyed and listed for sale at the federal land office in Mineral Point, which was still a few years away. In the meantime he had to wait somewhere for the government work to be done, so why not on the actual land itself? That was why he crossed the thin river ice at the start of the winter of 1838-39: to gauge the lay of the land when it was not obscured by high grass or trees in full leaf and to start logging the timber he knew he would need for fuel, fencing and the tiny log cabin that became the first Johnson family home. On all counts he was aided by the Indians, who tolerated his presence and who had facilitated land-looking by firing the bluffs, and perhaps the Prairie, only a few months previous.

The land Johnson selected for what would be the first farm on the Sauk Prairie was about as close to perfect as the site of a frontier homestead could be. In Otter Creek, it had clear, running water for the Johnson family and their livestock to drink until they could dig a bucket well to groundwater. On the bluffs, it had workable sandstone to quarry into building blocks and limestone to burn for the mortar to hold them together, plus woodland for pasture and free-for-the-labor lumber. On the Prairie, Johnson had pasture for his oxen richer than he could raise from seed and fine, ready-to-plow soil that could produce a marketable crop promising a cash return upon first harvest. So Uncle Billy came back in the spring of 1839, with his wife, Mahalia, their five sons and three daughters. Along with the usual dunnage of a frontier family on the move, they brought a hand-powered grain mill that would make them very popular with their Prairie neighbors, once they had some, a set of blacksmith tools, five yoke of oxen and "a large breaking plow."[2]

Uncle Billy left an entertaining account of his attempt to swim ten reluctant oxen across the spring-freshened Wisconsin, with two teenaged boys riding in a leaky canoe as hapless drovers. The animals went into the water opposite Prairie du Sac and eight of them swam dutifully across the river. The final pair was water shy and as soon as their snouts were dampened, tried to board the canoe. Terrified of drowning, the cowboys beat the spooked animals off, all but destroying their paddles and placing themselves, if not up the proverbial creek, then in the middle of the Wisconsin absent the means to reach shore. The oxen beached themselves on a sandbar and refused to move until Uncle Billy coaxed them across, but not until the current had carried them all past Sauk City. One beast was so traumatized by the experience that for the rest of its life the mere scent of the river would send it galloping, as much as an ox could gallop, for the hills. With all the excitement over the oxen's crossing of the Wisconsin, Uncle Billy neglected to relate how he completed the equally impressive feat of ferrying the "large breaking plow" across the river in a canoe. Like many events future observers would find suffused with historical significance, bringing the plow across the river was all in a day's work for the participants involved.

An 1830s breaking plow of the type Uncle Billy did not describe was a formidable, if primitive, piece of equipment. It had a heavy wooden beam, about fourteen feet long, attached to a pair of stout hardwood handles at the stern and two sturdy wooden cart wheels linked to a tongue for hitching up the oxen in front. It was assembled by a village blacksmith, or by a farmer like Uncle Billy, who had his own tools and was a jack of every frontier trade. The working parts where the cutting edge met the sod—plowshare, moldboard, landside—were made of wooden boards sheathed over with odd bits of flat iron pounded into as smooth a surface as the smith could bang out on his anvil. When gigged up and ready to plow, it could roll the prairie sod into a shallow furrow about eighteen to twenty-four inches wide. Johnson would have seen his father steer such a plow in Pennsylvania and he himself might have had one when he started farming on his own in Ohio. By the late 1820s and 1830s, especially if he were a man open to new ideas, Johnson would have progressed to a manufactured plow with a smooth cast iron share, moldboard and landside, and perhaps equipped with a knife-edged coulter mounted in front of the share to slice the sod before it was turned. To be sure, Uncle Billy's plow was not one of the razor-edged, self-scouring, steel, prairie killers manufactured by John Deere and a few other midwestern inventors, because these plows were not developed until the 1840s.[3]

No, Uncle Billy brought an old-fashioned plow across the Wisconsin, the kind of implement Thomas Jefferson would have recognized as similar to his own design and George Washington would have seen his slaves steer at Mt. Vernon. Antiquated it may have been, but Uncle Billy's plow was a powerful engine of change. With it he inaugurated an abrupt finale to 6,000 years of grassland evolution on the Sauk Prairie. Where humans once were participants in and stewards of the prairie/savanna ecosystem, they now had no more than temporary use for it; knew only that if they were to survive, it must be destroyed, so they did what had to be done.

Breaking was rugged work and depictions of prairie plowboys as hefty he-men were not far from the mark. Depending on the quality of the plow, the number of oxen under yoke (ten was the recommended minimum), the density and moisture content of the soil, the number of rocks therein, and the thickness of shrub and tree roots in the grass, a team of pioneer plowmen (one to steer the plow, one or two to guide the oxen) could break one to two acres a day. Uncle Billy and his boys, running a primitive plow through thick sod on loamy soil, plowed no more than sixty acres for themselves in 1839, "and as much more for their neighbors." Even if they started at sunup, the plowmen needed time to yoke the oxen, time to walk the ponderous beasts to the prospective field at the stately pace of two miles per hour and more time to survey the ground and decide how best to slice open the sod. The common practice was to work from the edge of the field into the middle, turning the oxen to the "haw" or left side while the sod rolled over to the "gee" or right. Since turning corners with a thirty-yard-long string of oxen and plow was difficult and troublesome, plowmen preferred to run straight furrows as far as the cattle could walk.

Since the sharper the plow, the easier the plowing, and since iron plows did not hold an edge very long, the breaking team was forced to stop often to sharpen the blade by tapering it with precise blows from a hammer. At midday, the oxen, if not the men, had to be fed and watered, then rested for about two hours "until they started to chew their cud." Then work could resume and continue until nightfall. The pace of the breaking slowed if the prairie sod was unusually thick, the soil sticky clay or if the field was salted with rocks or the tree roots known as "grubs." The Sauk Prairie itself was clear of rocks but they turned up to batter plowshares and slow the pace of settlement on the moraine in Merrimac. Grubs were also rare on the Prairie, "where scarcely a single root obstructed the plow on sixty and more square miles," but they waited with the rocks just beneath the surface of the hills and vales east of the moraine. Not even in Merrimac were the grubs as bothersome as those John Muir encountered on his farm, where the sod was "reinforced by the tap roots of oak and hickory," four inches thick and 100 years old.[4]

Improved equipment made it easier to master the grubs. John Muir reported that a newly sharpened steel plow would cut hickory and oak grubs as if they were "carrots and turnips." Oswald Ragatz, who migrated to Honey Creek with his family in 1843, teamed up with his brother Henry and became the "Ragatz breaking outfit." In 1844, they "had six pair [of oxen] and began breaking the prairie in earnest with a big thirty-inch plow which turned two feet of sod at a time." A plow of such size, powered by twelve stout oxen, could break three acres a day. The Ragatz brothers were soon hiring out at the rate of $2.25-$2.75 per acre, plus room and board, and "gained quite a reputation among newcomers." Uncle Billy Johnson was a self-sufficient pioneer, but few of his neighbors had the means or desire to purchase specialized breaking equipment or keep the small herd of oxen required to power it. Instead, they hired professional breakers like the Ragatz brothers. "Turning virgin prairie is back-breaking labor for man and beast," wrote Oswald, "and each night found us utterly fatigued. But we were exultant...we were converting sheer wilderness into a smiling habitable land."[5]

Befitting heroic effort bringing about change on a grand scale, pioneer lore is well-laden with prodigious feats of plowing. Uncle Billy Johnson is reputed to have broken, in addition to his sixty-acre field, a furrow around his entire original claim, into which he welcomed newcomers he favored and from which he ostracized those he did not. He is also said to have plowed a three-to-four-mile furrow straight east on the section line from his farm to the river above Prairie du Sac, so he could find his way home across the trackless Prairie in the dark of night. James Isaac Waterbury, who settled just south of the Johnsons in 1840, grabbed the handles of his prairie breaker and declared he would claim all the land inside the furrow marked by one day's plowing. His oxen gave out before he finished, but Waterbury claimed all the ground within a three-and-one-half-mile circumference anyway.

Fatigue was not the only hazard the prairie breakers faced. "This being virgin land we disturbed innumerable moles, field mice and snakes," wrote Oswald Ragatz. "The latter did not like to cross plowed land so always retreated into the dwindling grassy center piece.... They gave both oxen and us much trouble. It was only after I came out the victor in a fifteen-minute contest with an enormous blue racer, in which I had nothing but a three-foot stick to defend myself with that I overcame my fear of them. Thereafter snake killing was a great sport for me." Greater than the sport was the destruction of habitat. Ragatz killed more snakes and other grassland species with an hour of plowing than he could in a week of beatings with a three-foot stick.[6]

Once the prairie sod was broken and turned, the vegetation was left to die, with roots exposed to the sun and tops to wither beneath. Enterprising settlers, like Uncle Billy Johnson, dropped seed potatoes into the decaying sod and made their first Prairie harvest in the fall. The following spring, the field was cross-plowed to further break up the sod and loosen the soil. Only then was it planted to a field crop. Like most Wisconsin farmers in the years prior to the Civil War, Uncle Billy and his family planted the staff of life on the midwestern frontier, summer wheat. Since it could feed man, woman, child or beast, and put cash in the family purse, wheat was the ideal pioneer crop.

In the late summer of 1840, the Johnsons mowed their wheat and found that "their harvest was never better." The stalks were stacked into shocks to dry in the field. When they were ready, bundles from the shocks were carefully forked onto a clear patch of ground so the seed heads could be threshed off the stalks. With no mechanical thresher on hand, the pioneers let their oxen separate the grain from the stalks by trodding on them. The stalks were then raked away and the chaff winnowed out of the grain by tossing it in the air. Wheat not destined for the home farm was bagged and freighted by wagon to flour mills as far away as Milwaukee for a cash return. As the lumber industry developed up the Wisconsin River valley, it created a regular market for wheat and also for oats, potatoes, hogs, cattle, oxen and horses raised on the Sauk Prairie that lasted until well into the 1880s.[7]

The staff of life and the font of progress, wheat was also the founder of villages. In Sauk County, Baraboo, Reedsburg, Rock Springs, Loganville, Leland, Blackhawk and Plain all owe their birth to the presence of waterpower turning grindstones to mill wheat. Closer to the Sauk Prairie was the dam and mill Charles Bryant financed and Henry Staines constructed on Honey Creek early in the 1840s. Erasing a "few of the largest mounds there," including one carefully made of gray clay to contrast with the sandy soil beneath, and quarrying limestone from the blufftop, Staines erected a mill house, fifty by thirty-two feet with "four run of stone." Better known as Loddes mill, it served farmers in the Towns of Prairie du Sac, Honey Creek and Sumpter for over a century. By 1859, when wheat ground at the Honey Creek mill took first prize at the Wisconsin State Fair, the Sauk Prairie had been converted from native grassland to white flour breadbasket.[8]

Since a winter had to pass between prairie breaking and planting, the settler had time to perform a task as essential to farming as plowing. With farms surrounded by acres of grazable prairie, savanna and forest pasture, farm animals were left to roam at will. The free range enabled farmers to fatten livestock virtually cost-free, but the stock couldn't always distinguish between the unplowed bluestem it was encouraged to eat and the new and precious stalks of wheat it was not supposed to nibble. Fences were necessary to keep livestock out of the new fields and the fencing material nearest to the Prairie came in the form of wooden rails split out of the hardwood trees of the savanna and the bluffs. "It takes a good wagon load of heavy oak rails cut and split ten feet long to make a fence of 16 or 18 yards, but when done it is a perfect wall of wood seven feet high," wrote Mazomanie settler John Greening in 1847. He spent his winters logging the hills adjacent to his Dane County farm just as his neighbors on the Sauk Prairie cleared the bluffs encircling them. The "south shed of the Baraboo Bluffs," wrote William Canfield in 1861, "... bore a growth of white, red and black oak, some hickories, also a few burr oak and maple....Most of the oak was 'made' into rails by the farmers on the prairie before the ceremony of purchasing from the United States Government was performed."[9]

James Waterbury helped himself to enough of Uncle Sam's timber to fence off the entire three-and-one-half-mile perimeter of his claim, plus another mile where the road to Sauk City crossed the property. Using John Greening's estimate, it took about 100 wagon loads of rails to lay a mile of fencing, and about 450 wagon loads to enclose the Waterbury spread. Fencing, in addition to construction and heating, required an immense number of trees, and trees immense, in both size and number, were available. In 1889, for example, the Baraboo newspaper reported the cutting of a 275-year-old white oak on the J.W. Wood farm along Skillet Creek. It was news because trees of this age and size were already scarce in the Baraboo Hills. The tree measured "sixteen feet six inches in girth at the butt end, and over twelve feet at its smallest girth, just below the main fork," forty-seven feet above the butt. It was labor for a team of heavyweights to fell such an old-growth hardwood with axe and crosscut saw, slice the trunk and heavier branches into ten-foot logs, split them into rails with axe, maul and hickory wedges, often cracking their way around or, maybe, through knots as thick as a man's torso. Then all the loggers had to do was muscle the green rails up into the wagon bed or sled. The only easy part, if easy it was on a frigid winter day, was the ride back to the farm as fast as the oxen could trot, slip or slide downhill. In the years between 1840 and 1860, the old-growth hardwoods of the bluff forests and the savanna were converted to hundreds of miles of rail fences. Had a Civil War balloonist floated over it, the view of the Sauk Prairie would have resembled a checkerboard with the squares separated by zigzag lines of rail fences. The logging of the bluffs, much of which returned to forest, and the clear-cutting of the savanna woods, which did not, were alterations of the environment as profound as the plowing of the Prairie.[10]

As George Knapp, who rebuilt the fences on the Waterbury farm in the 1880s recalled, raising a rail fence was a "...humdrum task, which involved keeping the fence exactly on the line, making the angles proportional and placing the rails right end to and right side up, to make the finished fence look symmetrical. Criticism taught you to do it right, if pride in your work failed to do so." Prairie farmers continued to exercise their pride making rail fences for near a half-century. The original barriers James Waterbury and other pioneers assembled from oaks centuries old had rotted to duff by the time George Knapp came along thirty years later, but Knapp, whose family took over the Waterbury place in 1873, found new fencing material "east of [Otter Creek], on the slope rising to the prairie level." There stood oak trees, "largely six to eight inches in diameter, with straight trunks and straight-grained and because they were crowded, few limbs. They split easily and made excellent rails." [11]

Those easy-to-split, straight trunks were signs of another alteration the American settlers wrought on the grassland environment. With the Indians and their fires in exile, any patch of dry land not plowed for crops turned to woodland. As Aldo Leopold stated, "go count the rings in any set of stumps on any 'ridge' woodlot in southwest Wisconsin. All the trees except the oldest veterans date back to the 1850s..." These second-growth forests were not the same as the park-like woodlands the pioneers had logged off. As Knapp said, "they were crowded" and brushy, with much less sun reaching the ground and thinner grass than in the original savanna. The ground cover beneath the trees had also changed, since it had been worked over by farm animals, principally by the most energetic excavator of all, the frontier hog.[12]

As much feral as tamed, the pioneer hog was expected to fend for itself and whole herds ranged the Baraboo Hills, where they were as conscientious at rooting out the native vegetation as the breaker plow had been at eradicating the prairie sod. So ubiquitous were free-range hogs that registering the ear notch patterns that identified their owners was one of the first responsibilities of town government. At harvest time, neighbors organized drives to move the hogs out of the woods for slaughter that, had the settlers fired the woods first, would have resembled the deer drives of the Indians. Inevitably, some hogs went truly feral and had to be hunted out of the woods with dogs and rifles. The hogs' only enemies, other than the settlers, were the black bears and gray wolves that also ranged the hills. H.E. Cole tells the story of the hog that upset the travelers spending the night at the inn Prescott Brigham built on the East Sauk Road near Kern's Corners. Just as the guests were settling in for the night they heard an animal pounding over the ground outside with a pack of wolves snapping at its hindquarters. Soon a hog—panting and panic-stricken—burst through the inn door and parked beneath the table, apparently preferring to trust its fate to an inn full of hungry travelers than to a pack of hungrier wolves outside.[13]

With the prairie broken and converted to fields, with the savanna logged, plowed and planted, with the bluffs cleared of old-growth trees, with any untended patch of ground nosed over by livestock and turning to brushy woods, with native plants and animals replaced by introduced alien species, the original Sauk Prairie was as drastically reconstructed by the American settlers as the farmland appropriated for Badger Ordnance was reworked by the Army Corps of Engineers. The settlers needed about twenty years more than the Army personnel, who completed their work in about four years, but the result was similar. Little wonder that Edmund Rendtorff, who first came to the Prairie in 1840, declared in 1861, "What a change time has made here!"[14]

York Staters, Germans and Swiss

When federal surveyors started to stretch their measuring chains across the Sauk Prairie in 1842, they found about twenty families spread out along the arc of Otter Creek from the Johnson and Waterbury spreads in Section 15, north into Jameson's Hollow, where Highway 12 was later to run, and continuing east along the south face of the Hills into Merrimac. The collection of farms that came together on the south face was known as the Bluff or Bluffs and was the first recognizable community in the county not on the Wisconsin River. It began in December 1838, when Albert Jameson, Andrew Hodgett, Alexander Bills and Nelson Lathrop crossed the same Wisconsin River ice as had William and Roswell Johnson, then made their way north over the Prairie. At what was later called Pine Hollow they found an abundant supply of green boughs for bedding and camped under the stars on a frigid night bereft of woolen blankets but rich in youthful pioneer optimism. The next morning they rose to a splendid view of a Prairie sunrise and set about staking claims on new ground. Of the four, only Jameson and Hodgett returned to settle, Jameson in Section 34 of Sumpter, at what became the northwest corner of Badger Ordnance, Hodgett to land farther west. They were soon joined by Charles and Charlotte Teel, who built a farm on the tip of the ridge east of Jameson's claim; John Hoover, who claimed the entrance to Pine Hollow; and Hannah Brigham with her redoubtable husband, Prescott.

In 1840, at sixty years of age, Prescott Brigham migrated from Massachusetts via Blue Mounds and turned the Indian trail from Sauk City to Baraboo into a pioneer highway that carried people, freight and mail from the settled sections of southern Wisconsin to the frontier north of Baraboo. Brigham's highway was originally known as the "East Sauk Road," although it was sometimes called the "Middle Sauk Road," to distinguish it from another road that climbed the Hills east of Devils Lake.The confusion subsided somewhat after 1918 when Wisconsin numbered its state routes and lettered county trunks.The West Sauk Road became Highway 12 and the farthest "East Sauk Road" became Highway 113.The other "East" or "Middle Sauk Road" was designated County Trunk Z south of Kern's Corners and was lost inside the fence of the Badger Ordnance Works. Much of the right of way outside the fence was also abandoned, except that portion which World War II veterans named the "Burma Road."

A century earlier, Prescott Brigham established an inn on the "East Sauk Road" at a point where it started to climb the Hills east of Pine Hollow. It became the nucleus of the Bluff settlement and an important stopping point for travelers. Since the stage was carrying the United States mail up the road to Baraboo, it was logical for a post office to open at the Bluffs. Lewis Smith contracted to perform the service out of his nearby home in 1850.As a stagecoach driver, Brigham was well known in the county and a respected political elder who filed the plat for the courthouse square in downtown Baraboo, then persuaded eleven county commissioners to place the county seat there. By 1853, Sauk County must have become too tame for the seventy-three-year-old teamster so he joined Charles Teel and fellow Bluff settler Elhanon Shaw on an expedition in search of fame and fortune in California. Traveling by oxcart, they took six months to reach the Golden State and stayed just about long enough to turn around and come home, penniless, in 1855. Brigham survived the California trip to die peacefully in his rocking chair at his Bluff home after a final ride to Prairie du Sac in 1862. He lies buried in the Pioneer Cemetery on land taken for Badger Ordnance.

Another man who spent a lot of time traveling arrived at the Bluffs a year after Brigham. Reverend Thomas Fullerton, a young circuit rider following the Wesleyan tradition of carrying the Bible's message in the saddlebags of a horse bound for the frontier, began to preach regularly at the Bluffs in 1841. He delivered his first sermon at the Teel place on a midsummer morning, choosing St. Paul's text on a Christian love that is,"patient and kind...not jealous or boastful...not arrogant or rude" to inspire his tiny congregation. The Methodists continued to meet at the Teel house until 1862, when they built a church at King's Corners. It was known as the "Bluff church" until 1876 when the building was moved and installed on a new foundation at the southwest corner of Section 11, Sumpter, just inside the line of the property taken for Badger Ordnance. For a few years, the Methodists shared the church building with the local Baptist congregation. When the number of Baptists fell too low to support their own ministry, they merged with the Methodists, who were relatively few themselves. From 1876, until its demise in 1942, the Sumpter Methodist Church did not have a resident pastor but shared its minister with congregations in Baraboo, Merrimac or Prairie du Sac.

By the 1860s, the center of the Bluffs community had moved onto the Prairie to the crossroads at Kern's Corners, named for Charles and Charlotte Kern, who came to the Prairie in the latter half of the 1840s. The Bluff post office shifted among several locations along the road until it landed in the store opened by Luther Daniels in 1866. Three years later it moved to King's Corners. Also part of the Bluffs/Kern's Corners community, School District No. 1, Town of Sumpter, was already operating in 1863 when the voters agreed to build a frame schoolhouse on the southwest corner. Built with twelve-inch-wide pine lumber that may have come off a raft floated down the Wisconsin, set on a footing of stone quarried from the Hills nearby and eventually topped with a tall bell tower, the 1863 Kern's Corners school cost $1,300 to build. It remained in use until it was taken by the federal government, sold, dismantled and moved in 1942.

The early settlers of the Sauk Prairie, whether in the Bluff community or farther down Otter Creek, followed the typical frontier practice of settling first on the edge of a prairie. It was a practical decision but also a choice based on timidity. Easterners accustomed to the sheltering shadows of the woods were reluctant to expose themselves in the open prairie. These fears turned into preposterous panic during an "Indian Scare," as in 1845 when the Bluff settlers, fearing an Indian "attack," fled en masse to Prairie du Sac, organized an armed force led by Agoston Haraszthy and hunkered down to wait for an assault that never came. The inevitable exception to the rule of settling first on the edge of the prairie was bachelor farmer Isaac Gibbs. In 1839, he claimed 140 acres of land in Section 2, Sumpter, a good quarter mile away from the creek and a mile south of the Hills. He was there when the surveyors ran the metes and bounds and when Prescott Brigham brought his first stagecoach up the East Sauk Road, which ran along the Gibbs property line. The land he chose was fertile, flat and Gibbs did well in the fifty-five years he spent living on it. When the Kern's Corners school board called a meeting of the taxpayers to discuss building a new school in 1863, the trustees asked Gibbs if he would support it, since he was the "biggest" taxpayer in the district. He announced that he would vote "against anything short of a good one." Although he had no children of his own, Gibbs joined the building committee and helped plan, as well as pay for, a school that was "good" enough to remain in use for seventy-nine years.[15]

Like Isaac Gibbs, who hailed from Massachusetts, the majority of the original settlers on the Sauk Prairie were of Anglo-Saxon lineage, Protestant religion and from the New England, Middle Atlantic and eastern Great Lakes states, with New York the leading state of origin. They were soon joined by members of the first wave of German immigrants who would become the dominant European ethnic group of Sauk County. The 1860 census listed 959 residents in Kingston/Sumpter, with 394 from the eastern United States, 359 youngsters born in Wisconsin and 209 Europeans, 178 of whom were Germans who came from a score or more of the states in that disunited nation and the German-speaking cantons of Switzerland. Merrimac's population of 815 was similarly proportioned, with 354 from the eastern states and Canada, 240 Wisconsin-born and 221 from Europe, with 118 of them from Germany and Switzerland.

A general observation on the settlement of Sauk County, supported by a good deal of evidence, holds that the Yankees claimed the most fertile, easy-to-work land and left the rocky moraines, wooded hills and flooded lowlands to the late-coming Germans. After they skimmed off the initial fertility of the Prairie soil by repeatedly planting wheat or went bust when the hops bubble popped in 1869, the Yankees sold out and the Germans moved down from the hills. Ethnic stereotypes come into play here. Yankees were supposed to be financially shrewd, more willing to pull up stakes when the winds of chance blew fair or foul and less interested in hard manual labor than the stolid, thrifty sons and daughters of Germany, who accepted the drudgery of clearing stumps and rocks, saved their pennies to acquire the best farmland they could and diligently took up the daily grind that created America's Dairyland.

The pattern of settlement on the 10,500 acres taken for Badger Ordnance confirms the general observation, if not the stereotypes. On the plat maps of the 1850s, '60s and '70s, German names are hard to find on Prairie parcels, but they fill the squares on the moraine. Stephan Hofsteter bought forty acres on "Yankee Street" (Highway 12) in Section 10 of Sumpter in 1855 and two members of the Roicks family had forties on the East Sauk Road in Section 13, but the rest of the Prairie belongs to folks with names like Mather, Willis, Carpenter, Baxter and Brown. The Eugene and Maria Keller family are a special case that illustrates the limitations of historical generalizations. Eugene's grandfather Caspar left Germany for New York state in the 1700s and his children and grandchildren were born and raised in the Empire State. Although aware of their Germanic heritage and future members of the Sumpter German Lutheran church, Eugene and Maria Keller were about as American as second-generation York Staters could be when they moved to Section 23 of Sumpter in 1857.

By then the moraine in east Sumpter and Merrimac was thick with

Germans, nearly all of whom arrived in the late 1840s and early 1850s. In Sumpter, Section 1, by 1851, could be found Henry and Fredericka Steidtmann Gattwinkel, Carl Graf, John and Michael Weirich, with brother Henry Weirich across the road north in Section 36. In Section 12 were William and Christina Graf Siebecker, and John Frederick Steuber, while Section 13 was home to John and Mary Thilke, plus Johann Gottfried and Wilhelmina Steidtmann Schlag, who claimed their farms in 1847, followed shortly after by Wilhelmina's brothers John August and Christian. Section 13 was also the home of Jacob and Cecelia Burgy, whose daughter Anna married the farm boy next door, Henry Hill, even though he was a Yankee. It was one of the first Yankee-German marriages in Sumpter and testimony to the rapid rate of assimilation typical of German immigrants. Farther east in Merrimac, Section 31, lived Andreas Gross; in Section 6, John Gaber; in Section 7, Herman Kurtz and Charles and Rosalie Naffz; in Section 8, Adam and Barbara Eschenbach; in Section 18, Traugott Burckhardt and George Frese. These families, and others that followed, were part of a distinctive, yet internally diverse, German community that reflected the divisions in its homeland. For example, the Germans were more religiously divided than the Yankees, bringing their Evangelical and Lutheran creeds to Sumpter and Merrimac, as well as the humanist principles of the *Freie Gemeinde* or Free Association.

With its roots in German Methodism, the Sumpter Evangelical congregation was organized by John Thilke in 1860. The church building was erected and the cemetery established on land Thilke donated in 1863. In 1871, the Thilkes sold their farm to John's cousin, John Henry Thoelke, and his wife, Maria, who then found their name informally transferred to both the church and the cemetery. Sumpter Lutheran was organized by William Siebecker and George Frese in 1862. As the German population spread over the Prairie, the congregation grew. In 1877, for example, Carl Shimniok came out of the Hills in the Town of Freedom and, following the standard pattern of ethnic succession, purchased the William Sprague farm on the East Sauk Road, immediately south of the Kern's Corners school. Although they were a good three miles away, the Shimnioks, along with many other German Lutherans from the northern part of the Prairie, remained faithful members of Sumpter Lutheran until its demise. A victim of paved roads and family autos, the little country congregation lost it members to village churches in the 1930s. The building was no longer in use when it was taken by the federal government in 1942, sold and dismantled.

Sumpter Lutheran did not have its own cemetery, but made use of the Kingston Cemetery about one and one-quarter miles away to the east and north. Kingston cemetery took its name from one of the original towns of Sauk County, from which Merrimac was split in 1855. The remaining portion of Kingston kept the name until the onset of the Civil War prompted its citizens, who were better patriots than spellers, to rename it in honor of Fort

Sumter, at Charleston, South Carolina, where the first shots of the sad struggle were fired. The cemetery, which kept the name of Kingston even though it was in the Town of Merrimac, was part of a bequest from Charles and Rosalie Naffz, who migrated from Germany in 1849. Naffz was a genuine '48er, who participated in the liberal democratic movement in Germany that was crushed by the armies of the autocrats in the revolutionary year of 1848. He was also an entrepreneur who, shortly after he arrived in Sauk County, purchased — at a discount — warrants for federal land granted to veterans of the Mexican War. He traded the warrants for two 640-acre sections of land and sold them at a markup to other Germans, some of whom were also political exiles. Among the exiles were Naffz's sister Emilie and her husband, Carl Duerr, who was known as the "Latin farmer" because of his habit of reading works of classical literature while attempting to steer a plow. More philosopher than farmer, Duerr was a member of the German Free Association, also known as the "Freethinkers," who held that men and women should revere humanist values of reason, tolerance and lifelong learning. A Freethinker himself, Naffz built a hall across the road from the Kingston Cemetery where the Merrimac Free Association held meetings and musical programs. In the 1860s, Naffz and other members of the Merrimac group joined the Sauk City Association which Duerr had founded a decade earlier and remains active to this day. The Merrimac hall was sold to a local German singing society in 1878, became home to the Sumpter-Merrimac Band in the late 1880s, and was a venue for local musicians and dances into the 1900s. It succumbed to disuse and decay and was demolished in 1924.

While their churches, organizations, music and family links reminded the German farmers of their European roots, the circumstances of their new lives, as well as their own inclinations, encouraged them to adopt a new American identity. No institution was a greater force for assimilation than the two public schools within the German community. They were known as Joint District #4, Sumpter and Merrimac, and Joint District #5, Sumpter, until 1918, when the state legislature decreed that all country schools should have names. Then #4 became the Lakeside School, and #5, whose board also wanted "Lakeside," but was too slow on the draw, settled for Lakeview. As to the whereabouts of the lake, the Prairie du Sac dam had backed up the Wisconsin River into a flowage christened "Lake Wisconsin" only five years before the school-naming ordinance passed and lake consciousness was still high along its newly flooded shores. The Lakeside school was only about one mile away from the shore of the flowage and Lakeview about one-quarter mile away from Gruber's Bay, which was close enough for the school boards. Whether named, numbered or identified by the farm next door, the rural school was the place where the children of immigrants learned the English language, as well as the culture, customs and traditions of their new homeland. Like other country children in the northern and western United States, where rural schools were established in every town, they were not only literate, but among the best-educated rural people in the world; and their education enabled them to be not just better Americans, but better human beings. At school these educated, immigrant children learned to deal with other young people who were not German, among whom they found friends, enemies and future spouses as well. Tellingly, in 1906, the parents of children at Lakeview, nearly all of whom were first- or second-generation Americans with no more than a Sunday service knowledge of German themselves, voted to begin German language classes at the school. The following year, they voted German out and did not raise the question again.

The Germans assimilated into Americans as the century progressed and they continued to purchase Prairie ground from Yankee settlers who were content to retire to nearby villages. "Prairie du Sac as well as Baraboo, Lodi and a thousand other places are full of these superannuated hayseeds..." wrote Charles Abbott in 1898. He was a superannuated New Hampshire hayseed himself, who moved to Sumpter in 1854 and watched many of his contemporaries cash out their farms in the 1890s. The going price was about sixty dollars per "improved" acre, quite a bit more than the $1.25 per acre that Uncle Billy Johnson and other firsters paid the federal government for virgin ground a half-century earlier. The higher price must still have been right since names familiar on the moraine — Weirich, Waffenschmidt, Gattwinkel, Graf, Roick, Thoelke, Stedtmann, Schlag, Steuber — turn up with increasing frequency on Prairie plat maps in the 1890s. After the turn of the century, they are joined by newcomers with the surnames of Huber, Leutscher, Pretsch, Ehret, Ballweg, Gasser and Machler. Although they were in the majority, not all the moraine-to-Prairie migrants were German-Americans. Charles and Elizabeth Premo, whose heritage was French by way of New York, moved down the hill from a farm near Parfrey's Glen to land in Section 11, Sumpter, in 1878. Likewise, not all the newcomers came from the moraine. John Peter Kindschi moved to Section 23 of Sumpter in 1883, after leaving a blufftop farm in Section 20 that his family had worked since the 1840s. Kindschi was a German Swiss who purchased 160 acres on "Yankee Street" that had previously belonged to John Dennett, a York Stater who had lost his farm in the hops bust. When he bought the place, the hops barn was still standing, but Kindschi used it to stable his horses and concentrated on dairy farming. By 1942, the Kindschi family owned at least four parcels of land that were taken for Badger Ordnance, including the site of Badger Village.

In 1869, farmers in the Towns of Sumpter and Merrimac raised over 75,000 bushels of wheat. It was the last hurrah for the crop that had nurtured the pioneers. The line on the graph of wheat production would run steadily, sharply downward and fall below 6,000 bushels by 1900. A few oxen were still working in 1869, but barely enough to muster a breaking team of the size the Ragatz brothers proudly steered through the sod in the 1840s; nor were oxen needed, with the Prairie in fields and the woodland in pasture. Instead, the farmers of Sumpter and Merrimac tended more than 600 "milch cows" that produced about 46,000 pounds of the butter that was replacing wheat as their cash crop. By 1900, they would be tending 1,400 milk cows and expanding their herds. They were already raising considerably more corn and oats than wheat in 1869 and would steadily increase production of these crops as the years went by. The corn fed the cattle, as well as the hogs that were no longer allowed to range free, and the oats fed the 700 horses that worked the fields the oxen had opened. The last report on Uncle Billy Johnson's breaking plow dates from 1880 and says that "now it stands at the door of a tumble-down log cabin, a dilapidated memento of early days." In Sauk County the pioneer era had ended, the frontier had closed.[16]

"Every Neighbor Was My Co-Worker"

To be successful, a dairy farmer had to submit to the discipline of daily milking and take pains to preserve the purity of the milk after it left the cow. Since it spoiled easily and mechanical refrigeration was still years away from the farm, raw milk had to be processed into butter or cheese. At first, farmers made their own cheese and butter, which they sold to village storekeepers who then resold it locally or shipped it out of town. Home cheese or butter making was usually considered to be a female chore and many farm women were very good at it. Others produced inferior cheese and the low-grade butter scorned by eastern merchants as "western axle grease," because it was better applied to wheel hubs than dinner biscuits. Improvement required professionalization. In 1878, Harold Hill and cheese maker Peter Young started the Sumpter Cheese Factory on the Young farm which straddled the zigzag on old County Z at the corners of Sections 14, 13, 23 and 24. The factory reportedly produced 18,000 pounds of cheese in 1880, a substantial volume for an upstart, then went out of business after only a few years. Young then moved the operation to the cheese-making heartland of Richland County. The experience of cheese maker Addison Brown was similar. In the 1880s, he ran a cheese factory just east of Highway 12 in Section 10, Sumpter, but, lacking support from local farmers, he was forced to move to a site west of Baraboo by 1891.

As cheese makers Young and Brown learned, Sumpter and Merrimac farmers, like most dairy producers in the eastern towns of Sauk County, preferred to market their milk in the form of butter. They patronized the Bickford-Lawson Creamery, which opened in Prairie du Sac in 1885, and the Sumpter

Creamery, which farmers from the western part of the town incorporated in 1888, so they could hire professional butter makers to process their raw milk into a marketable product. The Sumpter Creamery was located a few hundred yards east of the Stone's Pocket school in Section 9 and received milk from as far away as Gruber's Bay in the southeast and from the Bluff farms north and east of Kern's Corners. It remained open until 1942, when the loss of production from the farms taken for Badger Ordnance, coupled with the national trend towards consolidation into larger marketing units, prompted its merger with the Sauk City Creamery. In the Sumpter Creamery's heyday at the turn of the 19th century, John Peter Kindschi drove the collection route, picking up the cream that farmers laboriously skimmed from their milk and bringing it to the creamery where it was processed into butter for the Chicago market. The Sumpter Creamery, like hundreds of other small dairy processing plants in Wisconsin, was a vital link between local farmers and the national economy. It added value to their raw milk by turning it into butter, which it traded nationally for cash that it returned to the local economy.

The butter, the bull calves, the fat hogs and the money they fetched traveled to and from Sauk City and Prairie du Sac by train. Local farmers could not earn a living feeding the growing populations of Milwaukee, Chicago and other urban markets to which they were inextricably linked without rail service. It did not start out that way for, in their rush to reach the Mississippi River by the fastest, easiest-to-build route, the first two railroads to cross Wisconsin all but bypassed Sauk County. The La Crosse and Milwaukee, which touched down in the county across the bridge from Wisconsin Dells in 1857, and the Milwaukee and Mississippi, which came through Spring Green in the same year, had little impact on Sumpter and Merrimac. Not until 1873, when the Chicago and North Western ran its line from Madison across the river to Merrimac and up the Baraboo Valley, did local farmers find a shipping point for livestock, butter and cheese. After 1881, much of the Merrimac trade shifted to Sauk City and Prairie du Sac when, fearing that the North Western might run a spur down the river valley from Merrimac, the Milwaukee Road, successor to both of the earlier railroads, extended its line from Mazomanie. By 1892, the Prairie du Sac creamery was buying and shipping one rail carload of butter and one of eggs every week. On one "Sumpter Stock Day" in 1907 five carloads of hogs and two of cattle left the Prairie du Sac yards. By the 1920s, as many as twenty carloads of livestock were shipped out weekly.

The railroads also brought greater personal mobility and improved communication to village and country people. The Milwaukee Road sold tickets for a "day in Madison" in which travelers could depart from either the Prairie du Sac or Sauk City depots early in the morning and return in the early evening. The village depots were also the point of departure for many longer journeys, with older people visiting family in the eastern states or Europe and younger people leaving the home farm to seek their fortunes in the cities or the western states. The railroad also brought the telegraph line, with its near instant delivery of news to the depot office and twice-daily postal service to the villages. Mail was then delivered by wagon to rural offices where country people could pick it up. It was slower than modern "snail mail" but faster than Prescott Brigham's stagecoach and it became much more convenient after 1900 when the post office began delivering mail directly to farms.

From the 1870s on, the production and marketing of milk for butter, and the raising of cattle and hogs for meat, defined farming on the land taken for Badger Ordnance, as they did throughout the southern half of Wisconsin. In Sumpter and Merrimac, some farmers supplemented the basics with orchard fruit, sawlogs, sorghum, maple syrup, honey, blooded horses or fancy fowl, but dairying and livestock raising remained the foundation of farming until 1942. They prescribed the amount of land farmed and its allocation to field or pasture, dictated the crops grown and equipment acquired to raise them, determined the number and breed of livestock and the design of farm buildings.

Prairie farms tended to be larger than other farms, with the Town of Sumpter spreads averaging the largest acreage in the county and surpassing 200 acres by 1942. Part of that acreage was in the Hills, where Prairie farmers customarily owned forty acres or more of woodlot pasture. Both the resulting woods and pasture were inferior to a managed woodlot and planted pasture, but the pioneer practice of keeping grazers in the woods was hard to break. Corn and oats were the leading field crops, with about as much ground planted to one as the other. Standard wisdom held that about one-quarter of cropland had to be reserved for oats, most of which fed the horses that worked the land and pulled the family buggy. By 1901, Sumpter farmers were averaging about twenty-four bushels of oats per acre and about twenty-one bushels of corn. Along with their prairie neighbors in the town of Troy, who had slightly lower yields, and in Honey Creek, with slightly higher, Sumpter farmers led field crop production in Sauk County.

Corn not cut green for silage was left to dry in the field, laboriously cut off the stalk and husked by hand, then stored on the cob in cribs. Oats required horse-drawn mowers and threshing equipment that was also powered first by horses then by lumbering steam "traction" engines that could move ponderously from farm to farm — but not in the fields — about as fast as a span of oxen could walk. By horse or by steam, threshing was labor-intensive and required cooperation among neighbors who shared the chore of processing each other's crop. The work began in early summer and continued until fall, as the oats matured and the schedule set by the owner-operator of the equipment. Between the 1880s and the 1920s, the Weirich, Heimlich and

Waffenschmidt families ran steam traction engines that powered corn shredders, clover hullers and silo loaders, as well as threshing equipment, on the land taken for Badger Ordnance. Threshing was an exciting time on the farm, with teams of horses hauling wagon loads of shocked oats to the thresher, gangs of men and boys loading the shocks into the hopper, stacking the discarded stalks and bagging the grain for storage, with the steam engine harumphing, the drive belts slapping, the machinery whirring, clanging and banging. Threshing days, including the gargantuan meals farm women laid out for the work crews, are much spoken of in farmstead memoirs. They have often been used to illustrate the friendly, cooperative nature of farming at the time. However, it was cooperation born more of necessity and less of enthusiasm. Robert Hatz, whose father Jacob ran a threshing ring for twenty years in western Sumpter, remembered that his Dad's first responsibility was to keep the machinery running smoothly. He also had another job, "more trying at times, that of keeping the ten farmers working together in harmony." When new equipment that enabled individual farmers to thresh for themselves became available in the 1920s, the neighborhood threshing rings dissolved, largely unlamented.[17]

Dairying and stock farming also required that the cowsheds and hop houses of the pioneers be replaced with multi-story barns fitted out with milking stalls, calf pens, feed bins and tall mows stuffed with a winter's worth of straw and hay. Starting in the 1880s and accelerating after 1900, the big dairy barn, with a silo alongside, became the definitive architectural feature of a Wisconsin farm. With a few exceptions, all the farms taken for Badger Ordnance had a dairy barn about fifty feet long and thirty wide that was built in the decades either side of 1900, or a newer one that replaced a barn built in those years. Silos, where cornstalks fermented into nutritious feed that increased milk production, also made their first appearances in the 1880s and soon became a symbol of progressive dairy farming. Another sign of progress was the disappearance of the "general" or "dual" purpose cow, a near mythical animal that traditionalists believed could be a top-quality producer of both milk and meat. Adequate in the 1870s, dual purpose cows could not compete in the increasingly competitive market of the 1900s and dairy farmers replaced them with selectively bred Jerseys and Guernseys, with their high butterfat milk, or with high-volume Holstein cattle.

By 1918, all three breeds had active associations in Sauk County, with many members on the land taken for Badger Ordnance. The development of dairy farming also brought with it the expansion of agricultural education programs. At the Farmer's Institutes which preceded the University of Wisconsin Extension work that began during World War I, and many of which were held at the Sumpter Town Hall, farmers were introduced to new tools and practices, such as the Babcock butterfat test, the principles of selective breeding and the necessity of testing for tuberculosis.

Between the 1890s and 1942, farming changed incrementally, but not by leaps and bounds as it did after World War II. Indeed, the farmers of 1940 had much more in common with the farmers of 1890 than with the farmers of 1990, or even 1970. The uniform plantings of hybrid corn, alfalfa and soybeans, the application of chemical fertilizers and pesticides, the general use of machine milking and bulk handling of whole milk on and off the farm, herd improvement by means of artificial insemination, even the universal replacement of horses by tractors, did not take place until after the shooting stopped in 1945. In this sense, farming as it is remembered on the land taken for Badger Ordnance is a time capsule of farming as it had been for fifty years.

Of the 1,339 people who lived in the Towns of Sumpter and Merrimac in 1900, and of the 1,111 who were there in 1940, virtually everyone lived and worked on a farm. A rural resident who did not farm, common in the year 2000, would have been a very queer duck sixty or 100 years earlier. Farming was how the people on the land taken for Badger Ordnance earned a living, but their lives were larger than their livelihoods. They were also parents and children, homemakers and neighbors, town officers and school district trustees, church members, lodge brothers, club sisters, growers of apples, makers of quilts. At the start of the 20th century they were still part of a community where people relied directly on each other for care, companionship, conflict and all the other interactions that went along with being neighbors in a culture where a neighbor was not merely the person who lived in the next house down the road.

In the 1920s and '30s, for example, it was a daily summer chore for young Robert or Wayne Hatz to walk the family cows to pasture about a mile down the town road from their home farm, which was located in Section 19, Sumpter. The road ran through the farms of John, Martin and Valentine Accola, and, boys left on their own being easily distracted, the cows sometimes got into the Accola cornfields. "Every neighbor was my co-worker when the cows got into their field," Wayne recalled. "When I asked [Valentine Accola's daughter] Lorene if her parents ever complained about the Hatzs' cows getting into their fields, her reply was, 'No, that was part of daily life, that neighbor helped neighbor.' " When Wayne asked Margaret Accola what she remembered about the Hatzs' cows raising the dust in front of her house on warm summer days, she recalled that her Mother did not complain but merely said, "Close the windows, here come the Hatz cows."[18]

This is the community of neighbors Erhart Mueller chronicled with such precision and affection in his five volumes on the Town of Sumpter and those parts of Merrimac taken for Badger Ordnance. It is where, as Mueller wrote, Nellie Shaw Premo and Nelli Hatz Stone started the Home Circle Club, first to simply take care of neglected cemeteries, then to conduct what are now called educational and community service programs. It is where the Sumpter

Methodist church, larger than the town hall next door, served as a community center, where the Grange, the Sauk Prairie Farmer's Club, the Farm Bureau and the Masons met, and where the country schools held graduation ceremonies, which, since most of the students were not going on to a village high school, were the only graduation most country kids would have. Mueller's community was also the home of the Rook Club for card players who met at the Urban Mather farm, of the Walter Welch family who welcomed picnickers to their picturesque grove at Pine Hollow, but never on Sunday, and the Elite Club for women and their families in east Sumpter who liked to drink tea in each other's parlors, picnic at Durward's Glen and Gruber's Grove in the summer and who sponsored tongue-in-cheek debates on questions of the day such as "Women's Suffrage, a good thing?" and "Bachelors are happier than married men?" It was a place where in 1913, when the number of students at Kern's Corners School declined to five and since "five children and the teacher cannot play a great variety of games," teacher Bessie Cole arranged a "Play-Day." A total of forty-nine students from other rural schools and Baraboo came to the school where they played baseball, hustle ball and other playground games that required more players than the Kern's Corners quintet could muster. The day was topped off with a late dinner cooked on a campfire before the children left for home. "There was an absence of speech making and perhaps that is the reason the children enjoyed it so much," reported the county *School Annual*. In typical fashion for the times, the six teachers in attendance agreed to form a teachers club, "to meet once a month for better acquaintance and the discussion of school problems."[19]

But change was coming. At one Farm Bureau gathering at the Sumpter Methodist Church near the turn of the 19th century, Nellie Luetscher Stone read a "theme" she had written on "Life and Home" that included the prescient lines, "Some people are always looking with suspicion at any change in customs. They sit in the corner and moan for 'the good old days' and say sadly that instead of progressing, we are slipping backwards fast." Even as Nellie Stone spoke, the community of neighbors that Erhart Mueller chronicled was "progressing." Like it or not, they were experiencing a technological revolution triggered by the automobile and electrical power that would alter their lives as severely in the thirty years prior to 1942 as the Army bulldozers remade their farms in 1942.[20]

Modern Times

The first automotive voyage across the Sauk Prairie went unrecorded but Frank Keller reportedly piloted a one-cylinder Oldsmobile down local roads in 1899. Ten years later, nine autos were listed on the Town of Sumpter's personal property tax roll. The new county highway commission was already at work on what would become the largest expansion of county government since the county farm was established in the 1870s. By 1921 Sauk County had 5,581 licensed autos and 400 trucks, with farmers much more likely to be auto owners than villagers. The highway commission had laid out a network of county trunk "farm to market" highways to serve farmers, was grading and graveling rights-of-way and erecting concrete bridges over rivers and creeks. It had also created the county's first full-time police force in the form of a highway patrol empowered to enforce newly enacted traffic regulations.

Agitation for improved rural roads started with bicyclists in the 1890s and accelerated when rural delivery of mail began in the 1900s, but it was the automobile that prompted the state legislature to fund highway construction and mandate that local governments build, improve and maintain rural roads. Since statehood, rural roads had been the responsibility of town boards. They divided their towns into road districts, supervised the election of a "Roadmaster" for each district and selected the date of the annual spring "road day" when landowners could reduce their town tax bill by working on the highways. Using their own horses, wagons and tools, for which they received additional credit, the men laid logs corduroy-fashion across low spots, filled potholes, dug drainage ditches and graded the right-of-way — if the Roadmaster kept them on task. If he was lax, the work crew could turn road day into an outing with the boys and still gain a tax credit. The result was a road system of uneven quality, or of no quality at all. Erhart Mueller tells the story of the traveler who asked the Prairie farm boy which was the better of the two roads heading up the Baraboo Bluffs. "You can take whichever you please," the boy replied, "and before you get there you will wish you had taken the other."[21]

In 1918, the state highway commission mapped the first system of state and federally funded trunk roads in Wisconsin. The Prairie du Sac-Merrimac road that became Highway 78 was not on it and would not become part of the state system for a few more years. The Prairie du Sac-Baraboo route, including the "West Sauk Road" through Jameson's Hollow, was designated Federal Highway 12 and earmarked for state and federal money in 1918, but saw little improvement until 1925. By then, state highway engineers had learned from previous experiences and resolved to make the new Route 12 a model of modern highway design. State contractors flattened the grade through the Hollow and increased the width of a single lane from nine to ten feet. They laid out a twenty-foot-wide, two-lane roadway with concrete curbs, gutters and culverts. The new road ran south to "Johnson's Corners," where it turned east to the town hall and Methodist church. Then it turned south again for three miles to the "Cemetery Corners" before heading east away from Sauk City and to Prairie du Sac. The route was indirect but it insured that motorists crossing the river on the Highway 12 bridge — new in 1922 — could not escape traveling through the business districts of both villages. On its other end, the road headed north to and through Baraboo where it connected with the already paved Lake Delton road, to become the first all-concrete road across Sauk County. As part of the federally funded highway connecting Chicago to the Twin Cities via Madison, Baraboo and Wisconsin Dells, the new route stimulated tourism at the Dells, especially at Delton where Chicago home builder Ralph Newman would soon dam Dell Creek into "Lake" Delton and surround it with a second-home subdivision. The new road also played a small role in the national military program since Wisconsin National Guard troops used it to reach Camp Douglas in Juneau County for summer training sessions. The Guard had been content to run its mule trains on the old dirt road since before World War I, but the modern motorized military of the 1920s preferred a paved route. In the summer of 1940, when the United States Army brought tens of thousands of Guardsmen to Fort McCoy in Monroe County, where they took part in the largest peacetime maneuvers the American military had yet to hold, Highway 12 bore a continuous caravan of military traffic.[22]

With representatives of the Guard and the American Legion present, the new Highway 12 was dedicated on Armistice Day, November 11, 1925. Governor James Blaine presided at the ribbon cutting, along with Big Bear, an Indian performer from Wisconsin Dells. Cradling a pipe in his arms, Big Bear struck a pose while standing on a platform bearing a sign that read, "Peace Day, A Pipe of Peace." So it was that a model federally funded, state highway that would only make the Sauk Prairie a more desirable location for an operation like the Badger Ordnance Works was completed in 1925 and was dedicated on a "Peace Day," set aside to commemorate the end of the war to end all wars.[23]

Even before modern Highway 12 was built, actually in less than a decade since 1909, a profound transformation had occurred in how farm people could live. A ten-mile trip to and from Kern's Corners to Prairie du Sac, once an all-day affair requiring a good measure of justification and planning, could now be made in minutes on a whim. There were a few conservatives, of course, who argued that just because you could get there faster didn't mean there was any more reason to go to Prairie du Sac, Sauk City or even Baraboo, but they were in the minority and growing ever fewer over time. By 1930, every farmer on the Sauk Prairie had a car. Factory-made trucks were less common, but that was because many farmers used their cars as if they were trucks or converted a second car — usually a Model T Ford — into a truck without bothering to relicense it.

The easy mobility that came with the automobile accelerated the decades-old movement of people from farms to cities. The population of Sumpter and Merrimac had peaked in the 1870s, when the children of large immigrant families were still at home. As they matured, these young people found less opportunity in their home townships and more in other places. Many who wanted to farm migrated to the frontier of the Great Plains states but many more selected a city where they found better-paying jobs, educational opportunities, a higher standard of living, livelier entertainment and escape from the low esteem in which farmers were customarily held. The family histories collected by Erhart Mueller contain numerous profiles of farm boys and girls who made good off the farm. Robert Siebecker was one of them. Born in 1854, he was one of eight children of William and Christiana Graf Siebecker, who had settled in Section 12 of Sumpter a few years earlier. Not much interested in farming, and with plenty of brothers and sisters who were, he stretched out his studies at the Lakeside school until he was seventeen years old, then moved to Madison and the university. There he befriended another farm-to-city migrant, Robert La Follette, who had left his own family farm in the Dane County town of Primrose. Siebecker also met and married La Follette's sister, Josephine, then settled in Madison to pursue a career as an attorney and a partner in La Follette's firm. In 1891, La Follette, smarting from the loss of his seat in Congress the previous November, was anxious to return to power, and Siebecker was a circuit judge in Dane County. One of the cases before Judge Siebecker involved Wisconsin's senior United States Senator, Philetus Sawyer, the lumber baron from Oshkosh and backroom boss of the state's Republican Party. Although each man had his own version of the story, the La Follette account—which attracted more press—says that Sawyer summoned the ex-Congressman to a private meeting at which the Senator produced a wad of cash which he offered in return for La Follette's promise to persuade brother-in-law Siebecker to rule in Sawyer's favor. La Follette refused and, shocked by Sawyer's presumption that both he and Siebecker could be bought, decided that a system that allowed rich men to buy political power and influence must be reformed. La Follette then broke with the regular Republicans to launch a decade-long campaign that carried him and his Progressive Republicans to power in Madison in 1901. Sawyer denied it all and, along with other "Stalwart" Republicans, remained an implacable foe of La Follette for the rest of his life. Robert Siebecker played only an offstage role in this story, but it is worth noting that a person who had great influence on the birth of the La Follette legend, if not La Follette Progressivism, was born and raised on land taken for the Badger Ordnance Works. Few of the other young people who left the farms of Sumpter and Merrimac in the years between 1870 and 1940 could claim a footnote in Wisconsin history books, but even if they merely became teachers in

Madison, shopkeepers in Milwaukee or factory workers in Janesville, they were equally able to illustrate the lack of opportunity on the farm in those years and the greater promise extended by city life. The population trend continued downward in Sumpter and Merrimac until the 1940 head count of 1,111 was no more than two-thirds that of 1870. In 1941, the school census of Sumpter and Merrimac fell to 235 students, including the Merrimac village school. A county committee on school reorganization recommended that Parfrey's Glen and Lakeview schools (average enrollment "14.8 students" each) be closed and the children moved to Seibecker's alma mater, Lakeside. The taking of the Lakeview School for the Badger Ordnance Works only expedited what the decline in rural population would have brought about anyway.[24]

Just as new roads and autos brought country people to town, they brought the town to the country. The construction of the Prairie du Sac dam in 1914 flooded wetlands on property owned by Theodore Weigand in Section 8, Merrimac. Although he didn't give up his farm, Weigand platted a fourteen-lot subdivision in 1912, even before the dam was completed, and started renting cottages for fishing vacations. Initially, Weigand's guests traveled by train to the Merrimac depot, where he would meet them and then carry them back to his cottages by water in a canopied, gasoline runabout. Weigand's became a popular place for family and club picnics and fishing outings for both local and out-of-town people. By the 1920s, anglers were driving their own cars to the fish camp and many had purchased a cottage in Weigand's growing development. By 1942, the subdivision contained forty-four lots owned by local farmers — Garth Premo, Robert Organ, William Eschenbach — and also by out-of-staters who lived as far away as Chicago and Rockford in Illinois and St. Petersburg, Florida. On Highway 78, near the entrance to Weigand's, stood William Jenewein's gasoline station and auto repair shop. It was a symbol of the young automobile age and, along with Weigand's camp store and resort operation, the only non-farm commercial property taken for Badger Ordnance.

By creating "Lake Wisconsin" the Prairie du Sac dam brought 20th-century style tourists to what had been strictly farm country. It also brought electricity to the farms of the Sauk Prairie at least fifteen years earlier than it might otherwise have arrived, and vividly demonstrated the ability of 20th-century industrial technology to reshape the landscape on a grand scale. The Wisconsin River valley had, since the 1890s, been subject to what one journalist diagnosed as "hydromania," or the compulsion to convert waterpower to electricity. By 1910, new dams and generating stations had been built at over twenty waterpower sites on the Wisconsin and a system of reservoirs to maintain a uniform flow of water to them had been created on the lakes and streams of the river's headwaters in Vilas and Oneida counties. Nearly all of these power plants were sited in or near small industrial cities like Wausau or Wisconsin Rapids and were designed to supply power to local industries, principally the papermakers that had replaced the lumber mills there. In the early 1900s, Magnus Swenson, a Madison entrepreneur who had made a fortune processing beet roots into table sugar, contracted hydromania. He organized the Southern Wisconsin Power Company and obtained a state charter to construct a dam and power plant in the middle of the Dells of the Wisconsin River, at what was then still known as the village of Kilbourn. With a population of 1,100 and no industry to speak of, Kilbourn could consume only a tiny slice of the 6,000 kilowatts of power the new plant began to generate in 1909. Swenson hoped to market the power in the nearest city of some size, Madison, but was forced to sell it at cost to run street railways in Milwaukee. Undaunted, he moved downriver to a point where the glacial moraine crosses the river with a ledge of solid rock only thirty-five feet below the sand. Here, just upstream of Prairie du Sac, Swenson's Southern Wisconsin Power Company started work on what was then the largest hydroelectric power installation in the United States outside of Niagara Falls, New York. Construction began in 1911, with a workforce of 400 men living in temporary camps near the site. American-born workers, who held skilled and supervisory positions, were housed in a $4.50-a-week bunkhouse on the Sauk County shore, while the common laborers, chiefly African-Americans, Italians and Eastern Europeans, had low-rent quarters under canvas across the river in Dane County. The work crew at the dam brought a flurry of ethnic diversity to the uniformly north European families of the Sauk Prairie. Clifford Schweppe, who grew up on 100 acres in Section 14 of Sumpter, remembered meeting a former dam worker who recalled that, "There were quite a few black people here then. There were Italians. There was a mixture of all kinds of people...If one had a goat the Italians wanted it. The Italians like goat meat. These Englishmen and Germans around here were more interested in beef and pork and chickens."[25]

The vast enterprise did more than introduce its neighbors to exotic Italian cuisine. For people who had never witnessed the construction of anything more substantial than a barn, the Prairie du Sac project was a wonder to behold: the steam pile driver pounding thousands of pilings into the river bottom, the coffer dams that diverted the river, the giant steam-powered mixers, the line of wheelbarrows snaking up the scaffolding to pour concrete for the dam, the penstocks wide as a wagon, the turbines that could create whatever it was that made light bulbs glow and, when it was all done, a 9,000-acre "lake" gradually filling the old riverbed. In the eyes of one correspondent to the Baraboo paper, it was comparable to only one other contemporary undertaking, the Panama Canal, and the lake above the Prairie du Sac dam should be named accordingly — Panama Lake. The power company did not take the suggestion to honor Panama in Wisconsin, nor did it accept Swenson, Tripp, Gay, Koshwago, Saucomac, Sauco or any of the other colorful titles offered by the locals, but opted for the obvious, Lake Wisconsin.[26]

Although it did not take as long to build as the Panama Canal, the Prairie du Sac power station did experience delays caused by floodwaters in October 1911 and by a barrage of ice floes in the spring of 1913. Not until the fall of 1914 was the 14,000-kilowatt station ready to generate power. Once again Southern Wisconsin Power had failed to find customers who would buy power at a profitable rate and, facing bankruptcy, Swenson sold the Prairie du Sac and Kilbourn stations to John Beggs, the Milwaukee streetcar tycoon who was already buying Kilbourn's power. He soon sold the plants to Chicago magnate Samuel Insull, who made them the property of the Wisconsin Power and Light Company, which he incorporated in 1923. WP&L would grow to supply power to about one-third of Wisconsin, but demand for electricity was still so limited that Kilbourn and Prairie du Sac would meet all the company's needs until World War II. Thanks to Magnus Swenson, an entrepreneur whose reach was greater than his grasp, the farmers of the Sauk Prairie found themselves adjacent to a reservoir large enough to spread one foot of water over 193,000 acres of ground and one of the largest underutilized power plants in the United States.

With more power than he knew what to do with, Grover Neff, manager of the Kilbourn and Prairie du Sac stations, set out to create a market. Wherever he could, he bought out the small locally owned hydro plants that had been installed throughout southern and central Wisconsin in villages like Tomah, Lake Geneva and Friendship and replaced them with power from his central plants. Farmers along and near the village-to-village transmission lines were welcome to connect if they were willing to wire their houses and barns, set poles and run distribution lines to the company's wires. In Sumpter and Merrimac, practically a stone's throw from the Prairie du Sac plant, farmers led by Fred Waffenschmidt, Henry Thoelke and Julius Steidtman organized a stock company that connected to Neff's power lines in 1918. A year later farmers in western Sumpter organized a company of their own and came on line. So anxious were the farmers to bring power to their homesteads that they often ran the lines through fields and cleared off trees planted decades earlier that might be in the way. "A row of great maples in particular, forty rods long, went into cordwood almost overnight," to make room for the power line for the Mather farm on Highway 12 in Section 14 of Sumpter. Farmers who did not have power company transmission lines running alongside their farms, or who did not organize their own mini-distribution companies, did not see electrical service until after Rural Electrification Administration cooperatives were organized in 1936. In 1930, less than one in ten American farms had electricity, nearly all of them on the outskirts of cities where lines could be easily extended. Sauk Prairie farmers, due to a favorable location and their own initiative, were among an elite, electrified minority.[27]

With electricity in their houses and barns, with improved roads and modern autos, most of the farmers on the land taken for Badger Ordnance enjoyed a higher standard of living than farmers throughout most of Wisconsin. Just as it had been on the divide between grassland and woodland, and on the edge of the glacial advance thousands of years earlier, the Sauk Prairie occupied the edge of an economic divide separating the more prosperous parts of Wisconsin from the less advantaged. The divide could be marked on a map by drawing a line from Green Bay to the triple corner where Iowa, Illinois and Wisconsin come together. Southeast of the line, with a few exceptions, most notably in northwestern Dane County, the average value of a farmhouse in 1930 was greater than or equal to a farmhouse on the Sauk Prairie. Northwest of the line, again with a few exceptions, the average value of a farmhouse was less than on the Prairie, where homes were worth $2,300 to $2,900. Sauk Prairie farmhouses were more valuable than, for example, farmhouses in the Town of Reedsburg or in all of Juneau County, because the farmers who built them could afford to build well and install such amenities as electric lights and appliances, central heating and bathtubs. A bathtub was a good measure of living standards because it presumed that the home also had an electric pump and a hot running water system, which meant that the people who lived in it had progressed beyond the stage of having to take a bath in a washtub in front of the kitchen stove, and only on Saturday night. Sauk Prairie farmers also had autos, radios and telephones, just like the state's most prosperous farmers in Kenosha or Waukesha County and unlike most of their less comfortable cousins in the Town of Dellona or Adams County. Sauk Prairie farming operations were also on the statewide economic divide. Gross farm income on the Prairie was about $2,600, and the value of farmland and buildings ranged from $80 to $120 per acre, equal to or slightly less than farms to the southeast and greater than nearly all of those to the northwest. If not the richest farms in Wisconsin, Sauk Prairie farms were comparable to the wealthiest in the state.[28]

That was in 1930, before all of American agriculture encountered the most serious crisis it had known. The farm economy had been in decline since the end of World War I, when the ability of American farmers to produce grain, meat and dairy products consistently outstripped the demands of the market. Farm numbers fell, with marginal operations like those struggling in the blowsand of Delton or the thin loam of Dellona the first to go out of business. After a few years in the late 1920s when the farm economy stabilized, prices and income started to fall again until they hit rock bottom in 1932. In 1933, the gross income for all farms in Sauk County had dropped more than fifty percent below the 1927 level, to average less than $1,200 per farm. Sauk Prairie farmers were a little better off since their incomes were higher than the county average to start and had declined by only one-third. They were better able to adjust as rural people have always adjusted to hard times. They hunkered down, cut expenses, spent time instead of cash and relied on themselves and their neighbors to feed and care for kith and kin. The progress they had made in the 1910s and early 1920s, progress that had enabled them to put a Ford in the shed, a chandelier above the dining room table and a Philco in the parlor, crunched to a halt. The desire to buy a "pick-up" truck, brand new in the mid-1930s, to install electric milking equipment or trade in the horses for a rubber-tired Allis tractor could not be gratified. Perhaps the most visible sign of the slowdown was found in the barn, which was still occupied by work horses. Sauk Prairie farmers owned just about as many horses in 1940 as they did in 1900 and would continue to rely on in-the-flesh instead of mechanical horsepower until after World War II.

The weather did not ease conditions for horses or farmers. A continental drought that began in the fall of 1933 continued through an open winter that killed off pasture grass and did not break in Wisconsin until the summer of 1934. The dry weather stifled the oats and stressed the young corn on the Sauk Prairie, and unleashed apocalyptic dust storms that carried Dakota soil over the Baraboo Hills and east to the Atlantic Ocean. The dust clouds were reminiscent of the rains of loess that had fallen on the Sauk Prairie in the immediate post-glacial period, but the Prairie soil itself, although light in texture, resisted erosion. Despite eighty years of plowing and, unlike the dunes near Spring Green or the rolling sand of Adams, Waushara, Waupaca and Portage counties, which blew into drifts three feet tall, the soil of the Sauk Prairie retained its natural integrity and withstood the wind. In 1935, the weather and harvests returned to normal, but in 1936 drought returned, this time bringing record summer heat that sucked dry the maturing corn and burned up the pastures. On July 13, 1936, the official thermometer in Wisconsin Dells read 114 degrees and set a record for heat in Wisconsin. It was just about as hot on the Sauk Prairie, maybe hotter in the middle of an unshaded field in the late afternoon after the sun had been incinerating the atmosphere since dawn. After the drought and heat came the flood. In September 1938, the Wisconsin River rose to a record height, pierced the levee at Portage, backed water into Weigand's and Gruber's Bays higher than in any living memory and roared through the floodgates at Prairie du Sac at the 100-year-flood rate of 80,000 cubic feet per second.

Even before the extreme weather visited the Sauk Prairie, farmers there had signed up for the agricultural relief programs instituted by the administration of Franklin Roosevelt. The economy had so soured by 1932 that even these most staunch of Republicans deserted their party and made Franklin Roosevelt the first Democratic presidential candidate to win a majority in Sumpter and Merrimac. Voters there had remained faithful Republicans in the bellwether gubernatorial election of 1890, when other Wisconsinites turned the "dairyman's friend," William Hoard, out of office and installed

Milwaukee Democrat George Peck. They had remained faithful Republicans in 1916, voting against incumbent Woodrow Wilson even though they supported his policy of keeping the United States out of World War I. They were able to remain anti-war and still vote Republican in 1918, when they returned Representative John Esch to Congress, since he, like other progressives, had voted against American entry in the war. By 1932, however, it was the economy, and Sauk Prairie voters were not stupid. They abandoned seven decades of tradition and opted to row the national boat with Roosevelt. They did it again in 1936 and in 1940, although with less enthusiasm than in 1932. Voting for a Democrat was about as radical as Sauk Prairie farmers would get. Erhart Mueller remembered Prairie farmers killing pigs and burying them because the cost of finishing and shipping them was higher than their market price, but direct action like dumping milk was a step they were not willing to take. Farming was their life's calling, not just a way to make a living, and pouring milk on the ground would have been a sacrilege. They did not take part in the Farm Holiday movement, which urged farmers to withhold products from the market in the summer of 1932, although Prairie du Sac's William O. Hart organized a visit and a rousing speech at the county fairgrounds by national Farm Holiday leader Milo Reno. They participated in the Milk Strike of February 1933 but only to turn out and protect the Sumpter Creamery, in which they owned stock, from more radical farmers who threatened to vandalize it.

A month after the Milk Strike, Franklin Roosevelt was inaugurated. Sauk Prairie farmers, with their power lines and parlor radios, were among the minority of American farmers able to hear his voice of hope in their own homes. New Deal farm programs began in 1934 after Sauk County, in order to be eligible for them, restored the position of county agricultural agent and hired Charles Karch to fill it. Karch, his successor David Williams and the county agricultural committee implemented aid programs based on the simple premise that farm prices and incomes would go up if the supply of farm products for sale went down. The federal government purchased "surplus" grain and livestock and instituted soil conservation programs that took land out of production. In addition to financial aid, the New Deal offered incentives to participate in farm improvement programs that supplied lime as a soil amendment, introduced "superphosphate" fertilizer to the Prairie and expanded the planting of alfalfa for hay, which would have far-reaching effects on the prairie environment after World War II.

By the late 1930s, the Depression was far from over, but the farm economy was showing some signs of improvement. It received an added boost in September 1939, when the German army invaded Poland. Although no one on the Sauk Prairie was likely to rejoice at the news, the outbreak of war in Europe meant increased demand and higher prices for American farm products. It was good news for a bad reason. The World War I years, 1914-1918, had been the most prosperous period in memory for Sauk Prairie farmers, and if the United States could stay out of the new war, there was reason for optimism. One of the optimists was Irvin Luetscher. His family had been among the first Swiss immigrants to join Barthomelew Ragatz in the Honey Creek settlement in the 1840s and the old Luetscher farm was still in the family. In 1916, Irvin and his wife Margaret moved to the Prairie by purchasing the old Isaac Gibbs place and, aided by wartime prosperity, got off to a good start on their own farm. Despite the slowdown in the 1920s and hard times of the 1930s, the Luetschers made plans for a better future. They had three sons coming up — Melvin, Leroy and Wilbur — and Irvin and Margaret talked about buying each one of them his own farm. In order to do so, the Luetschers had to improve their own operation. They started at the heart of the farm, by building a new barn, 112 feet long, forty high, twice the size of most of the other barns on the Prairie. They had a woodlot on eighty acres up the hill in Section 36 with plenty of white oak timber that Irvin, his sons and neighbors could fell and have milled for the building. It was affordable, doable and typical of farm life in good and bad times. The Luetschers starting logging the oaks and, in 1928, hired Otto Stoeffels to set up his portable sawmill in the pasture next to the woodlot. Irvin took the job of tail sawyer, catching and stacking each new board as it peeled off the rotary saw and piling the slabs for firewood. Nothing was wasted. The barn foundation was laid, block by block. The beams were notched, fitted and raised, purloins and rafters spiked into place. Board siding and roof sheathing were nailed down, asphalt shingles laid up, sliding doors and track hung. Calf pens and double stalls for eight horses were partitioned off inside the barn and pipes for an electric milking system run to new stanchions built for the family's twenty-two Milking Shorthorns. The barn was ready for the grain, hay, horses and cows that would make it the centerpiece of a working family farm, but the Luetschers hadn't completed their plans. In 1936, the year of the summer drought, they used more of their own lumber to build a new farmhouse, driving nails into rock-hard, white oak wall studs in between chores in the hot sun of that broiling summer. It was hard work, but they were building their home with lumber from their land, where they planned to live out their days. Once it was roofed and covered by the brick veneer the Luetschers chose over wood siding, the white oak house would stand about as long as any wooden house could.[29]

Five years later, in mid-October 1941, the county agriculture committee called a meeting for farmers in Baraboo. Agent Dave Williams announced that, instead of restricting production as it had been doing because of the Depression for the last few years, the government now wanted farmers to produce ten to twenty-five percent more hogs and eggs and corn and milk. Prices would be higher too, in 1942, because of the war in Europe. It was good news for the Luetschers, as it was for all the farmers on the Sauk Prairie.

The new barn that Irvin worked so hard to build, that looked like a white elephant when the Depression hit, would pay off handsomely, especially if the United States stayed out of the war. If the family's luck held, the three Luetscher boys would see better times than their parents, and the family farm on the Prairie would prosper with them.

III

Powder Plant

The farmers of Sauk Prairie have been dispossessed of their land, and a great industrial center has replaced this one time fertile agricultural domain. Where once a great agricultural harvest was reaped, a greater harvest will be reaped which will strengthen our defense and security.

David P. Mayer, U.S. Army Corps of Engineers, 1942[1]

The Fixer

Leo T. Crowley was a fixer, a salesman whose prime stock in trade was himself. Possessed of the fabled Irish gift for gab, he had a knack for bringing people together, smoothing over differences, making a deal on a handshake. He had relied on these innate skills to get a leg up in the world ever since his teenage years, when fate thrust a heavy burden of responsibility on his youthful shoulders. His mother, born Katherine Ryan of a God-fearing Irish Catholic family in Janesville, had raised Leo for the priesthood, but injuries to his father and two elder brothers — gruesome severings of limbs while they worked in the rail yards of the Milwaukee Road — had forced the aspiring cleric to give up the collar and find a job to support his family. The Crowleys left tiny Milton Junction and moved to the near west side of Madison, where all eleven of them crowded into a tiny house only ten feet away from the Milwaukee Road tracks. Leo made deliveries for a local grocery and turned his pay over to his mother, but carefully saved his tips to invest in his own future. He graduated from Madison Central High, but attending the university was out of the question. He had a family to support.

Young Leo put his energies into his career and rose fast. Six years after starting as a salesman for the General Paper Company in 1913, he became its sole owner. Building on 1920s prosperity and his own entrepreneurial skill, he expanded General Paper's operations, ventured into wholesale groceries, real estate, insurance and banking, creating desk jobs for his disabled brothers and buying a comfortable home on Edgewood Drive that he shared with his mother and unmarried sisters. He became prominent in Catholic circles, raising large sums to build St. Mary's Hospital and opposing the Ku Klux Klan when that hate group resurfaced in the mid-1920s and, with Jews and African-Americans sparse in Madison, focused its know-nothing venom on Italian-American Catholics in the Greenbush neighborhood. Politics came next, and even though 1920s Wisconsin was a one-party state where Progressive Republicans vied for power with Stalwart Republicans while the Democrats warmed the benches, Crowley joined the minority party. He worked hard to put his fellow Irish-American Catholic Al Smith in the White House in 1928 and, although Smith lost, the hard-charging "Leo the Lion" made a name for himself as the Democrats' man to watch in Wisconsin.

The onset of the Great Depression in 1929 caught the ambitious entrepreneur overextended. In a few short years, his paper, grocery and banking businesses were failing and his investments in stocks and real estate were near worthless. He was heavily in debt, with every asset, including the family home, mortgaged to the rooftop, yet he was unwilling to give up the comfort and status he had worked so hard to achieve for himself and his family. He engaged in questionable banking practices that his critics argued should have put him behind bars, but for which he was never prosecuted. Instead Crowley juggled his debts, exploited his connections, persuaded and cajoled his way out of the courtroom. He had developed a close friendship with Albert Schedeman, a Madison businessman and Democrat who had served in the administration of Woodrow Wilson. In 1930, Crowley engineered his friend's election to the mayor's office, which made Schedeman one of the most prominent Democrat officeholders in the state. In 1932, Crowley managed his campaign and, with help from the generous coattails of Franklin Roosevelt, Schedeman became the first Democrat elected governor of Wisconsin since George Peck in 1890. Crowley played so prominent a role in the Schedeman administration that the press referred to him as the real governor of Wisconsin, and Democrats, unaware of his compromised record, touted him as a candidate for governor in 1934. Schedeman/Crowley served one two-year term, until the resurgent Progressive Republicans led by Phillip La Follette sent the Democrats into exile from the Governor's Mansion for another twenty-five years.

By then Leo Crowley was on his way to Washington, D.C., where he had accepted an appointment as head of the Federal Deposit Insurance Corporation. A controversial experiment at first, the FDIC, with Crowley as chair for its first twelve years, became one of the most successful New Deal programs. By rescuing failing banks and eventually insuring deposits in virtually every bank in the country, the FDIC restored the public's confidence in the financial system. Crowley himself became a leading figure in the Roosevelt administration, known as the "president's man," even though he was roundly denounced—in private—by other officials who could not overlook his past. Crowley kept to himself and kept busy. A "nothing-a-year" man who accepted no pay for his government job, he supported himself and his extended family with lucrative private sector positions. "No conflict of interest," he said, when asked about his $50,000-a-year position with the Standard Gas Company. "Sir Leo," who had been knighted by Pope Pius XI, attended Mass every morning and continued his charitable and church work. He was dubbed the "most prominent Catholic layman in America" in 1940, when that title might have gone to another well-known Irish-American, Ambassador Joseph Kennedy.[2]

In November 1940, although he already had one full-time job at the FDIC, and another with Standard Gas, and despite bitter opposition from Secretary of the Treasury Hans Morgenthau, Crowley was appointed "custodian" of alien property. Crowley would supervise the assets of foreign companies frozen in the United States due to the outbreak of World War II in Europe, a job that expanded to include all German, Italian, Japanese and other enemy property after December 7, 1941. Even before he was named custodian, Crowley had started campaigning to keep Wisconsin in the Democratic camp and re-elect President Roosevelt for a controversial third term. He neutralized support for Iowa's Progressive governor Henry Wallace, helped persuade Senator Robert M. La Follette, Jr., to become the only nationally prominent Progressive to endorse the President, and brought Wisconsin home for Roosevelt.[3]

In this context, with his reputation as a deal maker who was both a friend of business and of President Roosevelt, the story Crowley told nearly thirty years later about his role in the siting of Badger Ordnance Works is credible. By modern standards, Leo Crowley was not accessible to the news media. Many public figures of his generation were equally close-mouthed, even those who held an elected position, which Crowley did not, and even if they were not under a cloud of suspicion like the one that had hovered over Crowley since the early 1930s. Not until 1969, when he was eighty years old and had been out of Washington for twenty-three years, would Leo Crowley agree to talk to the press, and then only on his own terms. He would be interviewed on tape by the Washington correspondent of the *Milwaukee Journal*, Laurence Ecklund, who had reported on Badger Ordnance for the Journal in 1941, but Crowley refused to share any notes, letters or personal papers with the reporter. During the interview, Ecklund was primarily interested in Crowley's feud with President Harry Truman over Crowley's order to discontinue Lend-Lease aid to the Soviet Union at the end of World War II in Europe, but he also asked about other matters. He began one question by quoting from an undated article in "the *Wall Street Journal*," that said, "the political support of Leo Crowley is a matter not taken lightly by members of Congress in other states as well as Wisconsin...Badger State members have not forgotten that it was he not they who saw to it that a $65 million powder plant was located in their state."

Ecklund: "Do you confirm that?"

Crowley: "I'll tell you about that. The President of the Chicago and North Western, Bud Williams, he came to see me and said they were making an investigation up there and wanted to know if I would help him. [I said] Yes,

I'd help him.

"I went over to George Marshall's office with him and introduced him and...[Ecklund interrupts] "then they had to get all that farmland. I got three-four real estate men to get options...They finally worked it out."

Ecklund: "Was that really the genesis of getting it located in Wisconsin?"

Crowley: "Yes, I think so."

Ecklund: "The fact that the railroad wanted more business?"

Crowley: "That's right."

Ecklund: "That story checks out, that it was Mr. Leo Crowley who got that powder plant for Wisconsin."

Crowley: "That's right."[4]

On the eve of World War II, the Chicago and North Western, like every railroad in the country, was reeling from a decade of hard knocks. The company had gone bankrupt and into receivership in the mid-1930s and was still scraping its way out of the red at the end of the decade. Locally, the North Western had been reducing service on its Merrimac-Baraboo Valley route since at least 1912, when it completed construction of a new main line across Marquette, Adams and Juneau counties and relocated its round house, repair shops and yards from Baraboo to the new village of Adams. When the North Western inaugurated its high speed "400" passenger service from Chicago to St. Paul in 1935, the fast trains roared across the flat, straight Adams line and not up and down the hilly Baraboo route, which was slowly slipping to milk run status. By 1939, four passenger trains stopped at Baraboo every day, but only one of them also stopped at all the other stations up the Valley. Two freight trains—one through, one local—also crossed the Merrimac bridge and steamed up the Baraboo Valley. The through freight was a non-stop, but the local train stopped at every depot to deliver coal, lumber and merchandise for Main Street stores and to haul out the products of area farms and industry: cattle from La Valle, dried milk from Union Center, woolen goods from Reedsburg and the milk by-product casein from the Hercules Chemical Company plant in Elroy.[5]

The Milwaukee Road's service to Sauk City and Prairie du Sac followed a similar, downward course. By the 1910s no more than one daily train brought freight to the twin depots and carried out Sumpter livestock, Prairie du Sac butter and peas canned in Sauk City. Mail delivery shifted to trucks in 1921; passenger service was reduced to one train in 1924, then restricted to the caboose of the freight train in 1937. The ultimate reduction came in 1940, when the railroad decided that it was frivolous to maintain two stations in two neighboring villages on a dead-end spur line. The Prairie du Sac depot was closed and the Sauk City depot moved to the Steidtmann lots midway between the two villages. Had it not been for the presence of the Badger Ordnance Works, the Sauk villages would most likely be on the list of communities that lost rail service in the 1940s.

For the Milwaukee Road, the siting of a large industrial plant anywhere in the vicinity would raise the Sauk Prairie spur from the dead. For the hard-pressed North Western, a plant would revitalize the underused Merrimac-Baraboo Valley line, while not adding traffic to the already busy main line through Adams. So it was only good business for the President of the C&NW to ask a friend in the Roosevelt administration for help in reviving rail traffic in Sauk County. Confirmation of Leo Crowley's conference with Roland "Bud" Williams, or of their meeting with Army Chief of Staff George C. Marshall, has yet to be found. Nor is there any indication that Marshall would have acted on Crowley's lobbying for Williams, nor if he was in the position to act on it, if he so desired. However, state politics and business lobbying had long been part of the process whenever federal tax money was spent and had played a well-documented role in the earlier siting of World War II ordnance facilities in Virginia, Illinois and Indiana. In August 1940, about one month after Congress made its first appropriation for Ordnance's $7.1-billion pre-Pearl Harbor buildup, political, business, labor and agricultural leaders from nine midwestern states met in Kansas City. Fearing that the Farm Belt might be left out of the rearmament program, they sent a resolution to the White House stating that, "The Middle West insists that...this area shall receive its just and proper share of the outlays of public funds. This is not a cry for pork...It is the voice of the Middle West asking for justice." A few months later, Minnesota pursued justice by hiring a full-time lobbyist to bring home defense contracts from Washington. The state of Wisconsin, with conservative Republican Julius Heil as governor, did not hire a lobbyist, nor did its Congressional delegation, made up of anti-war Progressives, anti-Roosevelt Republicans and Democrats of little influence, have sufficient clout to deliver a big slice of the swelling Ordnance Department pie. It did have Leo T. Crowley, and, if anyone from Wisconsin could bring influence to bear on the Army and deliver corporate welfare to a struggling railroad back home, it was "the President's man."[6]

Rearming for War

Even as Bud Williams and Leo Crowley were knocking on General Marshall's door, the Army Ordnance Department was working with the nation's railroads and munitions manufacturers to locate sites for new facilities. In July 1937, the Army's Ordnance Ammunition Production Office opened a liaison office in Wilmington, Delaware to coordinate plans with the munitions industry to rearm the United States. As the headquarters city of E.I. Du Pont de

Nemours & Company, the nation's largest manufacturer of explosives, and of its much smaller competitors, Hercules Powder Corporation and Atlas Powder Corporation, Wilmington was the big-bang capital of the United States. Nominal competitors, Atlas and Hercules had been fashioned out of Du Pont after the giant arms maker was found guilty of antitrust law violations in 1911. The corporate restructuring, which the federal court allowed Du Pont to draw up, left Du Pont family members and officials in control of all three companies. When Hercules and Atlas opened their own corporate offices, they were conveniently located directly across the street from Du Pont headquarters. All three munitions makers were equally well-positioned for profitable growth when World War I began in Europe in 1914, and they did very well, with Hercules' average earnings multiplying by a factor of six by war's end in 1918. Peace forced the companies to diversify. Du Pont, for example, marketed explosives from its TNT plant at Barksdale, Wisconsin, to farmers who used it to blast stumps and boulders out of northern farm fields. Hercules expanded its chemical division, one of whose products was casein derived from skim milk and used in paint, glue and coatings for glossy printing paper. By 1941, Hercules had three Wisconsin plants processing milk into casein, at Watertown, Durand and Elroy.

When Army Ordnance set up its liaison office in Wilmington in 1937, public sentiment in the United States had yet to abandon the pacifism and isolationism it had espoused since the end of World War I. In 1934, authors H.C. Engelbrecht and F.C. Hanighen published their best-selling indictment of the international arms industry, *Merchants of Death*. With a section devoted to American companies like Du Pont and Hercules, *Merchants of Death* was one of many factors registering in a 1936 poll that found four out of five Americans in favor of prohibiting "the manufacture of munitions for private profit." Over the next five years, world events prompted most Americans to rethink their stance. Adolf Hitler's prewar aggression in Austria and Czechoslovakia; the Fascist coup and overthrow of the republican government in Spain; Fascist Italy's attack on Ethiopia; the Japanese invasion of China; all became mileposts in the seemingly inevitable procession towards the world war that began with the German and Soviet invasions of Poland in September 1939. In less than a year, President Roosevelt declared a national state of emergency, Congress enacted the first American peacetime draft of men for the military and the War Department launched a massive rearmament program.[7]

The production of munitions was a key part of rearmament. It was obvious that the handful of government arsenals operating in the 1930s could not supply the billions of tons of explosives Army Ordnance projected it would need for a modern war, nor could the scaled down, peacetime facilities of commercial munitions makers like Du Pont and Hercules meet anticipated wartime needs. Converting similar industrial facilities, such as chemical or papermaking plants, to the production of explosives was much more difficult than, say, retooling an automobile plant to manufacture tanks. Consequently, the rearming of the United States would require the construction of dozens of new munitions manufactories. The free enterprise tradition prevented the United States from creating an arms establishment wholly owned and operated by the government, yet private industry was not willing to bear the burden on its own. American industrialists, even giants like Du Pont, were loathe to invest capital in huge new munitions plants with only a temporary life span. Although profitable while they last, wars eventually come to an end. Arms facilities must then be scrapped, idled or retooled to produce for peace, with an inevitably negative effect on the corporate balance sheet. To meet the demand to rearm and overcome corporate reluctance to invest in the national defense, the United States government shifted the investment cost and risk from industry to taxpayer. Munitions facilities would be government owned, but built and operated by contractors profitably reimbursed on a cost-plus-fixed-fee basis. The first government-owned, contractor-operated ordnance facility, or GOCO, built for World War II was authorized in 1938. Twenty-two GOCO facilities were under construction in 1940, with seventy-seven built by the end of the war. As defined by the Ordnance Department, a facility was a "plant" if "only fabrication or assembly" took place, as in the manufacture of bullets, rockets or bomb casings. It was a "works" if "basic materials were required for production," such as the ether, sulfuric acid and cellulose used to make the explosives that went into the assembled casings. Since it used "basic materials," Badger Ordnance was a works and remained a works—officially at least—until the definitions were changed in the early 1960s. To just about everyone in Sauk County, however, Badger was "the powder plant" from the day its coming was announced in October 1941.[8]

In selecting sites for its munitions facilities, the Ordnance Department used an evolving set of criteria based on defense strategy, natural resources, transportation, economics, agency rivalry and politics in the states and in Washington. Early on, Army Ordnance favored sites located near urban industrial centers with large resident workforces, and therefore included the populous states of the Eastern Seaboard as potential locations for new "works" and "plants." The Army also wanted its facilities to be close to sources of fuel—coal or natural gas—and, especially for ordnance works, to have access to plenty of fresh water. In 1940, after German armies had overrun France and were bombing Britain in preparation for a cross-Channel invasion, a War Department committee overruled Ordnance and declared that facilities should be located away from major centers of industry and separated from each other to minimize losses in the event of accident or military attack. They were also to be placed out of range of enemy aircraft in the heartland east of the Rocky Mountains, west of the Appalachians and 200 miles away from the Canadian and Mexican borders. The War Department also brought engineering factors under consideration: the topography, composition and drainage capability of the soil, the presence of gas or oil pipelines underground and high voltage power lines overhead. Technical factors, such as electric service, water resources and transportation, with the emphasis on railroads, were also important. After the War Department made its recommendations, a special presidential committee weighed in with the notion that local economic conditions should be considered. Munitions facilities were to help impoverished communities recover from the Depression and stimulate industrial development in the agricultural southern and western states. In September 1940, a disastrous explosion injured 100 people and killed fifty more at the munitions works operated by Hercules Powder at Kenvil, New Jersey. As a result the Ordnance Department doubled the required distance between production lines at works facilities, which meant that the already sprawling installations would gobble up even more land. At the end of 1940, the War Department concluded that many of the sites it had already chosen — including Radford, Virginia, and Charleston, Indiana, which served as models for Badger Ordnance — contradicted its own recommendations and that of the President. Radford, for example, was built on hilly ground that required extensive grading and leveling. As a result, the Ordnance Department and the Quartermaster Corps agreed to work together and create a War Department Facilities Board to tighten up the whole process of site selection. It would recommend sites only if they were 200 miles from the Canadian and Mexican borders and in between the Appalachian and the Rocky Mountains. Studies of transportation, labor, housing, fuel and other resources also had to be completed prior to site selection and, to keep the door open for political influence, the choice was not final until it was approved by the President.

Not In Massachusetts

Some time in the winter of 1940-'41, a man who identified himself as representing the North Western Railroad and the War Department visited the office of Sauk County Treasurer C. H. Wilcox in Baraboo and requested a detailed "plat of the bluffs between Merrimac and Devils Lake State Park." Wilcox didn't make much of the visit until the following October when a question from a reporter for the *Milwaukee Journal* stirred his memory. With the reporter's help, Wilcox also remembered "the inquiry he received from a Chicago engineering firm in May." The letter "requested detailed plats of some 60 square miles in the Towns of Bear Creek and Spring Green..." Then there was the visit only a few days previous from "two army men from Rock Island arsenal [who] obtained detailed plats of the town of Fairfield in the northeast corner of Sauk County."[9]

All of these inquiries went unreported in the local press, but the county treasurer's office was rarely a flash point for news and an out-of-towner asking for a plat map was hardly a stop-the-presses event anyway—at least not until October 29, 1941. Only then did the earlier visits by railroad and Army men, and the letter from the engineering firm in Chicago, set heads in the courthouse to nodding as if to affirm that they had known about it all along. October 29, 1941, became memorable in Baraboo and beyond because on that day the office of U. S. Representative William H. Stevenson announced the "Construction of a $65,000,000 powder and acid works by the Hercules Powder Co. ...at Merrimac, on the Wisconsin river in Sauk county..."[10]

It was customary then, and is today, that projects requiring a large expenditure of federal taxes be announced by the Representative of the Congressional district in which the project was located, regardless of party affiliation. A La Crosse resident and graduate of the local teachers college, Representative Stevenson put himself through the University of Wisconsin law school by teaching at Leo Crowley's alma mater, Madison Central High. After finishing law school in 1919, he joined the Republican Party, served as county attorney in Richland and La Crosse counties, and was elected to represent Wisconsin's Third Congressional District in November 1940. The arrival of a multi-million dollar defense installation in his district could have given a freshman Representative something to crow about, but, by his own admission, Stevenson had nothing to do with the siting of Badger Ordnance. The Ordnance Ammunition Production Office in Wilmington had already "built up a file of available sites throughout the country with the assistance of the various railroads..." and, if Treasurer Wilcox is to believed, was collecting information on the Merrimac area either before or very early in Representative Stevenson's first term, which did not begin until January 1941. Six months later, in June of 1941, the Chief of the Construction Division of the Army Quartermaster General, General Brehon Somervell, had wind of a Wisconsin project and asked the Ordnance Department for information. Somervell was interested because until mid-December 1941, the Quartermaster, not the Corps of Engineers, held responsibility for the construction of new munitions facilities. In November, the General was informed that the new project would be "in the vicinity of Merrimac, Wisconsin...has been approved by the President, the Under Secretary of War and the Office of Production Management... [and] is designated as the Badger Ordnance Works."[11]

In the vicinity of Merrimac, Wisconsin, Stevenson's announcement launched successive and conflicting waves of shock, disbelief, anger, acceptance, greed, opposition and glee. With the actual site of the new plant yet to be specified, public attention focused on Merrimac. On the weekend of November 1-2 the tiny village of 234 souls was all but overrun by job seekers, real estate speculators and other "city slickers" who told ferry pilot Herman Wartner they were "going to Merrimac to buy the place out." Merrimac was not for sale, but Governor Julius Heil was sufficiently perturbed to send three presumably worldly Madisonians to warn the country folk about shady deal makers. He needn't have worried. As ferryman Wartner said, "If anyone is going to make money on my place, that fellow is me."[12]

The typical reaction to the startling news was exhibited by Merrimac storekeepers Grace and Harry Johnson, who are also candidates for the honor of being the first residents of Sauk County to hear about the new facility. Grace answered the phone the morning after Representative Stevenson made his announcement and got an earful from a reporter for the *Milwaukee Journal*.

"Did you ever in your life! They say Merrimac is going to have a $65,000,000 powder plant and that it's going to get started right away," said Grace to Harry, with whom she shared duties as postmaster.

"Don't believe it," Harry replied. "Must be Merrimac, Mass. Somebody's always getting us mixed up."

The mix-up continued for about another three weeks, but not over Massachusetts. Speculation was centered on Merrimac, Wisconsin, which just happened to be on the Chicago and North Western Railroad and had a post office. As for where the actual plant would be located, several nearby sites were mentioned. The hills and valleys northeast of the village and on into the Columbia County Town of Caledonia were talked up, so were the flat but marshy grasslands of Fairfield farther north, the Sauk Prairie directly west and the sandy flats of Spring Green Township to the southwest. About a week into November, an airplane appeared over the Sauk Prairie and, starting at King's Corners and using Highway 12 as a perpendicular baseline, cast cold shadows over the harvested fields as it flew precise patterns back and forth from the road east to the river. By the fall of 1941 the airplane had become the herald of war, even in Wisconsin. Japanese Mitsubishi fighters over China in 1938, German Stuka divebombers over Poland in 1939 and over Russia in June of 1941 had carried combat to a countryside at peace. Combat did not come to the peaceful Sauk Prairie in 1941, but the war did, and its herald was an airplane armed not with a gun but with a camera.

Events moved quickly now, with important action taken not just in Sauk County, but also in Washington, D.C., and Wilmington, Delaware. At Hercules Powder, engineers had been working on preliminary designs for the "Merrimac plant" based on their Radford Ordnance Works since mid-October. On November 10, the Army issued a letter of intent to hire Hercules to design, engineer, build and equip the plant. Soon the aerial photos arrived from Wisconsin and Hercules drafters began to line out topographical maps based on them.[13]

On November 14, officials from the federal office of "defense health and welfare," accompanied by Milwaukee reporter Laurence Ecklund, visited Baraboo, Portage, Sauk City and Prairie du Sac to assess their ability to adjust to the "impact of a defense boom." They concluded that Baraboo and Portage were in the best shape to absorb the shock and that Baraboo, for example, could provide housing for up to 1,500 new residents. Portage Mayor William Mohr touted his city's hospital, its new ordinance regulating trailer camps and it streets, which were "cleaner than any other streets in the state." When the federal officials asked about vice, Mohr reported that "Portage hasn't had a sporting house for 30 years." In Prairie du Sac, village president Walter Doll said that the board was about to enact a trailer camp ordinance, was thinking about hiring a second police officer and that the lock-up had two big cells that could handle up to four prisoners at once. In Sauk City, village president John Schneider reported on a rumor he had heard saying the Milwaukee Road was about to extend its spur to Merrimac and that he figured the new defense plant would require "extra provisions" for schools. The federal people encouraged the locals to organize an area defense council, similar to the county Council of Defense created for World War I, and were pleased when Stewart told them that a meeting for that purpose would soon be held in Baraboo. None of the local people — the boastful Portage Mayor, the diligent clerk from Baraboo, the well-meaning elderly men from the Sauk villages — had any idea of what was about to take place and how it would slam into their communities like a runaway bus through a highway billboard.[14]

On November 17, seventy representatives from cities, villages and towns in Sauk, Columbia and Dane counties met in Baraboo under the aegis of the League of Wisconsin Municipalities. Founded at the turn of the 19th century, with Baraboo Mayor Frank Avery as one of its first board members, the League was, and is, an educational and lobbying organization for cities and villages in Wisconsin. Madison Mayor James R. Law was the League's president in 1941 and represented the largest city in the region, so he chaired the meeting. The group resolved to work together to manage the "defense boom" and Law named a steering committee that included Baraboo City Attorney Robert H. Gollmar, Portage Mayor Mohr, Merrimac President Frank Schulz, Prairie du Sac trustee R. J. Schwarz and Sumpter Town Chair Garth Premo. Although well-intentioned, the area council was quickly knocked into irrelevance by the accelerating pace of events and crippled by internal rivalry for the spoils of the boom. Mayor Law, who seemed to excel at inserting his foot in his mouth just prior to speaking to the press, dealt the area defense council a major blow early on when he bad-mouthed the ability of Baraboo to supply needed public services for the flood tide of new residents the powder plant would attract, compared to Madison, which the Mayor was confident had everything any crowd of newcomers could want and need. In response, the Baraboo Chamber of Commerce, prickly over Law's dismissal of St. Mary's-Ringling Hospital as a "sub-station" where the doctors "treat little more than whooping cough," threatened to boycott Madison wholesalers and persuade businesses throughout Columbia and Sauk counties to join in shunning the capital city. This shabby squabbling for spoils among the cities stood in sharp contrast to what was happening on the Sauk Prairie, where government people were already doing difficult work in which they could take little joy and farm families were facing the loss of their homes.[15]

On November 18, Captain Walter G. Hoar, U.S. Army Quartermaster Corps, who had just finished his assignment to appraise and acquire land for Fort McCoy in Monroe County, told reporters that earnest work would now begin in Sauk County. He also announced that the Merrimac plant would be built "farther west than believed originally...about 8,000 acres of land will be appraised and acquired...it will be largely in the Town of Sumpter...The area which is a rich farming community is known as the Sauk Prairie." Although the final boundaries had yet to be drawn, Hoar was the first government official to name the Sauk Prairie as the site for the plant. He opened a "land office" at 108 Third Street in Baraboo on November 24, and Federal Land Bank appraisers on loan to the Army would soon be visiting farms in Sumpter and Merrimac. The link is not clear, but it is likely that Hoar's "farther west" statement was based on the findings of H.N. Wendle, an engineer for Hercules Powder who visited "property near Merrimac" earlier in the month and filed a report in Wilmington on November 20. It was Thanksgiving Day in Wisconsin, but, had they seen it, the farmers of Sauk Prairie would have found little to be grateful for in the report of the Hercules engineer. In describing the property, Wendle found that "the eastern portion toward the north is rough and uneven but can be utilized for sidings, storage...[and] is very like the original ground in the double base and nitroglycerin area at Radford; but in traveling west...the topography is ideal for the construction of a powder plant...The western portion of the site is almost perfect..."[16]

On November 19, President Franklin Roosevelt signed the orders establishing the Badger Ordnance Works in Sauk County. He authorized the expenditure of sixty-five million dollars to build a works that would have three production lines to manufacture 300,000 pounds of smokeless powder and 3,000 pounds of diphenylamine per day, plus as much sulfuric acid as these processes required. Making the announcement for the President in Washington was Representative Theodore Wasielewski of Milwaukee, the state's senior Congressional Democrat. He had already asked the War Department for preferential treatment in the bidding on Badger for Wisconsin contractors, as had Republican Senator Alexander Wiley. State Democratic party chair and national committeeman Thomas King, who held no elected office but exhibited great interest in Badger, not only wanted to put Wisconsin contractors at the head of the line but also tried to spread the money around as much as possible. He proposed that the Badger project be bid out in $500,000 slices, so that many small contractors could have a helping of the pie. No one in Wisconsin, the politicians or the contractors, had ever had a crack at such a colossal chunk of public largesse and it is not surprising that Wiley, King and Wasielewski wanted to spread it where it would do the most good, for them. Badger was big for 1941, even though the initial $65 million estimate was only about one half the final bill toted up in 1945. Badger was a larger and more demanding construction job than the $22.8 million expansion of Fort McCoy. It would employ more people than the new Navy radio school in Madison, which would eventually train 40,000 men and require the expansion of the Truax Field airport. Badger also dwarfed other Wisconsin pre-war defense spending champions, easily surpassing the $26.5 million awarded to Milwaukee's A.O. Smith company to build bomb casings and the $35 million allotted to Manitowoc shipbuilders to produce submarines.[17]

Word of the President's action had yet to reach Baraboo on November 19 when the Sauk County Board passed a resolution to "go on record as being opposed to the erection of the proposed powder plant," by a vote of twenty-seven yeas, eleven nays. The measure was submitted by Sumpter Supervisor Garth Premo, whose land was likely being inspected by Captain Hoar's men even as he made the motion. Co-sponsors were Earl Cramer of Merrimac village, Nean Shanks of Merrimac town and Conrad Nolden of Prairie du Sac town. Premo said that he was less concerned about the economic effects of the plant's opening than of its inevitable closing. "If they build that plant here the result will be that it will reduce the total valuation of the county after the plant is no longer needed. It will leave us with a lot of buildings on the land and a lot of people the county will have to feed." Premo raised what was likely to be the most persuasive argument he could make to a county board that had been dealing with swelling welfare roles for the better part of a decade. He did not mention the loss of productive farms or the pain inflicted on the farmers. He probably did not think it was necessary since he was speaking to a board made up of farmers and their neighbors from villages dependent on the farm trade. Of the eleven supervisors who voted against the resolution and, implicitly, in favor of the plant, D.F. Murphy was from Prairie du Sac village, George Gavol from Sauk City, George M. Haase from Spring Green village, Arthur Cheek and George MacFarland from Baraboo city, all places with businesses likely to benefit from Badger. Conrad Nolden of Prairie du Sac town and Roy Gasser of Honey Creek also voted nay, even though their towns bordered Sumpter, while the others who voted not to oppose Badger represented towns in the northern part of the county where the impact of the powder plant would be much lighter than in Sumpter and Merrimac.[18]

The county board's vote added another voice to the chorus of letters and telegrams individual citizens had been sending to their representatives. On November 21, August Derleth published the first of several open letters he would send to Congress and the President. Nationally known for his short stories and novels, Derleth was the county's resident celebrity. More meaningful than his fame was his choice of subject matter. He wrote of his home, of the Sauk villages, of the Prairie at his doorstep, and viewed their alteration by the powder plant as a painter would look on the irreparable destruction of his canvases. To President Roosevelt he wrote, "It is remotely possible that you may be familiar with the Sac Prairie...its beautiful earth, its villages and people — some of us who represent, like myself, a fourth generation on this land, in this village; you can understand with what apprehension we look forward to the influx of elements which may work blindly, unwittingly, unintentionally, perhaps, but nevertheless inevitably, to bring about such change here that not even the inexorability of time can ever again rectify the errors that may be done."[19]

Many other individuals of less fame and literary renown, but with feelings as heartfelt, wrote to political leaders, farm organizations and the press. Among the most prolific letter writers were Albert Mueller, his wife, Dora Accola Mueller, and their son Erhart. They were Sumpter farmers whose land was not threatened by the plant, but who felt impelled to stand with their neighbors in defense of their homes. Erhart Mueller was so marked by the experience that, after retiring from farming in 1973, he made it his life's work to preserve in his archives and re-create in the pages of his books the community that Badger Ordnance destroyed. It is yet another irony of the Badger Ordnance story that the Sumpter he had known as a boy and young man had to die at the hands of the government so the elderly Erhart Mueller could bring it back to life forty years later.

On November 22, the Muellers probably attended, along with August Derleth and about 250 other opponents of the powder plant,a meeting at the Prairie du Sac high school. Since they were the focus of events, farmers whose land was threatened were asked by Chairman Garth Premo to sit in the center seats, and about eighty farm couples came together. Premo then turned the meeting over to Emil Jorgenson, supervisor of county farm agents for the University of Wisconsin Extension, who raised the crowd's spirit by announcing that the United States Department of Agriculture had protested the siting of the plant on the Sauk Prairie. He went on to say that, "There are reports that the farmers are opposed to selling their land even if they could get a good price. The purpose of the meeting is to point out how you feel about it." The farmers weren't very talkative but they responded by approving the text of a telegram to be sent to Wisconsin's Senators and Congressional Representatives, as well as the National Farm Bureau, the National Farmers Union and the National Grange. It read: "Preliminary survey seems to indicate that the proposed government powder plant to be located in Sauk County, Wisconsin, will include a large area of the best farming land to be found in the county, state or nation. In view of the fact that there is so much land suitable for such a plant and of small value for farming purposes, we, the undersigned citizens of the Sumpter-Merrimac area, vigorously protest the inclusion of this high value land in the area to be designated for the plant. We petition you to use your influence to correct what seems to us a gross injustice. Any effort you may be willing to put forth in our behalf will be greatly appreciated."[20]

The telegram was signed by all eighty of the farmers and dispatched the next day. Also riding the Western Union wire was a message to the White House from William Evjue, editor and publisher of the Madison *Capital Times* newspaper. It said, "I believe that it will be unpopular if the rich farming land in the Sauk Prairie area is appropriated for the new Merrimac powder mill project...The people of southern Wisconsin feel that land less desirable and less productive should be used for such a purpose." The message was transferred to the editorial page of the *Capital Times* on November 24. Two days earlier, the Wisconsin Agricultural Conservation Committee, state agent of President Roosevelt's Agricultural Adjustment Administration, telegraphed Agriculture Secretary Claude Wickard to protest the siting of the powder plant "in a large area of the most productive agricultural land in Wisconsin....This section of Sauk County is one of the heavy producing areas of milk, poultry and pork products needed for national defense." The group had high hopes for help from Secretary Wickard. He and his wife had visited the Sauk Prairie in the summer of 1939, stayed on the H.K. Hill farm, and visited the Irvin Luetscher and Swain Mather farms, before wetting a line to fish in Weigand's Bay.[21]

The barrage of opposition continued, with the progressive Wisconsin Council of Agriculture and the conservative Wisconsin Farm Bureau coming out against the powder plant. "When civilian authorities surrender their rights to the military, the stage has been set for despotism," wrote the Council's executive secretary, Milo K. Swanton, to Dora Mueller. The Council represented thirty-eight Wisconsin farm organizations, "federated together...doing all we can in this effort to put the powder plant on sub-marginal land." At its annual convention in Madison on November 28, the Farm Bureau, with Swain Mather, John Litscher and Emma Kindschi attending, met with former Indiana Governor Clifford Townsend, head of the USDA's Office of Agricultural Defense Relations. After the meeting Townsend wired the War Department, saying, "It is criminal to destroy such a garden of Eden. It would seem that a more suitable location for the plant could be found in Wisconsin."[22]

Orland Loomis thought he had one in Juneau County. A Mauston native, and former attorney general, "Spike" Loomis would become the last

Progressive to be elected Governor of Wisconsin in 1942. He helped the supervisors of the Juneau County Board formulate a petition to the federal government to build the plant in Juneau, upstream of Wisconsin Dells. George Engebretson, mayor of the City of Stanley, in eastern Chippewa County, wanted to see the plant built on the Eau Claire River in Clark or Eau Claire counties because the federal government held title to extensive acreage there. "Eight thousand, 10,000 acres, it's there waiting for it—and it wouldn't cost a dime," he told reporters. Gus Johnson, the mayor of Ashland, and Congressman B.J. Gehrmann thought they had the trump card on site selection. Instead of building a new facility, they urged the War Department to modernize the Du Pont TNT plant at Barksdale near Washburn, Wisconsin. Built in 1906, it was the only commercial plant in the United States producing TNT in 1941. Barksdale had a workforce of 600 people experienced in making explosives, while Bayfield, Ashland and neighboring counties had many unemployed miners, ore dock and lumber mill workers, plus underutilized housing in the city of Ashland, which was also served by six railroads. Abundant land not in farms surrounded the site and the biggest reservoir of water in North America was just down the hill in Lake Superior. On the negative side, Barksdale did not meet War Department site standards because it was less than 100 air miles from Canada and potentially in range of enemy bombers. That an alternative to the Sauk Prairie should be found also struck a chord at the other end of the state. On the same day the Wisconsin AAA wired Agriculture Secretary Wickard, the *Milwaukee Journal* asked in its front page opinion column, "why a powder plant need be erected on rich agricultural lands in a state that has sand wastes, marginal lands and cutover areas not suitable to agriculture?"[23]

On November 24, the day the federal land office opened in Baraboo, Emil Jorgenson was out talking to farmers on the Sauk Prairie. He stayed so long that he was late for a meeting he had promised to attend about forty miles north at the Adams County courthouse in Friendship. As he drove up Highway 13 from Wisconsin Dells, he traveled over a flat, sandy plain about half in farms, half in wooded jack pine "barrens." As it had in Sauk, the Green Bay glacier had halted in Adams County. It had left outwash plains that were later submerged under a meltwater lake and covered with a dusting of loess. The similarities ended there, for the soil that evolved to cover most of Adams County was characterized as the "poorest agricultural soil" in Wisconsin. Here were the "sand wastes," the "marginal lands," the "areas not suitable to agriculture" much in the minds of those who did not want to see the powder plant on the Sauk Prairie. Jorgenson knew this country well. He had been one of the Extension agents who started Adams County farmers in their planting of hundreds of miles of pine windbreaks to anchor their blowsand back in 1935. On the evening of November 24, he sat down with George Wood, chairman of the Adams County Board, the chairs of the county Republican,

Democratic and Progressive parties, plus a handful of other officials, and crafted a telegram to send to President Roosevelt. It said:

"You may likely have wondered why the good lord created Adams county in the first place, but now we know it was for the location of this plant. We believe it is a natural.

"Adams County has all the advantages attributed to Sauk County—location on the Wisconsin River, proximity to the same power sources and to centers of population such as Wisconsin Rapids, Stevens Point and Marshfield

"Adams County has better railroad facilities, better road building materials, better sanitation possibilities, cheaper land and no serious problems of rehabilitating landowners."

Wood topped off the effort by saying that "We've got everything they have just 40 miles below us—plus the fact that we want this project."[24]

The Adams County telegram was the culmination of two weeks of feverish activity to preserve the farms of the Sauk Prairie by agricultural people at the local, state and national level. Two days later, November 26, Senator Robert La Follette responded to a letter he had received from Albert Mueller. "I called on Under Secretary of War Robert F. Patterson today to present the protests that I have received from you and other people in the Sauk Prairie country..." he wrote, then went on to say that Patterson, "would make certain that all of the surveys of other sites on less valuable land...are rechecked to determine whether or not there may be other sites that will serve the Army's needs just as well." The news appeared in the press as well. After four weeks of unsettling rumors, followed by a week in which the Army appraisers appearing in their farmyards made the worst of those rumors real, the farmers of Sauk Prairie had some reason for hope.[25]

On November 28, the *Baraboo News-Republic* announced that "instructions had been issued by the War Department to proceed at once to investigate three other sites than the Merrimac area." The location of these sites was not made public but Adams County, the popular alternative, was a likely candidate. Adams had extensive stretches of level ground stretching from the banks of the Wisconsin River on the west to the terminal moraine of the glacier, ten to twenty miles east. The Dell Prairie, for example, just north of the city of Wisconsin Dells, spread over the better part of two townships. It was just as close to the impoundment and hydropower created by the Wisconsin Dells dam as the Sauk Prairie was to Lake Wisconsin and the Prairie du Sac turbines. It also sat above an aquifer that could supply millions of additional gallons of groundwater. The Dell Prairie was only a few miles northeast of the Milwaukee Road tracks in Wisconsin Dells, eight to ten miles southwest of the Chicago and North Western line across Adams County and about ten to twelve miles west of the Plover-to-Portage "P" Line of the Wisconsin Central Railroad. The population of the town of Dell Prairie was about one-half that of Sumpter and its agricultural output a fraction of Sauk Prairie's. Locating a plant there would displace fewer people and have a much smaller impact on farming. Workers and housing were a different matter. The population of all Adams County was 8,449, about one-third that of Sauk County and only about 2,000 more than the city of Baraboo, so it could not furnish as many workers as Sauk. On the other hand, War Department planners did not expect Sauk County to supply the entire Badger workforce and Badger was plagued by a shortage of workers throughout World War II anyway. As H.N. Wendle said of the Sauk Prairie, "The labor question is bad and the housing question is in the same classification." Men and women from a fifty-mile radius commuted to Badger in Sauk County by bus or car and they might just as easily have commuted to a Badger in Adams County. Housing was short in both places as well and would have required the construction of accommodations for workers in Adams County, as it did in Sauk. All these arguments worked for Dell Prairie, but they would also apply to Strong's Prairie upriver in west central Adams County and to the barely inhabited barrens of the Town of Rome farther north.[26]

The stated desire of other communities for the powder plant, the suitability and proximity to the Prairie of other sites and the recheck promised by Senator La Follette gave hope to those who wanted to keep the plant off the Sauk Prairie. Other interests, both national and local, worked to bring it there, as state newspapers reported in late November and early December:

• Survey work continued on the Prairie and after only two days the ground was so dotted with flags on stakes that "observers said it looked like a golf course." (*Milwaukee Journal*, November 24)

• Appraisal work also continued, with Federal Land Bank appraisers valuing farms for purchase by the government. They were preceded by "negotiator" Silas Knudsen, whose job was to secure landowners' permission for the appraisers and surveyors to enter their land. (*Baraboo News-Republic*, November 24)

• A steady stream of powder plant job seekers stopped at the land office on Third Street in Baraboo, but were told to come back when the construction company opened its office in town. Police Chief John Nietz reported that the job seekers "seem to be mostly well dressed and prosperous looking," and the police had not yet found it necessary to jail any of them. As the number of inquiries increased, the Chamber of Commerce set up a register in its office where job seekers could sign up to be contacted when work was available. (*Baraboo News-Republic*, November 27 and 29)

• Baraboo's Brittingham & Hixon lumber company announced plans for a new lumberyard at Merrimac and hired contractor George Isenberg to build it. (*Milwaukee Journal*, November 24)

• The Columbia County Board telegraphed word to Washington that it had adopted a resolution, "welcoming the location of such a large industry [as the powder plant] in or near Columbia County. (*Capital Times*, November 26)

• The Madison and Wisconsin Foundation, a business organization in the capital, wired War Secretary Patterson that "when the government decides that the Merrimac location is the proper site for the powder plant we here accept that decision in a wholehearted spirit of cooperation..." (*Capital Times*, November 26)

• The Baraboo City Council went on record as favoring the plant and Mayor Carlton Hornung named a citizens committee to further the effort. The city's Chamber of Commerce was even more active. Local attorney and county Democratic Party leader William Hayes was dispatched to Washington to visit the War Department and Congress and deliver the message that Baraboo business wanted the plant in Sauk County. Hayes later reported that the powder plant would trigger an economic boom unlike Sauk County had ever seen. Everyone would benefit, even the fish in the Wisconsin River, since the plant would draw sixty million gallons of water a day out of the Wisconsin and return it "as pure as spring water." Chamber members Clark Wilkinson and John Lange traveled to Joliet, Illinois, to gather information on the impact of the Elwood Ordnance Plant, which occupied 37,000 acres of prime farmland nearby. They reported that the "businessmen and other residents of Joliet are well satisfied." The farmers protested but "all got a very fair price and all were well satisfied, with most of them resettling on farms within a radius of 200 miles." (*Capital Times*, November 30, *Sauk County News*, December 4, and *Baraboo News-Republic*, November 29 and December 4)

• Sauk City was more reticent than Baraboo. The village board refused to pass a resolution presented by downtown businessmen in outright favor of the plant, but pledged full cooperation in the event the plant was built on the Prairie. Officials in Madison, Portage and Lodi went to the other extreme and urged the federal government to not only build the plant on the Sauk Prairie, but disregard all protests therefrom. (*Capital Times*, November 30, 1941)

• Thomas King, the Democratic Party chief, was the guest of the Baraboo Chamber of Commerce at a luncheon in the Silver Room of the Warren Hotel on November 25. Under normal circumstances, a Democrat politician at a business lunch in Baraboo would have been as welcome as bad fish, but mutual financial interests make enemies friends and King told the Main Streeters what they wanted to hear. He rhapsodized that the "area was due for a wonderful change...the powder plant would be like a large park...there would be no unsightly buildings. The ground will be sodded...trees planted and with the use of camouflage which hides the area as a defense measure, it would all be a thing of beauty." He also said that "only two spots in the United States were deemed suitable for this plant...here in Sauk County and another one in Minnesota." He also insured his listeners that " The Democratic party will not as a political party mix up anywhere in the defense program," thereby raising, but not answering, the question of why — if not for politics — he had come to Baraboo. (*Baraboo News-Republic*, November 25)

• A week later, Congressman William Stevenson jumped, if not into bed, then on the bandwagon with King. In November he had said, "it is a mistake to use the very fine fertile land in the Sumpter area...for a powder plant when there are so many locations near there that are sandy and not of any particular use." On December 2, he visited Baraboo and delivered a press release that said, "I am firmly convinced that the powder plant should be located in the site chosen in the Sumpter-Merrimac area. In these times of crisis, everyone will be forced to make sacrifices...Farmers will be treated very well by the government."[27] (*Baraboo News-Republic*, December 2)

• The next day, Thomas King called Mayor Hornung and said that "plans for the powder plant in the Merrimac vicinity are going ahead." King had contacted Washington, "this morning, talking to Leo Crowley...at the capital." This is the sole public reference made to Leo Crowley and Badger Ordnance during this period. That it was made by the head of the state Democratic Party is no surprise.

After his conversation with Crowley, King was able to tell Hornung that, "I can assure you definitely that plant is set for the Merrimac vicinity and you can let your committee know that this is definite." (*Baraboo News-Republic*, December 3, 1941)

• On the outskirts of Prairie du Sac, surveyors from the Milwaukee Road were staking out a right-of-way to lay new tracks across the Weynand, Wyttenbach, Belsky, Moely, Colby, Gruber and Kindschi properties. Captain Hoar reported that he had obtained permission from about eighty percent of landowners on the prospective plant site to survey and make soil borings on their land. (*Sauk County News*, December 4)

• On December 5, Major Wayne Hauck of the Quartermaster Corps arrived in Baraboo. He been ordered to leave his post as construction quartermaster at the Indiana Ordnance Works, near the small city of Charleston on the Ohio River. Hauck had started construction of 400 brick and steel, plus 200 wood frame buildings at Indiana in October 1940, and completed enough of the works for production to start in April 1941. Having seen an ordnance works through from start of construction through eight months of operation, Hauck was one of only a handful of experienced ordnance construction managers — military or civilian — in the United States. He made several trips in and out of Baraboo in December, always close-mouthed about his duties. As he said, his orders were to "report to Merrimac and await further instructions." No further information was "obtainable from the army officer, relative to his plans, exact site location, when employment will begin, when, where or how...." (*Baraboo News-Republic*, December 6 and 16, 1941)

And so it went, as the calendar marked the passing of days to Sunday, December 7, 1941.[28]

Patriotic, All of Them

Garth Premo was born and raised on the Sauk Prairie. The son of country school teacher Grace Kellogg, who grew up in a log settler's cabin in the Town of Honey Creek, and Steve Premo, whose family moved from New York to a Merrimac farm in 1849, Garth enjoyed the idyllic life of an all-American farm boy at the turn of the last century. He was born in 1889, the same year as Leo T. Crowley, and grew up with his brother, Ken, two years older, as companion and partner in farm boy mischief. The Premo boys helped work the farm in Section 11, Sumpter, which their father had acquired from his father and which would eventually pass to Garth. They finished eight grades at the one-room Waterbury School, then, at their mother's insistence, drove a buggy into Prairie du Sac every school term for four years so they could be in the minority of farmers in their day who had graduated from high school. A self-educated and cultured rural woman, a pillar of the Sumpter Methodist Church who enjoyed poetry and music, Grace Premo created a sense of the world beyond the fence line in her children, and the farm provided sufficient means to enjoy it. The two-story-plus-dormered attic house the Premos built in 1912 had central heat, and was rich in bedrooms, porches and plumbing, with bathrooms on the first floor, the second floor and in the basement so the family could wash off the grit of the farm before they entered the house. When Garth married Nora Steidtman in 1916, the young couple spent their honeymoon in New Orleans. When they returned they moved into a bungalow Grace and Steve had built next to the big house, both of which had been decorated with hand-painted murals by an artist Grace had hired from Chicago. In time, and according to farm family custom, as Garth's and Nora's family grew with the addition of two sons, while Grace and Steve edged into retirement, the couples swapped homes.

Garth developed into a tall, sturdy man and was remembered by Erhart Mueller as a "seven horse farmer," because he had means and ability beyond the range of an ordinary farmer for whom four horses were enough. The Premo farm was, like other Sauk Prairie spreads, above average. It allowed Garth to buy a lot and build a cabin at Weigand's Bay—only minutes away from the farm and its twice-daily milkings—where he could fish and hunt, and to whose cool shores the whole family could retreat on hot summer nights. Owning a lakeshore cabin, as if Garth were a lawyer, doctor or prosperous downtown merchant, was unusual for a farmer, but within the means of the Premos and of other Prairie farmers as well. The farm also allowed Premo to be active in farm organizations and local government. He served as town clerk, keeping the minutes, supervising the polls, conducting correspondence, from 1914 to 1929. He moved up to town chair and county supervisor in 1930 and, as was customary, held both positions until Badger Ordnance forced him to leave Sumpter in 1942. As a farmer, Premo was appointed to the county agricultural committee and, in that position, witnessed at first hand the New Deal expansion of the federal government into virtually every aspect of farming from the baiting of grasshoppers with arsenic to the setting of prices for hundredweights of milk. He had seen plenty of directives from Washington to Sauk County farmers. None of them prepared him for the whirlwind that fell upon his home, farm and community in November 1941.

It started with phone calls from neighbors asking if the Premos knew anything that they didn't about this-powder-plant-everyone-is-talking-about. It picked up with calls from reporters from all over the state also asking for information the Premos didn't have. Next came invitations to meetings like that of Mayor Law's area defense council, which named the Sumpter Supervisor to its executive committee, but by its very nature clashed with the interests any Sumpter representative held most dear. After he submitted the resolution to the county board opposing the siting of a powder plant anywhere in Sauk County, and won the vote handily, Premo became the acknowledged leader of the local opposition. He presided over the meetings of what was variously known as the "Sauk Prairie Farmers Committee" or the "Sauk Prairie Landowners Committee." When they could catch him, Premo spoke to reporters, but he was not as media conscious as even county supervisors would be a few decades later. He is conspicuously absent from the first big splash of publicity directed at the Sauk Prairie farmers, a Sunday spread in the *Capital Times*, parts of which appeared in other state papers, on November 23.

Nora filled in for him and gave voice to the sentiments felt by all the Prairie farmers. "We think we should have been consulted before they went this far," she said. "We can't understand why the government wants to take the best land in the state for a powder factory, when there is so much other land available." At this time, prior to Pearl Harbor and the completion of the appraisal process, what the farmers resented most was that they, although citizens of the United States, had no voice in decisions that so profoundly affected their lives. They had all been taught Lincoln's Gettysburg Address in the cradle of citizenship, the country school, all believed government was supposed to be for the people. When it wasn't for the people, actually seemed to turn against them, their sense of loss and betrayal was real. A century earlier, Yellow Thunder must have felt the same way. He faithfully deposited individual grains of corn in a leather poke to keep track of the tax dollars he paid on his forty acres in Delton because he believed a paid-up American landowner could not be evicted from his home. Yet he was loaded on the train west along with any other Ho Chunk man, woman and child the federal posses could catch. The Prairie farmers were not loaded on a train, but they were evicted from their land. The sense of powerlessness they experienced before a government that was supposed to be their own accompanied and aggravated all the other insults and injuries they would face in the coming months and would not fade from their memories for many years more, if ever.

They were also farmers, dirt farmers in the best sense of the term, who worked the land with their own hands, watered it with their sweat, knew it, cared for it. They could not understand why land that was so obviously meant to raise fine green growing things to sustain life should be deliberately choked off from doing what it was most suited to do. "Maybe we're acting selfishly," said Nora Premo, but we can't understand why the government wants to take the best land in the state for a powder plant..." Invited to speak on a WHA radio broadcast, Garth Premo pointed out that "eighty farmers" whose land was threatened had previously agreed to raise 5.3 million pounds of milk, 175 million eggs and 1.5 million pounds of pork more in 1942 than they had in 1941. It was silly, sinful to take out of cultivation farms capable of so increasing their production. On the Prairie it made as much sense as shooting a blue ribbon plow horse and hitching up a busted plug, or, in terms city folk could understand, gagging Kate Smith just as she was about to sing "God Bless America." But it was still 1941, and American farmers, on and off the Sauk Prairie, had yet to see suburban sprawl gobble up square miles of 200-bushel-an-acre corn ground for cloverleafs, ranch houses and parking lots. The countryside they knew was strictly rural. It was, as it ought to be, land in farms.[29]

Taking the Sauk Prairie out of agriculture also flew in the face of what the government itself had been telling farmers for nearly a decade: that agriculture was in crisis because too many farmers were working marginal or submarginal land and dragging down farmers working good land. That was why there was a Resettlement Administration moving farmers, sometimes forcing farmers, off of wetlands and cutover ground in central and northern Wisconsin. Some of them were even relocated all the way to Alaska. Yet the powder plant was coming to the Sauk Prairie instead of a place like Adams County, whose bumper crop was the sand burr.

Hurt and puzzled, the Sauk Prairie farmers were also shocked. "This is such a sudden blow that we are really stunned," said Lillian Kindschi, "it seems like the bottom has just dropped out of everything for us." The Kindschis and their neighbors were not mobile people, accustomed as their World War II-generation children and their baby boomer grandchildren would be to living in six, ten or more different places in a life time. Nora Premo had been born and raised in her parents' home in east Sumpter, where Steidtmans had lived since the 1840s. She moved to the Premo bungalow as a newlywed, then to the main Premo farmhouse soon after. If her life went as it was supposed to, she and Garth would move back to the bungalow and live out their days with children and grandchildren next door. This was hardly moving at all, but farm people did not move, could not move. They made that choice when they took up farming, when they could have left the land for the cities, as so many of their brothers, sisters and cousins had. Their lives and livelihood were tied to the land and they did not leave it lightly. The thought of giving it all up, of not watching the morning sun filter through the curtains of the familiar wavy-glassed kitchen window, of not finding the coffee cup in the cupboard made from oak milled by your father's hands, of not smelling Grandmother's lilacs in May, of not seeing the neighbor's black-and-whites in the pasture across the road, of not smelling, not hearing, not feeling the familiar, the known and the comfortable was too hard to bear. "I've lived on this same farm ever since I was born — 1889," said Swain Mather. "If I have to go now, I'll never start digging dirt again."[30]

Yet there was resignation, too, at least on the surface and even before Pearl Harbor. "We all have to sacrifice for our country," said Emma Dahnke. "Lots of boys from around here are in the army and navy. Maybe leaving our farms is our way to sacrifice."

"If somebody could show us that it was absolutely necessary for us to move, in order to provide for defense," said Irvin Luetscher, "then we'd move — and move quick. We're patriotic — all of us..." They were patriotic, even when it was not easy, even when they thought their government was not worthy of their patriotism.

On the evening of the day Thomas King lunched with the chamber of commerce in Baraboo, over 100 farm men and their families crowded into the Sumpter town hall for what the Baraboo paper said was a private session. Emil Jorgenson was there, Garth Premo was not. He was in Iowa, along with Christ Gruber, and the Kindschi brothers, Glen and Harry. They had traveled to the outskirts of Burlington, Iowa, where the Army had recently finished building the Iowa Ammunition Plant. It was a shell-loading facility, not a works, that had required the displacement of farmers from land as fertile as that of the Sauk Prairie. Knowing that their neighbors were gathering at the town hall, the men stopped in Davenport, Iowa, to place a long-distance call home. Premo reported that the Iowa farmers had received a "pretty good" price for their land and that they had not been rushed off of it. If the Prairie farmers had to go, at least they would get a good price for their land and the government would conduct the eviction with common sense and compassion. A few weeks later it was revealed that the Army had not intended to pay the Iowa farmers such a "pretty good price." The local real estate broker who had appraised the farms was working on commission and the higher the total value he placed on the property, the larger his paycheck. He appraised the farms at $4 million, twice as much as the Army had budgeted for the acquisition. Warned that the farmers might riot if the appraisals were lowered, the Army sent the chief of its Real Estate Branch to Iowa. A soldier-diplomat, he was able to keep the peace and lower the appraisals, but the Army still had to pay about eighty percent more for the Iowa farms than it had anticipated. It was a situation the Army real estate people vowed never to place themselves in again.[31]

Premo and his friends returned to Sauk County in time to hear Agricultural Agent Dave Williams report on a phone call he had received from the Office of Agricultural Defense Relations in Washington. According to Williams, Washington wanted the Sauk Prairie because of its railroads, water supply, electric power and level topography, which would ease plant construction and that of an airfield that might be built nearby. The Prairie soil was solid enough to support footings for heavy machinery and Sauk County received less snowfall than areas farther north. The call came on the day after Senator La Follette's recheck was supposed to begin and, if it was a result of it, then the recheck went quickly and took place on the desktops of officers and engineers in Washington and Wilmington. No other report that might be read as the findings of the recheck was made public, although Williams did request that Washington send someone out to talk, maybe listen, to the farmers. No one came. In an interview thirty years later, Captain Hoar indicated that he had taken part in a re-evaluation of the powder plant siting. "In analyzing it," he said, "we felt the housing conditions, the rail facilities, the water facilities, were such that we couldn't afford to vacate [the Sauk Prairie] and seek isolated land that didn't have those facilities. It would have meant building a huge housing unit...so we stayed with the selection."[32]

The matter was settled without any formal announcement. After the Pearl Harbor attack it hardly seemed necessary. On December 8, feelings were so strong that the Sauk County Board did not even take the time to come to Baraboo and meet in a special session to change its position on the powder plant. The board's action was reported in the newspapers, but no official record of a meeting was kept. Apparently, once a majority of the supervisors agreed by phone, a "petition" was sent via air mail to Washington. It rescinded the November 19 resolution opposing the powder plant and said "...this is an advantageous location and necessary for the defense of the country."

The powder plant would, "furnish supplies for those boys who are now prepared to make greater sacrifices than anyone will be called upon to make because of these necessary displacements," and concluded irrefutably with "...the liberties that we enjoy come first."[33]

They were patriotic — all of them, including the Supervisor of the Town of Sumpter, Garth Premo.

Condemned

The month of December 1941 is justifiably remembered as a time of tension, shock and sorrow in the United States. On the Sauk Prairie the atmosphere was tense and sad, but the process of establishing the powder plant continued according to the Ordnance Department's prewar plans. Money to build the powder plant would not be available until 1942 and the War Department's letter of intent to Hercules Powder Company authorized no more than $20,000 for the aerial survey, topographic maps, soil investigations and other planning work in 1941. The government schedule, with the announcement in October followed by construction the following spring, gave the landowners time to clear out, find new land and start farming again without missing a growing season. If there was such a thing as a down time for a Wisconsin farmer, a good time to shop for a new farm and to move, it fell between November and February. After the United States entered the war, the pace of rearmament accelerated and the Army paid less attention to the needs of farmers. Landowners evicted for the 11,700-acre Gopher Ordnance Works in Rosemount, Minnesota, were told in March of 1942 that they had several weeks—in spring—to leave their farms. On the Sauk Prairie in December 1941, temperatures were dropping, snow was falling, the ground was freezing. Outdoor construction was suspended, but necessary preliminary work was underway so that ground could be broken as soon as winter ended in 1942.

Land acquisition was the first order of business and the first instance of person-to-person contact between land owners and land takers. Six Federal Land Bank appraisers from Minneapolis, Minnesota, had been assessing property on the plant site since Thanksgiving week in November. The Land Bank had been established in response to the 1930s banking crisis to help credit-starved farmers obtain financing and its farm appraisers were about as experienced as could be found in the Upper Midwest. They were also salaried federal employees, which meant they had less incentive to inflate property values than, say, the commission broker in Iowa who had busted the Army's budget and feathered his own nest by paying better than top dollar to the farmers who lost their land. The Army also presumed that government appraisers with rural experience might be more aware of the sensibilities of the country people with whom they were dealing than contract brokers. Earlier in 1941, farmers near Elwood, Illinois, where Clark Wilkinson and John Lange visited, had so bombarded Washington with protests on the treatment they had received from the Chicago real estate agents the Army had hired to appraise their land that officers were dispatched from Washington to conduct an inquiry. As a result of the discontent in Illinois, the high appraisals in Iowa and controversy in other midwestern states, the Army stopped hiring real estate agents from the private sector and replaced them with appraisers from government agencies like the Federal Land Bank. Badger Ordnance was one of the first places where the new policy was implemented.

Appraiser R.R. Smith, a member of the Land Bank crew who worked on the Sauk Prairie, later described the appraisal procedure he followed while testifying in the condemnation suit the United States filed against the Sauk Prairie landowners. He said that in the course of working for the Land Bank, he had valued property in all but four of Wisconsin's seventy-two counties. Before appraising the Sauk Prairie farms, he had examined recent sales of comparable property in the area and in adjoining counties. He made careful examinations of each property he had appraised, inspecting and measuring each building, consulting soil surveys and assessing the farm's earning power. Smith may have been as thorough as he said, and maybe the other appraisers assigned to Badger Ordnance were too, but it was also revealed in testimony, as reported by Erhart Mueller, that no appraiser had inspected the Swain Mather farm. The appraisal of this 120-acre farm, although it was handily located on Highway 12, was compiled from abstracts on file in Baraboo. The Mather appraisal was also discovered to be one of the farthest off the mark.

When the appraisers did inspect a farm, it could not have been a pleasant meeting for host or guest. On the farmer's side the attitude would have been, by golly, I don't want to sell this place, but if I do, I'm gonna get top dollar. Customary country cordiality was probably not exercised when the appraiser's car rolled into the farmyard, nor was the give-and-take bargaining that usually accompanied the sale of rural real estate. This deal was different, far different from any stockyard dickering or horse trading the farmer had known. The whole farm was at stake, for one. The buyer was setting the price and the seller had no right of refusal. On his side, the appraiser knew that the Army had announced a budget of one million dollars to acquire land for Badger. Cost overruns and confrontations with landowners, as had occurred in Iowa, were not acceptable. Time also mattered, even before December 7, and the work had to been completed as quickly as possible. Six men had about thirty days — counting Sundays — to inspect, assess, make a judgement and file a report on 138 individual parcels. They had to deal with wary, unfriendly farmers, impatient supervisors in Baraboo, Chicago and Washington, plus confusing messages that circulated down from the Army Chief of Real Estate, who told the appraisers to "satisfy the owners but to remember that Uncle Sam was not Santa Claus." Considering the number of Sauk Prairie land owners the appraisers left dissatisfied, no one mistook them for Santa's elves. "It was a very, very trying job," said Captain Hoar, "in almost every case there were strenuous objections." [34]

Hoar continued to deal with those objections until the middle of December. The appraisal work at Badger was just about finished and the

Army required his experience on other projects. Before he left he thanked the farmers and said that, "the government appreciates the whole-hearted cooperation extended to our appraisers in this work." Hoar and other land acquisition officers consistently praised the Sauk Prairie landowners for allowing them on their land. The farmers did cooperate, although they clearly did not like it.[35]

Major Hoar's successor in charge of land acquisition was a civilian employee of the Corps of Engineers who was no stranger to the area or the job, Silas J. Knudsen. Born in Chicago, Knudsen had farmed in Adams County in the 1910s and '20s and managed the Oak Villa Resort in Wisconsin Dells in the early 1930s. In the mid-1930s, he became project manager for the federal Resettlement Administration in central Wisconsin. The job of the "Resettlement," as it was known by its clients, was to identify impoverished shack dwellers attempting to eke out a living on land that could not be developed into viable farms. These operations were of so little value that they could not be even minimally assessed to pay for a one-room school, a town road or any other local government services. After struggling mightily with the problem for many years, federal and state officials determined that the best way to relieve these zones of poverty was to buy the property, move the landowners who still wanted to farm to better ground and take the vacated land for conservation purposes. While the Resettlement Administration did most of its work in northern Wisconsin, its reach extended to the equally submarginal blowsand and wetland sections of Jackson, Wood, Monroe and Juneau counties. Working out of an office in Black River Falls, Knudsen supervised the acquisition of tens of thousands of acres for the Necedah National Wildlife Refuge and the Central Wisconsin Conservation Area. When he signed on with the Corps of Engineers in 1941, Silas Knudsen had more experience in the business of land-taking and farm family relocation than just about anyone in Wisconsin. His title was "Negotiator" or "Optioner". He was the front man whose job called for him to make the first official visit to a farm in order to persuade the landowners to allow appraisers, surveyors and soil testers onto the place, then to accept the government's first offer to purchase. It was a hard task and, despite his experience, Knudsen was not quite prepared for it. Instead of working with impoverished, isolated landowners, most of whom were elated to have anyone offer to pay them for their land, Knudsen now had to deal with successful farm business men and women who expected to be treated as equals, if not with deference. As they saw it, the Sauk Prairie farmers were helping their government in its time of need. The government was not helping them and they should be treated accordingly. That misunderstanding and ill-will should result is no surprise.

Abrupt and businesslike, the negotiator followed orders and did his job, as did other government men in 1941, many of them to a much worse end. He is sadly and best remembered as the man who told the Sauk Prairie farmers that they had been living too well in their fine houses, with electric lights, indoor plumbing and central heat, when an oil lamp, outhouse and pot-bellied stove should be good enough. The story has been retold and embellished many times, but had enough of the truth in it in 1941 that Knudsen's commander felt compelled to officially deny it. Major Hauck placed a memo in the files at Badger Ordnance stating that at a meeting on December 29, 1941, he had not heard Knudsen tell the farmers that "Your standard of living is too high. You people don't need furnaces or plumbing. Get a heater, warm your stomach and then turn around and warm your back."

What Knudsen had said, according to Hauck, is that Knudsen had grown up without central heat and indoor plumbing and was none the less for it. His advice to farmers was that, "the first thing they should look for in securing another farm was good land, and that they might temporarily have to live in a home where they would not have the conveniences they now have in their present homes." The farmers listening to Knudsen did not hear it that way. They were grown men and women who managed successful farm businesses, and most of whom had not made a midnight trip to the back house in twenty years. They had progressed beyond it, took pride in having so progressed and were not about to step back at the behest of an arrogant government man. None of the Sauk Prairie farmers mistook Uncle Sam for Santa Claus, nor did they see the federal appraisers as Santa's elves. Instead, they would remember Silas Knudsen as the Grinch who stole the Christmas of 1941. He himself remembered how difficult the job was and how awful he felt about taking the land.[36]

The Land Bank appraisers finished their work a few days before Christmas, about one month after they visited their first farm. Considering the amount of work they had to do — measuring buildings inside and out, listing features and improvements, drawing site plans, photographing the properties, researching soil surveys, walking the woodlot, looking up tax records and sales of comparable land — a month was surely not too long a time. When they finished they had appraised 138 tracts totalling 10,565 acres, almost all in farms and woodlots, but also one gas station, one resort store, one town hall, three schools, three churches, three cemeteries and twenty-one vacation lots. The entire parcel was valued at $1.1 million, a tolerable — for the Army — ten percent above the budget estimate. Per acre prices for the farms varied according to the quality of the land and the value of improvements. For example, in Section 23 of Sumpter, Bernard and Eleanor Oelke owned eighty acres of prairie ground next to Edward and Maude Mockler's all but identical eighty. The Oelkes' farm was appraised at $5,200 or $65 per acre, the Mocklers' at $9,300, or $116 per acre. Valuations were equally varied across the tract, with a sample of appraised values, running roughly counterclockwise east from the Mockler place on the southern boundary, as follows: Jacob and Emma Gasser, 144 acres at the head of Gruber's Bay, $11,800;

Emilia Waffenschmidt, 237 acres in three parcels, one on Gruber's Bay, one on the river and a third at the foot of the Bluffs in Section 36, $19,350; Alma and Harold Thoelke, 139 acres adjacent to the Thoelke Cemetery, plus eighty acres in the Bluffs, $19,825; William Burkhardt, 172 acres adjacent to the Lakeside School, $19,000; Lucille and Charles Herr, 154 acres on the moraine north of the junction of County ZZ and Highway 78, $11,550; Celia and Ralph Kindschi, 210 acres on the East Sauk Road in the Bluffs, $14,000; Alex Steuber, 249 acres in the northwestern corner of the tract, $22,000; Lyman and Essie Kindschi, 140 acres on Highway 12, just south of Steuber's, $16,500; Paul and Anna Shimniok, eighty-two acres on County ZZ west of Kern's Corners, $14,300; Irvin and Margaret Luetscher, 160 acres just south of Shimniok's, $31,600; Garth and Nora Premo, 159 acres on Highway 12 at the town hall, $21,900; Swain and Cora Mather, 120 acres on Highway 12, $18,700. In every one of these cases, the landowners disagreed with the appraisers and, although it would take two years of struggle to prove it, in every case the landowners were correct, and not by just a few dollars.[37]

In appraising the properties, the Land Bank men were obliged to consider only the market value of real estate and improvements. The Army had paid "disturbance damages" to farmers for some of its earlier land takings, but discontinued the practice by the time its people came to the Sauk Prairie. There would be no payment for housing a displaced family might rent if they had to wait until spring to find a new farm; no compensation for fall plowing already completed, for storage of equipment or animals, for losses due to the forced sale of equipment or stock, for decreased milk production while the cows adjusted to new surroundings; no reimbursement for moving expenses or for the labor involved; nothing for time lost, grief suffered, stress inflicted. Intangibles were not valued, only real estate and only market value. Such a minimal standard was common and would have been acceptable had the market value set by the appraisers not been so inaccurate so often. The farmers were willing to give up home and land for their country. They were not willing to be cheated for it, and they would fight.

Not until the appraisals were finished did the Army make known the actual borders of the property. As indicated in the report of Hercules engineer H.N. Wendle, the property taken was selected because it satisfied the most important requirements for an ordnance works, access to abundant water and level land suitable for building. The moraine ground in the east and northeastern section of the property was taken because it was adjacent to the river, and the powder plant water system, with its five-foot main line, would cross it on its way from the pumping station in Weigand's Bay to the foot of the Bluffs. The moraine area was not considered prime building ground and, initially, none of the production facilities were located there. Rumors that the Glen and Emma Kindschi farm and additional property on the south edge of the property would be taken for an airfield proved to be just that, rumors. Instead, drainage and sewerage disposal dictated where the southern boundary line would fall. Powder plant operations were concentrated on the prairie and designed to run downhill from the Bluffs. The drains and ditches that accompanied operations would also run downhill to discharge into the river via the natural watercourse running south and east into Gruber's Bay. Wendle wondered if Gruber's Bay might not be the best discharge point since it was upstream of Prairie du Sac and Sauk City. If so, then sewers and drains should be directed to run into Otter Creek, which meets the Wisconsin below Sauk City. Considering the rudimentary sewerage treatment facilities used by cities and industry up the Wisconsin, the water flowing by the Sauk villages was less than pristine anyway and, by the standards of the day, the Army's handling of wastewater would be as good as or better than any system on the river. However, had the engineers in Wilmington followed up on Wendle's suggestion, the boundaries of the property might have been extended farther west to facilitate discharge into Otter Creek. Wendle also recommended, and the Army agreed, that a jog be taken out of the boundary line to square the northwest corner with Highway 12, even though it meant the purchase of an additional 300 acres, most of which came from the Alex Steuber family.[38]

The extension of the railroad lines into the property required the additional purchase of land on the south and northeastern ends of the property. In December, the Milwaukee Road repurchased right of way it had sold to Sauk City and Prairie du Sac in 1940, then proceeded to acquire access to the plant by buying a strip of ground crossing the farms of George Weynand, Fred Wyttenbach, Ed Belsky, John Moely, Earl Colby, Chris Gruber and Glen Kindschi. The Chicago and North Western waited to acquire land in Merrimac until after the 10,565-acre tract was identified and condemnation proceedings were underway in January. Major Hauck then dispatched his new public relations person, the engineer Alf "Barney" Oldfield, to help a railroad agent persuade landowners on the northeast border to option their land for sale to the North Western. Consisting of 584 acres in Section 32 of Merrimac, this property belonged to Roy Leiser (160 acres), Christ Huber (144 acres), Frank Roznos (120 acres) and Pauline Euker Spiegel (160 acres), and was purchased in time for the North Western to start work on its spur line into the powder plant in February 1942. Born and raised on the family farm, Pauline Euker had married into the Spiegel catalog family and was the only Badger Ordnance property owner with an address on Park Avenue, New York, New York.[39]

The War Department had been reluctant to publicize the final borders of the property until the appraisals were completed in order to prevent undue speculation on the surrounding real estate. The secrecy was intended to hold down the prices on new farms the displaced landowners might try to buy and to prevent commercial speculators from buying land for non-agricultural purposes. The Sauk County Board of Supervisors helped out on the second point by enacting a zoning ordnance that placed limits on non-agricultural development in rural areas at its November 1941 session. Badger Ordnance and the town of Sumpter would not be plagued by a strip of taverns, pawn shops, dance halls, or worse, directly outside its gates as were many other World War II industrial and military installations.

Although a dozen or so landowners were willing to sell, none accepted the government's first appraisal and offer to purchase. On December 29, Secretary of War Henry Stimson made their refusals superfluous by asking the Attorney General to order that all the property be condemned and taken prior to final purchase. John Boyle, United States Attorney for western Wisconsin, filed the petition in the federal district court in Madison on Saturday, January 10, 1942. The next day the first maps depicting the borders of the 10,565-acre tract were published in state newspapers. The landowners could not appeal, hinder or delay the taking and were obliged to vacate their land by March 1.

As part of the process, presiding Judge Patrick T. Stone would appoint a Condemnation Commission which would conduct hearings at which the government would defend its appraisals and the landowners attempt to overturn them. The Commission would then reappraise the property. If they were unsatisfied with the findings of the Condemnation Commission, the landowners had the right to a trial before a federal judge and jury, at their own expense. In turn, the United States could contest the Commission's findings and sue the landowners before a jury in the federal court. In the meantime, the condemnation order prevented the landowners from delaying the construction of the powder plant through court action. It enabled the Army to act, while the lawyers and landowners argued. The farmers could fight for more money, but they could not halt the military juggernaut from taking their homes.

Only the Dead were Undisturbed

The opening months of 1942 were among the darkest in American history. The Japanese had followed up their victory at Pearl Harbor by invading the Philippine Islands, capturing Manila and pushing American and Philippine forces into the besieged strongholds of Corregidor and Bataan. Imperial forces marched on to occupy the British colonies of Hong Kong and Malaya, capture American-held islands throughout the Pacific and threaten an assault on Australia. On the other side of the world, one German invasion force was poised to capture Egypt and the Suez Canal while another continued its advance into the heartland of the Soviet Union. In German-occupied Europe, the Nazis, no longer concerned whether their actions would bring the United States into the war, implemented their pernicious "final solution to the Jewish question" and prepared to send millions of additional victims to the death camps already in operation.

In the United States, people were troubled, puzzled by the setbacks in the Pacific and fearful for the future. Prayer services had already been held for young men who had died in combat, including Baraboo, where a memorial was conducted at First Methodist for young Leroy Church, who died while on duty with the Army Air Force at Hickham Field, Hawaii, on December 7. Undeterred but nonetheless anxious, young men and women volunteered for the military and lists of men drafted for service became a regular feature in the local papers. Defense work on the farm and in the factory, the mobilization of resources for victory, the need for hard work, dedication and sacrifice were emphasized from the pulpit, the radio and the newspaper page. In this atmosphere, President Roosevelt's Executive Order of February 2, 1942, which ordered the seizure of the property and the internment of 110,000 Japanese-Americans, went unnoticed and unprotested.

On the Sauk Prairie, eighty-some families had little time to notice events in faraway places. They had been given exactly seven weeks to move to new homes and relocate their businesses. They had to put aside their grief, their anger, and focus on the practical aspects of building a new life and livelihood. Those who planned to continue farming had to find suitable land and infrastructure where they could operate efficiently and profitably. They needed good barns, silos and milk houses for dairying; houses to breed, farrow and finish hogs; stock barns for horses, hen houses, sheep sheds, calf pens, chicken coops and fenced yards. They needed machine sheds, good wells and pumps, electricity and telephones. They needed pasture and woodlots—with fences standing—and cropland as fertile as they were leaving. Finally, all of it had to be located on a good road and not too far from the local stockyard and market village. In short, they needed a Sauk Prairie farm not on the Sauk Prairie.

The war presented a great opportunity to do well in farming, one that American farmers had not seen since 1918. It was as frustrating as it was challenging for the displaced farmers. Just when the depressed farm economy started to swing upwards, they'd lost their farms. Yet a displaced farmer who could find a new place, get a crop in the ground in the spring of '42, start milking cows and shipping hogs, could still come out ahead in the wartime boom. While they might be sad about the past they were losing, and angry at their present situation, the farmers had to try to think clearly and coolly about the future. "We tried to look ahead and see what our options might be," recalled Evelyn Hoppe. She and her husband Alvin, with two sons in grade school and two daughters too young for school, worked 120 acres on the Prairie.[40]

Money was part of the equation. The minority of landowners who accepted the government's appraisal could expect to see some government money in the spring and summer of 1942, but it would arrive too late to help buy a farm prior to the eviction deadline of March 1. The first direct purchase was authorized in early February from John Stempke for his twenty-acre woodlot tucked in the bluffs of Section 31 of Merrimac. Most of the summer cottage owners at Weigand's Bay also accepted early payment and a few moved their cabins just off the property to the Idlewild subdivision and would soon rent them to powder plant workers. Throughout late January and February, agents from the Corps of Engineers offices in St. Louis revisited the Prairie farms, promising higher appraisals and urging all to sign options authorizing the government buyout. On February 12, 1942, when Charles Eschenbach, "a member of the executive landowners committee," agreed to sell his ninety-one acres, the "deadlock" was supposedly broken. "I believe that within the next three weeks that practically everyone in the area will sign their option," said J.D. Barber, one of the new agents on the scene. He and "assistant land negotiator" Wesley White had come from St. Louis to smooth over the ill feelings generated by Silas Knudsen and the low appraisals. If they could get a majority of the landowners to sign off, so they said, the government would halt condemnation proceedings. While the new agents failed to obtain the majority of signatures they needed, they did succeed in dividing the landowners and stimulating ill feeling among them. Farmers who held out for a better price felt let down by those who signed off, while many who did sign off had second thoughts about it later. Most of the landowners who stuck out the condemnation process and the trials that followed did eventually come out ahead but they did not receive even a partial payment for their property until May of 1942 and many were not paid in full until the final months of 1944.[41]

"They gave us a dollar bill to make it legal and that's all some of the farmers had when they left Sumpter," remembered Raymond Jessie. Even John Stempke, who agreed to sell on February 5, was not paid until March 18, almost three weeks after the eviction deadline. So even if a majority of the landowners had accepted the government's first offer and signed the option to purchase, they would not have been paid in time to use the money to buy a new farm or merely to meet daily living expenses. In January and February, farmers looking for new places to live and work had to rely on their own assets and on credit, with mortgages backed by the new property they were purchasing and the government's promise to make at least a partial payment for their old farms.[42]

Despite these adverse conditions, they went shopping for new farms as good as the old — in a seller's market. Pete Litscher, who grew up on the family farm west of Kern's Corners, recalled looking for farms with his father, John. "We'd look at a farm and when the sellers saw we were interested, they'd come back in a few days and had the price raised." Alvin Hoppe "spent whole days looking at this farm and that," said his wife Evelyn. Like other dairy farmers, Hoppe had to drive over winter roads to a prospective farm, look it over, talk to the owner or realtor, then drive home — all in between morning and evening milkings.[43]

Mindful of the quality of the farms they required and the difficulties of moving a farm operation, and with only fifty days to clear out, the displaced farmers focused on property in Sauk, Columbia and Dane counties. Of sixty or so landowners on a list published in March 1942, no one showed interest in moving north out of the county or west beyond the towns of Freedom, Troy and Honey Creek.

To Columbia County went Lyman Hatz (Portage), William Burkhardt (Arlington), Raymond Jessie (Lodi), Paul Kirner (Caledonia), J. P. Steidtman, Louis Steidtman, Garth Premo, Swain Mather, Alvin and Evelyn Hoppe, Paul and Katherine Patrick (Columbus).

To Dane County went Harold Thoelke (Oregon), Ben Gasser, Ed Schroder (De Forest), Paul Shimniok (Verona), Fred Kirner (Middleton), Lowell Eschenbach, Roy Harrison (West Point), Addison De Weese (Morrisonville), Louis Frese (Stoughton), Emma Dahnke (Cottage Grove).

Remaining in Sauk County were Fred Waffenschmidt, Roy Mather, Frank Reusch (Merrimac), Richard Heimlich, Harwood Page, Roy Kindschi, Lorin Colby (Prairie du Sac), Edwin Schweppe, Irvin Luetscher (Honey Creek), Floyd K. Bass (Troy), John Shimniok, Ed Shimniok (Freedom), John Litscher (Greenfield), Louis Eschenbach, Charles Eschenbach and William Eschenbach (Baraboo). Those who moved farthest from the Sauk Prairie were Howard Ruesch (Watertown) and John Ehret, who looked at land in four states before settling for a place near Janesville.

Although all of these people were farming in 1942, not all of them bought farms that year. Fred Waffenschmidt, Lorin Colby, Ed Schroeder, Fred Kirner, Richard Heimlich, Roy Harrison, Roy Kindschi, Addison De Weese and Louis Frese all rented farms to work or live on.

Of the sixty landowners on the list, about one-third chose to quit or retire and move to Baraboo, Madison, Lodi or the Sauk villages. Among them were Charles Herr, William Heimlich, Charles Kunz, William Simon, Fred Kurtz, Chris Huber, Arthur Weirich, Walter Kindschi, Carl Kirner, William Reusch, Jake Gasser, Ed Mockler, Clara Schlag, Alex Steuber, Emma and Chris Waffenschmidt, Robert Waffenschmidt, Ray Ederer, Lila Welch and Theodore Weigand. Many of these men and women were already retired or had reached retirement age, but a few were in the their fifties, yet too wrung out to start over again.[44]

The month of February saw an auction sale — sometimes two or three —

nearly every day. It was so busy that iron-throated auctioneer Carl Swalheim, who worked for the Thorp Agency that handled most of the farm sales, lost his voice. Knowing that the sale could not stop and that there was another one tomorrow, Swalheim had an assistant call for bids, while he mimed his way through the sale. Nor was bad weather reason enough to cancel a sale. On the day Ray and Frantje Jessie had their auction, the thermometer read -16°F. At midday everyone took a long break and crowded into the kitchen for a cooked lunch and plenty of fresh, hot coffee — courtesy of the Jessies. Warm and full of food, the buyers were ready for their customary after-dinner nap, but out they went into the cold to bid and buy. The outcome of the auctions varied, of course. When Alex Steuber counted up a gross sale of $13,000, some of his neighbors, forgetting how much he had paid for the goods sold and that his life's work had just been placed on the block, said he was "rich." On the other hand, Wayne Hatz can remember his father buying auction items that he could not necessarily afford because he felt, "he had to help out those farmers."[45]

Farmers who were quitting liquidated everything and those who were moving took the opportunity to sell what they couldn't or wouldn't take with them. The gavel fell at Chris Huber's, the largest tract taken, on February 11. Twenty-eight Holstein cattle, seventeen "fresh and milking heavy," headed the sale bill. A "matched span of sorrel geldings" was listed, heavy-hooved, reddish-brown workhorses, advertised at 3,400 pounds, powerful enough to pull a plow from Weigand's Bay to King's Corners and not be winded. There were twelve bred sows, due to deliver anywhere from fifty to 100 pigs in April so they could be fattened over the summer and marketed in fall, and 300 hens that would start laying over 200 eggs a day when the sunlight increased as winter turned to spring. To feed them, Huber had 600 bushels of oats, 1,500 bushels of ear corn, fifteen tons of clover and alfalfa hay, plus silage. He was also selling a six-month-old McCormick-Deering tractor with rubber tires, a grain drill, grain binder, side rake, hay loader, walking plow, sulky cultivator, manure spreader, wagons, a motorized DeLaval cream separator, ten-gallon milk cans, and on and on.

At his auction Alex Steuber listed thirty head of Holsteins, a Universal Milking machine and 300 white oak fence posts, sharpened. Ed Mockler had twenty head of "good Guernsey cows," a new McCormick-Deering mower that had "cut ten acres" and, of course, "a lot of small tools and goods too numerous to mention." Harry Weirich, the old thresherman, had a field full of old equipment, parts and metal scrap — twenty tons of which were sold and recycled for the war effort. All the auctions were similar sales of the goods and gear of farms with one foot in the old days of horse plowing, hand milking and oat threshing, and the other foot in the modern times of alfalfa hay, power takeoffs, electric fencing and pickup trucks. Each auction was also unique in that it laid before public scrutiny the assets, operations, bits and pieces of one family's life on the land.

Although bargain prices could be found on goods they might need, the people who were moving farms had no time to attend auction sales. They faced a hectic round of packing and loading, calculating how to move a shed full of tools, a crib full of ear corn, a mow full of loose hay, plus heavy machinery, nursing calves, brood sows, piglets and pullets. So it was easier to handle, loose hay was baled by a hired operator who parked a cumbersome wire-baling machine next to a barn while the farmer forked tons of dried grass into the hopper. Dairy cattle had to be moved between milkings, with the morning's milk handled on the old farm and the evening's on the new. Since no one had a truck much larger than a pickup, no more than two or three cows could be moved at once. "We should have counted how many times we drove back and forth," said Mel Luetscher, "seems like that's all we did for days." The Luetschers had a relatively short move of a few miles to Honey Creek, and four men and one woman to do it. How much longer did it take the Paul Kirners to move to Caledonia, the William Burkhardts to Arlington or the Paul Shimnioks to Verona?[46]

Trucks had to be borrowed or hired, family and friends enlisted to help. Jack Eberhardt, a trucker from Prairie du Sac who had hauled livestock for most of the farmers, now had all the work he could handle and could name his own price. He is remembered, however, for working hard and long and not taking advantage of the situation in which circumstances had placed him and his neighbors. Some folks didn't have to hire a trucker. Since their home was west of Highway 12, the Lyman Kindschi family was among the last to move. On the appointed day, fourteen trucks owned by friends and family rolled in the yard and completed the move to a new place east of Baraboo in one day. Farmers could not drive their trucks on state highways unless they upgraded their license and paid an additional fee. Upon the request of Major Hauck, the State Highway Commission waived the license upgrade for the Sauk farmers — temporarily. It was the only instance of special consideration the displaced farmers received from government at any level, federal, state or local.

The relocation was stressful for farm families. Roscoe Premo, a young man in 1942, said it took him six months to recover from the ordeal. It was even harder for older people. Roscoe's grandfather, Julius Steidtman, could not bring himself to leave his home and had to be literally carried to the car by his sons. It was said that he was never himself again, even though he moved to a good farm near Columbus with his extended family nearby. Moving was especially hectic for farm women, who managed home and family and shared farm chores. Not only did the women have to pack the family's clothes, furniture and household goods, they also had to clear out and pack cellars full of home-processed food. Fruit, vegetables and meat in breakable glass jars and crocks, a hundredweight or two of potatoes, carrots, red beets or apples, sausage in lard, pork in salt, cabbage in brine, wine in bottles, all had to be packed, loaded and — since it was winter — unloaded and stored before it was damaged by the cold. Most farm women also tended poultry which had to be taken care of, even if the birds weren't going to be moved, just slaughtered, dressed out and served for supper. To a greater or lesser degree, and depending on the farm, women also shared the regular chores of milking the cows, loading the full milk cans on the hauler's truck, unloading and washing the empties, cleaning the barn and feeding any animal that was hungry, which was all of them.

"Many nights the boys and I would have finished the chores of feeding and milking before Alvin would get home with the moving truck," recalled Evelyn Hoppe. "Alvin was beside himself. We had to milk and feed the cattle in the old barn in the morning, load and haul them to the new barn to feed and milk them at night. I had to give the movers directions, prepare food for the noon meal and food to take to the new house, get the girls dressed and in the car and myself ready to drive them out to our new farm. My mother was able to come along with us and stay for a few days or I don't think I could have managed."[47]

Most families were multi-generational and help crossed the generational lines in both directions. Nora Premo and her sister Lena Kurtz moved their farms and also helped their parents, Julius and Agnes Steidtman, move theirs. The Bass, Waffenschmidt, Thoelke, Kirner, Kindschi, Reusch, Roick, Eschenbach, Knipschield, Steuber and Weirich families were also multi-generational, with children, grandchildren, parents and grandparents all forced to move. If not a parent or grandchild, there were brothers, sisters or cousins packing, moving and in need of aid. Arthur Luetscher, a bachelor, agreed to sell their parents' home farm to his brother Irvin and his family, while he himself bought another, smaller place nearby. Like most rural communities of the time, the Sauk Prairie was a place of which it was said, "Everyone's related, or going to be." There was hardly a person in Sumpter, northern Prairie du Sac, eastern Honey Creek, western Merrimac and the Prairie villages not related to a family displaced for the powder plant.

What to move was a question, what could be moved was another. To a farmer a manure pile was an asset, but probably not enough of an asset to fork into a wagon and haul ten miles down the road. At first, the Army real estate people told the landowners they could not remove permanent fixtures such as well pipes, water pumps, cow stanchions and drinking cups, then changed their minds and said the farmers could remove any equipment they had installed. Items built into the houses, such as full-length mirrors, kitchen cabinets, bathroom fixtures, lights and special woodwork had to remain. Often the owners could not see any sense in these restrictions. They had built these houses, designed these barns, installed appliances and fixtures which would be of use on their new farms. Now the government would not pay their full worth and would not let the owners take them. It is no surprise that after turning over his property to the government, more than one farmer indulged in a midnight visit back home to remove an article he felt was wrongfully taken from him.

In February, the town roads and county trunks were jammed with cars and trucks, trailers and wagons. Prairie farmers moving out, shoppers cruising for sales, Army surveyors staking boundaries, real estate men pressuring landowners to sign options, contractors looking for work, laborers out for jobs, the idle, the curious and the criminal just nosing around the action—all contributed to the general din filling the air of the gray winter days. Then on the appointed day, the farm women took a final look around the houses they had made into homes. Many would clean the counter, sweep the floor, one last time. Outside the men glanced over the yard, the sheds, the fields as familiar as the grainy calluses on their hands. The kids, more excited than sad, shoved the cow dog in the truck and jumped in themselves. Doors slammed, engines turned, the vehicles bumped out of the drive onto the county trunk one last time, and it was done. A few were allowed to stay a week or two past the deadline, but eighty-some farm families had pulled out. At dawn on March 1, the Prairie was silent. The sun rose over 800 deserted sheds, empty barns and vacant homes, shells with the life sucked out of them. For the first time since Uncle Billy Johnson swam his oxen across the Wisconsin in 1839, lowing cattle did not greet the Sauk Prairie dawn.

Only the dead were undisturbed. Confronted by opposition the customary calls for patriotic sacrifice could not overcome, and having adopted a more conciliatory policy after the condemnation process began, the Army abandoned its original plans to dig up and move the graves in the Pioneer, Thoelke and Kingston cemeteries. "...it is proposed to purchase the fee title on the land in the cemeteries," wrote Real Estate Manager Knudsen to his commanding officer on January 21. The Army would maintain the graves but no more burials would be allowed and access to the cemeteries would be restricted. The "next of kin" of the deceased would be granted special permits to enter the grounds to visit the graves accompanied by a guard. The flesh and blood of the farm community was gone, and the farm structures would soon follow. Only the remains of earlier generations who had built them both would stay to witness all that was yet to come.[48]

The Table of Explosive Distances

On February 25, 1942, six men drove up the town road that branched off from County Trunk ZZ and ran in front of the Bluffs by way of the Waffenschmidt, Weirich, Roznos and Pulvermacher farms. When they arrived at their destination, they parked, and, having obtained permission of the landowners, scuffed their way over the inch or two of snow on the ground to a surveyor's marker staked down just off the line dividing the Waffenschmidt and Weirich properties. One of them carried a fence hole auger, another a shovel, a third had the camera. When they found the right spot, with the bluff rising behind them and the snowy field beneath their feet, five of the six lined up for a photograph. Major Wayne Hauck—twenty-year veteran of the United States Army Quartermaster, newly assigned to the Corps of Engineers, in khaki, pants creases sharp, coat belted—drilled the auger into the ground, pulled up some earth and spread it like a dark cloak on the surface of the snow. To his right, leaning on a shovel, sheepskin coat collar turned up, cigarette between his lips, stood Horace M. Buck, project manager for the Mason and Hanger Construction Company. At Hauck's left was Alfred A. Oldfield, inevitably nicknamed "Barney," in suit, overcoat and snap-brim hat, an engineer who had worked on the Prairie du Sac dam twenty-five years earlier, now employed by the Corps, in charge of public relations for the powder plant. At his side was George Roberts, overcoat buttoned to throat, hat pulled down against the wind, civilian Resident Engineer and Operations Manager for the Corps. Next to him and looking the part in knee-high, lace-up boots, wool jacket and work gloves, was Ed Belcher, construction superintendent for Mason and Hanger. When it appeared in the Milwaukee newspaper the next day, the photo caption said that the officials were ceremonially breaking ground for the 75,000 feet of "cyclone" fencing that would encircle 7,500 acres of the 10,500-acre tract taken by the Army for the Badger Ordnance Works. It was a curious, quaint, peaceful ceremony. They could have been breaking ground for a new school, park or church, except no audience of well-wishers was present, no workers drafted for the occasion, no Army band, no politicians blowing hard on hot air, only the photographer to capture the image and announce the event to the world.[49]

The low-key nature of the groundbreaking was appropriate since the fence was only a minor pop in a construction blitz that would fashion the equivalent of a small industrial city on the farm fields of the ancient Sauk Prairie. In the words of its own historians, the Ordnance Department "was asked to perform an industrial miracle" for World War II, and anyone who visited Badger Ordnance after March 1, 1942, would testify that a miraculous change was underway. Yet Badger was only one of 149 projects making up the Ordnance Department's industrial "miracle," which also included the gigantic Detroit Tank Arsenal that produced one-quarter of all the tanks used by the United States in World War II, and the tiny Eau Claire, Wisconsin, Ordnance Plant that fabricated .30 caliber ammunition for M-1 rifles shouldered by American infantrymen.[50]

Badger fell somewhere in between Detroit and Eau Claire, neither the largest nor the smallest of Ordnance facilities, but still an impressive creation. The statistics describing Badger are always in the thousands, hundreds of thousands or millions, whether the item listed is earth dug, gravel dumped, concrete poured, water pumped, steam piped, sewerage drained, track tied, lights powered, sills laid, studs raised, rafters stepped, doors hung, roofs covered, machinery installed, money spent, people employed and, ultimately, "powder" produced. There never seemed to be only one of anything at Badger Ordnance, except for the American flag which flew from Anna Magli's windmill in the spring of 1942 until it was moved to the ninety-five-foot-tall cedar flagpole the Army erected in front of the new administration building.

Badger was designed and built under the direction of two Army agencies, the Ordnance Department and the Corps of Engineers, and two civilian contractors, Hercules Powder Corporation and the Mason and Hanger Construction Company.

Here is the mission they chose to accept:

- Build a facility to produce millions of tons of two extremely dangerous explosives on thousands of acres of expropriated farmland with no industrial infrastructure nearby except a two-lane concrete highway, a hydroelectric plant and two rail lines.
- Provide the means to supply the millions of gallons of clean water vital to the process, fabricate from raw materials some of the chemical compounds necessary for production and dispose of all the wastewater and chemicals in a reasonably safe manner.
- Hire personnel, including skilled tradespeople, to build the plant even though there are not enough workers available—skilled or not—within a 100-mile radius of the site.
- Hire personnel to operate the plant even though there are not enough workers to do so within a fifty-mile radius.
- Since housing is in laughably short supply, find or build shelter for over 12,000 construction workers and their families for about six months and then for 4,000-8,000 production workers and families for the duration of the war.
- Since workers usually come with children, build a school, a recreation center and a facility to care for children when parents are at work.
- Provide health care, transportation, job training, safety training and security.
- Do all the above while facing shortages of critical materials

including structural timber and everything made of metal, especially structural steel, piping and rails; while the gasoline and tires necessary for workers to travel to the plant are rationed; and while thousands of other equally vital defense construction projects are underway and making the same demands.

- Do all the above even though the demands of war will change, along with orders from Washington and designs from Wilmington, which means that additions, subtractions, alterations and construction will not stop until the war ends.
- Do it fast, there is a war on; the sooner the plant is producing, the sooner victory will come and the fewer innocent lives will be lost.
- Do it safely, there is a war on; accidents claim the lives of workers, delay production and prolong the war, which means more innocent lives will be lost.

Hercules Powder and Mason and Hanger Construction were no strangers to demanding, high profile projects. M&H had been hired to build and Hercules to design and operate the flagship powder plant of prewar rearmament, the Radford Ordnance Works on the New River in western Virginia. Ground had been broken for Radford in October 1940, and, with the contractors pushing hard, smokeless powder production had begun in March 1941. By the end of the year, M&H had also completed the New River Ordnance Plant where powder made at Radford was loaded into "bags" for heavy naval guns and shells for smaller artillery. In January 1942, both M&H and Hercules were ready to start work at Badger and both had personnel on the ground by the middle of the month.

Mason and Hanger was the kind of contractor the Army favored to bring about its industrial miracle. Although more numerous than arms makers like Du Pont and Hercules, companies capable of executing multi-million dollar federal contracts were in limited supply. Consequently, and despite pressure from politicians like Thomas King and Theodore Wasielewski, who wanted Wisconsin contractors to run Wisconsin projects, the big federal money went to big national contractors no matter where their home office was located. Based in New York in 1942, Mason and Hanger traced its roots to an antebellum Virginia railroad contractor. The company become nationally known in the 1890s for its role in reversing the flow of the Chicago River from Lake Michigan to the Illinois River, later dubbed as one of the "seven engineering wonders of the world." In the 1920s and '30s M&H was building subway tunnels in Manhattan and three tubes of the Lincoln Tunnel beneath the Hudson River from New York to New Jersey. In 1935, the company began work on the Grand Coulee Dam on a remote stretch of the Columbia River in the mountains of eastern Washington. A mile long, 550 feet tall, with its base alone covering thirty-five acres, and requiring fifteen million cubic yards of concrete, the Grand Coulee was the largest concrete structure on earth or, as

Woody Guthrie sang, "the mightiest thing ever built by a man." In addition to the dam, M&H also built "Mason City" to house 6,000 workers and their families for the seven years it took to complete the dam and power station. Compared to the Grand Coulee, a $65 million powder plant in Wisconsin was a home handyman's job.[51]

The original contract signed by Hercules and the Army on January 1, 1942, and upon which initial design, construction and contracting was based, called for an ordnance works capable of daily production of 300,000 pounds of "smokeless powder" for cannon and small arms, plus 3,000 pounds of the solvent diphenylamine necessary to process the powder. Plans and contracts changed early and often. About one month after Major Hauck augered his photo-op posthole, plans called for four smokeless powder production lines, one wood pulp and one cotton linter nitrocellulose processing line, two convertible wood pulp and cotton linter nitrocellulose processing lines, one acid plant, one plant to fabricate the nitrating agent Oleum and five lines to manufacture the high explosive TNT, plus infrastructure, maintenance, safety, security, personnel and administration structures. Following a relentlessly optimistic schedule, it was all to be completed and the plant making powder by January 1, 1943. Plans continued to change, starting with the cancellation of the diphenylamine works early in 1942. With construction going full bore, and thirty-nine structures completed, Badger's TNT lines were cancelled on August 1. Finally, in February 1943, "a small addition" of twenty-four structures was announced for the manufacture of ethyl centralite, or "E.C." powder, for use in hand grenades, tear gas canisters and blank cartridges.[52]

Despite the diversions, the Army and the contractors did not lose sight of their original mission, which was to design and build a works to produce the nitrocellulose explosive known as "smokeless powder." First developed in the 1870s, smokeless powder was a relatively new addition to the ordnance arsenal, which had for centuries relied on black gunpowder. Over the years, black powder charges had been crammed down musket, rifle or cannon barrels, followed by simple but lethal projectiles — smooth stones, metal scraps, lead slugs, iron balls. When the charge was ignited the projectile blasted out of the barrel at sufficient velocity to harm property, animal or person. In time, artillerymen refined the art of making projectiles and produced hollow iron balls or shells filled with explosives that would not detonate until they hit a target. This invention was called the bomb and now, instead of merely firing a smooth stone or solid iron ball to slowly knock down a castle wall, the cannoneers could blast it to smithereens.

Other inventors, instead of placing black powder in a heavy but reusable cannon barrel, packed it in a lightweight, disposable tube along with a non-explosive filler. The filler moderated and prolonged the explosion of the powder so the tube itself was propelled into the air, thereby creating the sky rocket. In short time, pyrotechnicians were loading the noses of rocket tubes with small charges of powder and other substances that would burst into colorful displays in the sky, i.e., fireworks. In the early 1800s, British inventor William Congreve filled the nose of his rockets with explosives that would detonate on impact like a bomb and wreak equivalent damage. Congreve's rockets provided the "red glare" to illuminate the "star-spangled banner" at Fort McHenry which inspired Francis Scott Key to pen the American national anthem.

In the late 1800s, inventors combined elements of the bomb and the rocket to produce the modern artillery shell. It was a two-part projectile, with a high explosive bomb in the nose and a less explosive propellant in the tail. When the propellant was fired in the gun barrel it launched the explosive-bearing nose of the shell, which would not detonate until the nose hit its target. During World War I, the mother of all artillery wars until World War II, shells with propellant in the tail and high explosives in the nose were fired from guns ranging in size from small trench mortars to the "Big Bertha" railroad gun with a range of forty miles. Although many Europeans, including the celebrated Alfred Nobel, had developed propellant formulas, its grimly superior performance in the guns of World War I had revealed that the finest propellant made was the smokeless powder patented about a decade earlier by the American firm of Du Pont. It was Du Pont's smokeless powder which the Hercules Company was licensed to manufacture at Badger Ordnance.

Although smokeless powder makes much less smoke than black powder, it is not really smokeless, nor is it a powder. Put simply, smokeless powder of the type made at Badger consists of cotton lint or wood pulp impregnated with an explosive compound of nitrogen. "Smokeless," as Badger workers called it, was adaptable. It could be processed into various shapes, sizes and strengths, with its explosive power toned down to propel a rifle bullet for 1,500 yards, or souped up to launch a 150 mm artillery shell fifteen miles. It is a dry, but pliable substance, whose individual "grains" are made in the shape of miniature cylinders. Grains of smokeless varied in width or length from a few millimeters to a few inches. The explosive power of each grain was regulated by the ingredients in its recipe but also by its size and shape. It could be precisely perforated to resemble a tiny length of pipe with one or more round, rectangular or star-shaped holes running end to end, or grooved on the outside. Badger Ordnance Works was initially designed and built to house all the processes of mixing, purifying, drying, compressing, sorting, testing, storing and shipping smokeless powder. In addition to direct manufacture of smokeless, Badger was also designed and built to fabricate many of the ingredients making up the powder and the chemicals used to process it, such as Oleum, diphenylamine, ethyl ether, nitric acid and sulfuric acid.

Powder plant layout was based on the assembly line principle which had been the industrial norm for decades. Each step in production took place at its own work station with the product rolling down the line from station to station as each stage was completed. The conventions of industrial efficiency and economy dictated that work stations be as close together as possible and assembly lines be housed under one roof. During World War II military contractors routinely erected massive buildings, roofing over acreages equivalent to whole production areas at Badger, but powder plants were not like other defense facilities. The explosive nature of their product required extraordinary safety measures that started with the overall design of the plant.

Space was the first component of safe powder plant design. Rural locations put safe space between the plants and large population centers. Space also played a role at the site, with production lines, transportation facilities and storage areas set well back from the plant boundaries and separated from each other by buffer zones. What looks like wasteful sprawl at Badger is actually a simple and economical safety precaution. Space was also used as a safety measure on the production lines, with individual work stations occupying their own "houses" with plenty of bare ground in between. That an explosion would take place was always the greatest fear. Using space to isolate an explosion first to a single structure, then to a single area, then to the plant itself created concentric rings of safety for the whole operation.

Utilizing space for safety meant distributing hundreds of buildings over thousands of acres of ground according to the specifications laid down in the Hercules Powder "Table of Interplant Distances for High Explosives." The smokeless powder finishing area, for example, was over one mile long by three-quarter miles wide and covered about 650 acres. Within the finishing area, buildings were separated by as much as 700 feet of space. Although the specifics varied from area to area, the principle of utilizing space for safety was followed throughout the facility. Erecting hundreds of small buildings instead of a few large ones meant hundreds of individual sites had to located, surveyed, staked out and prepared for construction. On any given day in 1942, as many as twenty-five survey crews were working at Badger to identify sites for buildings and infrastructure. Work progressed concurrently on fifty, eighty, one hundred, two hundred or more, production buildings plus the power plant, water pumping station, fences, access roads, rail lines, electrical service, drinking and process water mains, sanitary sewers and surface drainage systems.[53]

Safety, coupled with the demands of the war, also required special architecture and materials. Structural steel was in short supply and its use was limited to the power house and the nitrating and oxidation houses. Brick construction, although commonly used for exterior walls at the Radford, the Indiana and other Ordnance works before the war, was deemed too slow and costly after the war began for all but special purposes, such as the three-story nitrating houses at Badger. The extensive use of brick or block construction was also considered unnecessary, since the Ordnance Department assumed that most, if not all, of its new facilities would be dismantled when the war ended. Temporary construction standards for buildings that did not have to last longer than five years were adequate for all but a few specialized structures. Foundations consisted of an eight-inch concrete wall extending from below the frost line to six inches above floor level, but the basements that might have gone with them were not in the plans. The ground floor was a concrete slab left unfinished except where it was necessary to cover it with a non-skid, non-sparking or acid-resistant surface. Wood frame construction was the rule, with walls, joists and rafters or trusses built on site. Frame construction was fast, cheap and required a minimum of skilled labor. It also readily lent itself to one of the special practices of powder plant construction whereby production equipment was set in place on the slab floor, then the walls and roof raised around it. The press houses, for example, were built around the bulky hydraulic presses they were intended to shelter. To support the roofs of wider buildings like machine and maintenance shops, while maintaining open space and conserving material, roof trusses were supported not by load-bearing interior walls but by wooden posts set on piers. Plywood and other types of manufactured sheathing were not as commonly used in the 1940s as they would be in later years, and were earmarked for higher priority uses anyway, so all but a few buildings at Badger were sheathed, sided and roofed with boards. Roofing was rolled "Whitetop" asphalt composition material, reliable if not decorative, colored light gray.[54]

To minimize damage from explosions and insert another ring of safety into the plant, the designers of Badger made extensive use of "blow out construction." Large buildings, like the poaching and blending houses, and not-so-large buildings where sufficiently hazardous work was done — mix houses and block press and dehydration press houses — were divided cross-wise by interior brick or concrete firewalls that extended three feet above the roof and three feet beyond the exterior walls. The firewalls would deflect an explosion and prevent fires from spreading through the building. Identical firewalls were also attached to the east-facing ends of the main administration and office buildings to deflect explosions and contain them inside the production areas. Other press and mix houses had end walls, rear sidewalls and interior dividing walls made of concrete, with the front sidewall made of wood. An explosion would blow out the front wall and the roof, but be contained by the concrete interior divider and by the three exterior walls. The air dry and solvent recovery houses had a similar design, with concrete end walls and frame side walls, plus twelve-inch-wide wood-framed dividing walls with the space between wall studs filled with "poured monolithic gypsum," i.e., plaster.[55]

Last came the barricades. In addition to the blow-out design, buildings with especially dangerous operations were surrounded with free-standing, timber-framed, wood-sheathed, earth-filled walls known as "Repauno" barricades. Should one of the buildings tucked behind a set of Repaunos explode, the force of the blast would be directed upwards. It would blow the roof off and destroy the building, but the barricades would prevent shock waves and fire from spreading to nearby buildings. When the rocket powder line was built in 1944, lumber was in such short supply that buildings less than sixteen feet high were barricaded on three sides with an earth berm, with a Repauno barricade only on the fourth side facing the roadway.

Other special design features included powerful ventilation systems to remove hazardous gases from the workplace, or to recover fumes, such as nitric acid vapors, for recycling. Telephones, electric lines, lights and switches were covered to isolate sparks that could cause an explosion or, in the case of electric lights, mounted outside the building near a window which let a few dim beams of light inside. Elevators, with their sparky electric motors, were isolated in their own square silos and connected to drying houses by catwalk. Their use limited by tight deadlines and shortages of piping, automatic sprinkler systems were installed in only about forty structures in Badger M-1 as the original construction project was identified. In the rocket area buildings had interiors lined with "paraffin impregnated transite," which was a form of cement in sheet form, or with sheet asbestos paneling. Mixing houses, where nitroglycerin was combined with nitrocellulose, had floors lined with acid-, fire- and spark-resistant lead. Finishing houses, where the rocket paste was pressed, rolled and cut, were air-conditioned and protected by automatic sprinklers. Concrete was used more extensively in rocket finishing houses than in smokeless buildings. It was, for example, poured into two-story walls topped with a breakout roof to shield X-ray scanning equipment and into foot-thick, ten-foot-diameter arches, thirty-six feet long, to house extrusion presses.[56]

Perhaps the most conspicuous and unusual design element of Badger buildings was the fast-exit ramp which sprouted from workstations throughout the plant. The ramps were especially arresting on the three-story ether stills, where they looked like misplaced toboggan runs. Made of smooth steel shaped into a trough and heavier than a playground slide, the ramps enabled workers to run, skid, slide or otherwise locomote themselves out of a building as quickly as possible. Ramp or no, the fastest runner on earth could not outpace an explosion in his or her immediate work area, but if an accident seemed imminent, or if the tocsin sounded to warn the workers of an accident nearby, they could get down the ramp faster than a stairway and hit the ground running for a safe spot. Whether intended or not, the ramps were also vivid, omnipresent warning signs announcing the hazardous nature of Badger's operations.

Safety in construction and operation was not just a humanitarian or ethical consideration at Badger Ordnance. It was a macabre necessity. Material, equipment and trained people lost to accidental explosions were lost for production, and the plant was built to produce powder to win a war. If safety measures preserved materials, machinery and people for production, they were good. In addition, were it perceived that the plant was not safe, workers would not sign on. There were plenty of other high-paying, seemingly safer jobs to be had in the war-pinched labor market. Yet, without workers no plant could function, so safe design and operation were not mere frills or fringe benefits, they were a necessity. Although any untimely death is a tragedy, Badger's safety record was as good or better than that of other World War II ordnance facilities, some of which were wracked by explosions causing dozens of injuries and deaths. If on-the-job fatalities are an accurate measure, then building Badger Ordnance during World War II was almost as dangerous as operating the plant during the war, with at least four construction workers dying of injuries suffered while building Badger M-1 in 1942 and 1943.

The pace of construction was brisk, when it was not delayed by bad weather, paperwork snafus, transportation gridlock, shortages of labor, materials and equipment. Orders to "Hurry up and wait," well-remembered by veterans of all military endeavors, often seemed to be in force at Badger Ordnance. Yet, by any standard, an amazing amount of good work was done in very short time. If surveying, drafting, planning and management conferences are included, the work began in December 1941. Railroad-grading and brush-clearing crews hit the unthawed ground in February, work was in full swing as soon as the mud dried in April and peaked in August of 1942. From the first day of March, construction at Badger was not lineal, but simultaneous. Workers were active all over the site, from the water-pumping station in the river to the Oleum plant on the high ground, to the administration buildings on Highway 12. Work started on almost everything at once: laborers placing forms for the footings; linemen running power cables up from Prairie du Sac; steel haulers unloading I-beams for the power plant; section hands spiking down rail track to ties; dump trucks laying gravel on raw, soupy roadways. Everywhere bulldozers, power shovels and graders pushing, digging, scraping, filling, levelling and reshaping the landscape from farm field to industrial plant.

There is so much piping at Badger that the whole operation can be viewed as one vast, intricate plumbing job where the busiest tool was the adjustable pipe wrench. Liquid acids, steam, sewerage, drainage, nitrocellulose slurry, pressurized air, coolants, hydraulic fluid, inert gas, recovered liquid acids and fumes, alcohol, ether, nitroglycerin and water for drinking, cleaning, manufacturing, fire fighting, heating, cooling and waste removal all ran in their own plumbing systems. Pipes and fittings were made of steel, iron, brass, bronze, lead, clay or concrete. All were heavy to lift or set in trenches and, depending on their use, had to be cut, treaded, bolted, soldered, welded or cemented into place by hand. Although not plumbing, the plant's miles of electrical lines also flowed through pipes. Badger was a powder plant but, with only a fraction of its plumbing underground, it looked like a pipe farm.

Construction was labor intensive in 1942. Although hundreds of pieces of machinery were on hand, much of the digging, spreading, unloading, hauling and lifting, as well as all of the nailing, sawing, brick laying, pipe treading and wire pulling, was done by hand. Men worked in gangs, and instead of one or two carpenters with power-nailers sheathing a wall or roof with 4 x 8-foot plywood panels as they would in 2002, dozens of men placed single boards on studs and held them in place by hammering nail after nail, one at a time. At the height of construction in the summer of 1942, the roar of heavy equipment that enveloped the site was punctuated by the blam-blam, ping-ping of hundreds of hammers driving nails. An observer with only a small dose of imagination could transplant Badger to ancient Egypt at the time of pyramid building, or to Panama when the canal was dug — thousands of men working on a thousand jobs in a thousand different places.

To manage all these workers, Mason and Hanger shared the Army's preference for dealing with large contractors. It hired fifty-seven subcontractors from at least nine states to work at Badger, including the Pitt Construction Co., Pittsburgh, Harlan Electric, Detroit, and McCarthy Improvement, Davenport, Iowa. Contractors also came from St. Paul and Duluth in Minnesota, and Superior, Appleton, Watertown, Milwaukee, Madison and Stoughton in Wisconsin. Subcontractors receiving the biggest dollar amounts were: J.J. Duffy, Chicago, $816,737 for the water filtration plant; J.H. Findorff & Son, Madison, $1.1 million for general concrete work and eighty buildings, including two pulp dry houses and the cotton linter warehouse; James A. Norton, $1.2 million for the on-site concrete plant; Bowen & McLaughlin-Birch, Great Falls, Montana, $1.2 million for excavating, roads and railroads; Armco Railroad Sales, Middletown, Ohio, $1.3 million for outside water lines; E. M. Roche, Freeport Illinois, $1.3 million for plumbing and heating; Bentley Construction, Milwaukee, $1.4 million for the administration and shop buildings, hospital and fire station, two mess halls, eight change houses, sixty magazines and ninety-eight shipping houses. The contractor pulling down the largest dollar amount was the Starrett Brothers and Pontarelli & Sons, Chicago, paid $4.4 million to erect 308 buildings. On a list of slightly out-of-the-ordinary contracts would appear the $38,412 awarded to the Carrier Corporation, Chicago, for air-conditioning in the ballistic lab, the $97,720 paid to Daum and Zingen, Wauwatosa, to build the staff houses, the $111,000 awarded to Wolfes-Jensen, Chicago, for forty-six sentry towers and the $1.1 million to Walsh & Steenburg, St. Paul, for 129 high barricades. Unless Madison companies are included, local contractors are all but absent from the list. L.C. Welch, Baraboo, received about $11,000 for plumbing work and Central States Construction, Sauk City, earned $119,000 to supply earth fill for barricades and other uses. Edward Kraemer & Sons of Plain organized a consortium of contractors, including H. Turner, Boscobel, and Arthur Overgaard, Cambridge, to excavate, process and load 30,000 tons of gravel for ballast onto railcars from an expanded quarry at Okee.[57]

As the operator, Hercules Powder also contracted with specialty builders and installers to set up the $29 million worth of equipment it purchased. The acid concentrator plant cost $1.1 million, the Oleum plant $820,000, the ether and alcohol units another $290,000. Hercules was also responsible for installing the steam-generating units in the power plant and the equipment to handle the coal that fired the boilers. Automatic sprinkler systems were installed in forty-two buildings. Eighty-three elevators were erected to move raw materials, also forty-nine steel storage tanks, plus cooling towers, pneumatic conveyors and heavy-duty ducting for ventilating systems. In total, Hercules spent over $4.2 million on plant construction, not counting manufacturing equipment.

The scope and progress of the work during that first, chaotic year or so can be best charted by a construction chronology.[58]

A Construction Chronology

January, 1942

Major George Griffith of the Ordnance Department is named Commanding Officer. He joins Major Hauck, in charge of construction for the Corps of Engineers, H.V. Chase of Hercules Powder and Horace Buck of Mason and Hangar as managers of the project. Offices shared by all four men and their staffs open on the second floor of the Berkley Chevrolet Garage across Third Avenue from the courthouse in Baraboo. Three young woman, arguably the first wage earners of Badger Ordnance Works, are hired for secretarial duty: Mae Moersch, Shirley Gibson, Ruthann Frenz. Another early hire, for Mason and Hangar, is Baraboo teenager John Geoghegan, who becomes Badger Ordnance's first "mailman." He is assigned a new company Ford car and, day after day, stuffs it with mail sacks full of construction plans sent registered mail from Radford. "I had to sign for them all and deliver them to the offices." As the work expands, additional office space is leased in Baraboo at the offices of Wisconsin Power and Light, the basement of the First Methodist Church, the Ringling Mansion/Elks Club and the Knights of Pythias Hall.

At a meeting at the Corps of Engineers division office in Milwaukee, Major Hauck and Horace Buck propose a labor agreement with the American Federation of Labor building trades unions. They will not agree to union organizing on the site, nor to union restrictions on hiring or firing, nor to allow union members to refuse to work alongside non-union people. State labor representatives do not accept the terms, and M&H then agrees to operate on a closed-shop basis and extend the policy of employing only union members to its subcontractors. The AF of L accepts these terms and charters seven new building trades union locals in Sauk County, with John O. McGlashan named temporary secretary of the county labor council.

Construction workers will be hired out of the Building Trades Council office and the Carpenters and Joiners Union hall in Madison or in Baraboo at the Hod Carriers and Common Laborers hall and the Mason and Hanger employment office at 124 Fourth St. To supervise hiring for plant operations, the Federal Employment Service sets up a branch office at the county courthouse.

Construction wages are based on a United States Department of Labor study of prevailing wages in Sauk County. Top wages of $1.65 per hour go to operators of cranes, draglines, trenchers and other heavy equipment. Electricians, elevator constructors, steam fitters, stone masons, asbestos workers and other skilled tradesmen earn $1.37 per hour. Journeymen carpenters, linoleum layers and roofers make $1.15, and painters, $1.10. Plumber's helpers, truck drivers, mason tenders and laborers make $.75 per hour, while beginning carpenters earn the lowest rate of $.62 per hour. Construction workers also receive first aid and other health care benefits as part of their compensation package.

After touring the site the managers select property straddling the Anna Magli and Lyman Kindschi farms off Highway 12 for a two-story administration and a single-story general office building, which, along with the railroad spurs from Merrimac and Prairie du Sac, will be the builders' first priority.

February

Landowners on the northern and western sections of the property agree to allow construction crews to enter their lands to work on the railroad spur and administration buildings. The North Western and the Milwaukee Road railroads begin work on spur lines into the plant. After the photo-op groundbreaking, the Cyclone Fence Company, Waukegan, Illinois, starts erecting 75,000 feet of perimeter fence.

Hercules Powder organizes administration for operating departments: Administrative and Service, Accounts and Audit, Materials and Transportation, Safety, Security, Mail and Records, Personnel, Labor Relations, Contracts, Property, Timekeeping and Payroll, Engineering and Operations.

The Corps of Engineers and Hercules timekeeping and payroll staff set up an office in the former Civilian Conservation Corps camp at Devils Lake State Park. After a few days Hercules moves to a "farmhouse on the site" and sets up its first "brass house" time shack. Its first employees are hired here and many are sent to Radford Ordnance Works for training.

The fire department is organized and by the end of the year will have eighty-two firemen split among three shifts around the clock. Initial equipment includes one 750-gallon pumper, two 500-gallon pumpers, two half-ton trucks and 29,000 feet of hose. The fire alarm system consists of forty-five phones placed in boxes around the plant.

The first members of the guard force are hired by Captain Otto Cronce, a veteran of the Green Bay Police Department. By the end of the year there will be 252 guards and four lieutenants manning forty-eight guard posts and ten gates. The guards occupy their first headquarters in the former Alex Steuber farm house near Gate 2.

The Sauk County Highway Commission begins acquiring right-of-way for the "boulevard" expansion of Highway 12 to four lanes adjacent to the powder plant. The federally funded $650,000 project will include upgrading of the town road running directly south to the Sauk City bridge, which means Highway 12 will bypass Prairie du Sac.

March

Local and state newspapers report what proves to be an erroneous announcement from Congressman Stevenson stating that Badger Ordnance, "will be doubled in size and doubled in cost" with "the acquisition and use of another 10,000 acres of land." The misconception is understandable, since many other Ordnance Works have shell loading or other related plants nearby.

Except for about eighty structures retained by the Army, sealed-bid sales of the approximately 800 houses, barns, silos, sheds and other structures left on the property are held on four days throughout the month. Sales prices are low, with the Swain Mather house going for only $200. Dismantling and moving can double or triple the actual cost to the purchaser, yet many buildings are resold almost immediately. Even though the ground is thawing and it is mud time on the prairie, all buildings must be demolished or moved within weeks of sale. Buyers are hustling. Kermit Neale of Montello, who buys two barns, two hog houses and a double corncrib, brings twenty-five men to tear down, load and move everything in two days. The miles of wire fencing and posts on the property are divided into forty-six lots and sold by sealed bids on March 27 and must be removed by April 10. While most of the materials and buildings, like most of the farmers who formerly owned them, end up in Sauk, Dane or Columbia counties, a few move farther away. The Roy Kindschi home goes to a farm east of Lake Winnebago near Stockbridge in Calumet County.

General construction work begins on the main administration and office buildings, the power house, water wells, rail lines, contractors' workshops and warehouses, plus a temporary concrete-mixing plant off of Highway 78.

By the end of the month, building contractors are employing 1,611 workers on the site.

On March 20, the Chicago and North Western delivers the first carload of materials directly to Badger. The flatcar of lumber sits unloaded for several days because the materials yard is "a sea of mud."

Work begins on the power house designed to house four coal- or oil-fired steam boilers, each capable of generating 175,000 pounds of steam per hour. The plant will burn four-and-one-half tons of coal per hour and consume 300 gallons of water a minute. The superstructure will require 777 tons of steel and the foundation more than 4,000 cubic yards of concrete.

Work begins on the North Service Road, which runs east-west across the site. After grading, the road is surfaced with gravel from the quarry on the Kehl property across the river from Prairie du Sac. The North Service Road is the first of twenty-one miles of formerly public roads inside the plant to be improved. In addition, a thirteen-mile road is built around the fence perimeter.

Work begins on the water system, starting with the pumping station at Weigand's Bay, which can take up to forty million gallons of water a day out of the river and send it 3.5 miles to the six-million-gallon reservoir at the foot of the bluffs or directly to the filtration plant. After filtering, the water is stored in a four-million-gallon-capacity reservoir before being circulated through the plant. Until this water system can be completed, water for construction and fire protection is supplied by five deep wells distributed around the property, each capable of delivering at least fifty gallons per minute with pressure maintained by five 15,000-gallon elevated wooden tanks.

Work begins on the air compressor house and pipelines that will deliver compressed air to the press houses and other buildings. Pipe is also being laid on the surface for hydraulic pressure lines and asbestos-wrapped, steam-heating lines from the central power plant.

Work begins on the sewerage system, which includes three separate systems. Acid sewers deliver waste acids to neutralization plants. Sanitary sewers carry non-acidic wastewater to the treatment plant to be purified before discharge into the Wisconsin River via Gruber's Bay. The general purpose sewers collect water from road drains, parking lots, eaves troughs and other surface sources and discharge it untreated into the Wisconsin River via Gruber's Bay.

The Corps of Engineers sets up a temporary office for its Property Section on the former Leutscher farm and starts to keep track of hundreds of pieces of construction, maintenance and office equipment.

Railroad contractor Bowen & McLaughlin Birch sets up a temporary office and yard for equipment on the former Walter Schlag farm.

Dr. Howard L. Reed leaves his practice in New York City to set up the Badger clinic and hospital, temporarily located in the Anna Magli farmhouse.

On March 12, all roads leading into the plant are closed with gates manned by guards.

April

Construction workers employed at end of month: 2,977.

Message from Colonel Hauck, area engineer: "The area engineer has been assigned the mission of constructing the Badger Ordnance Works in the shortest possible time. He will not tolerate obstructionists of any nature...Remember that no special credit is given for finishing the job on schedule; our objective is not to meet the schedule but to beat the schedule."

Despite Hauck's determination and the demanding schedule, the Corps of Engineers' centralized procedures for approving contracts are so slow that contractors are forced to lay off workers while they wait for papers to move from the Corps' in-box to the out-box. Not until the process is changed to give on-site managers more discretion is progress made.

The Norton Company's temporary concrete plant on Highway 78 is operating and "six huge trucks," each capable of carrying six cubic yards, are delivering concrete to building sites. The plant at the Kehl gravel pit across the river is making 2,500 yards of gravel a day.

Work begins on Smokeless Powder Line C, the first of four production lines, optimistically scheduled to start making powder in October 1942. The B-Line is scheduled to start production in November, the D-Line in December and the E-Line in January, 1943. (There is no A-Line.) Each line will consist of one dehydrating press house, one alcohol pumping and accumulator house with alcohol tanks, one diphenylamine mix house with three storage tanks, two mix houses with refrigeration unit, three block press houses, four horizontal press houses, two loading platforms, one hydraulic refrigerator, one solvent recovery cooling tower and pump house, one powder line shop and office, one activated carbon recovery building, three material store houses, fifteen solvent recovery houses with twenty-seven-foot-high Repauno barricades plus safety chutes and platforms, fifteen water dry houses, six air dry houses, six rest houses, one preliminary blending house, plus beater and elevator houses. Each line requires a total of sixty-five buildings plus storage tanks and piping for chemicals, pressurized air, hydraulic fluid, chemical recovery, water, steam heat and sewerage. In addition, all four lines will make use of a total of ninety-eight shipping houses.

Work begins on fifty-four of the 120 buildings planned for the cotton and wood pulp lines, on the twenty-one buildings and storage tanks making up the nitrocotton acid area; and the first forty-foot, concrete-lined tunnel into the bluff of the cannon proving grounds.

The Farm Security Administration purchases 105 acres on the edge of Prairie du Sac from Raymond Heintz to erect a trailer camp to house plant workers.

The former Jennie Johnson Spoors house on Highway 12 is remodeled into a headquarters building for the Wisconsin State Police.

The former George Carpenter house, on land farmed by Raymond Jessie, is converted into a field office for Hercules Powder Co.

May

Construction workers employed at end of month: 5,555.

The Norton Company's on-site concrete plant is operating and keeping an expanded fleet of eighteen trucks delivering six cubic yards of concrete non-stop.

Construction work is delayed this month, and a labor crunch averted, when 3.75 inches of rain fall, causing an estimated loss of 74,000 man-hours of labor.

Despite the rain, the administration and general office buildings are finished enough for Mason and Hanger, Hercules Powder and the Corps of Engineers to move in. Dr. Reed is also able to move the hospital out of the Magli farmhouse into temporary quarters on the North Service Road.

Work begins on adding two lanes to Highway 12 from the Bluff south to Sauk City. The road is competed by the end of August.

Work begins on five H-shaped barracks west of Highway 12. Identified as the "Administration Barracks," they each have sixty rooms, and can house 104 people. A mess hall which can serve up to 200 people adjoins the barracks and serves food twenty-four hours a day.

The pace of work is indicated by the time required to excavate, pour footings and floor, raise walls and roof, then finish the interior of a 4,000-square-foot guard house. Started on May 8, it was completed on May 24.

Industrial Food Systems contracts to operate a cafeteria and kitchen at the former CCC camp at Devils Lake. A temporary cafeteria opens at the plant with food for 1,200 meals prepared daily at Devils Lake and delivered to the cafeteria and sixteen canteens throughout the site. Diners are urged to eat quickly and clear the seating for others, instead of loitering, smoking and visiting.

In time for Memorial Day visitors, the Thoelke, Pioneer and Kingston cemeteries are enclosed by white picket fences, with the graves weeded, trees trimmed and grass mowed.

June

Construction workers employed at end of month: 8,664.

On the Mason and Hanger office staff is Lillian Scott, identified as the "first woman registered architect in the state of Wisconsin." Scott is a graduate of the University of Michigan School of Engineering and worked in Milwaukee before M&H assigned her to Badger.

With over 250 structures underway, work is delayed by shortages of drum and boiler steel at the power house, and in general by shortages of structural timber and railcars to haul cement.

A large colorful sign identifying the site as "Badger Ordnance Works — An Arsenal of Democracy" is erected near the entrance gate off Highway 12.

The river pump house is excavated to water level and sheathing is prepared to line the hole and dig deeper.

Work begins on the solvent area with nine buildings and a tank farm for strong alcohol.

July

Construction workers employed at end of month: 11,689.

A.H. Thayer of the Baraboo American Legion Post presents a fourteen by twenty-six foot flag to Badger Ordnance. The flag is first flown from the new official flagpole on July 2.

Work begins on one 200- and one 400-man mess hall. Completed in September, they serve lunch to 1,800 workers every day.

The Hercules Broadcasting Department, with fifty-two speakers located at strategic places on the site, makes announcements, takes phone messages and pages up to 100 people a day to receive them.

Excavation at the river pump house reaches twelve feet below river level, erection of steel on the power plant is completed and twenty vertical presses have been installed on slabs on the production lines.

The main transformer bank is energized with the capacity of 45,400 volts, and contractors begin erecting 3,000 poles with 5,000 cross arms and 300 miles of wiring.

The perimeter fence is completed and lit with 556 pole-mounted fixtures. From the moment on July 24, 1942, when the fence lights are first switched on, night on the Sauk Prairie is no longer completely dark.

August

Construction workers employed at end of month: 11,744, a decline from the high of 12,188 at work in the first week of the month. In addition, approximately 1,000 people are working for Hercules Powder and its contractors.

On orders from Washington, work abruptly halts on the power plant, infrastructure and 100 buildings planned for the TNT lines. The cancellation prompts rumors that the whole powder plant will be shut down and jobs lost, but workers are quickly transferred to other projects in the plant.

Work begins on laying the fifty-eight-inch water main, in clearing the cannon proving grounds and installing equipment in the tubs and beater houses.

By the end of the month the new hospital is finished enough for the medical and clinical staff to move in, but the steel shortage delays work on the three-story nitrating houses.

Fitzhugh Davidson Charlton is the first worker to die as a result of an accident at Badger Ordnance. He is struck in the chest and fatally injured by an unloading hook when the rig breaks while offloading pipe.

September

Construction workers employed at end of month: 9,538.

Labor Day is not a holiday for construction workers at Badger Ordnance Works.

A severe storm on September 18 drops 1.34 inches of rain in twelve minutes and floods the site. About one-half mile of water lines, laid but not backfilled, and including the fifty-eight inch diameter concrete main from the river, are afloat in their trenches and must be relaid and reconnected. Five days later, the first snow of the approaching winter falls.

With steel on hand, work resumes on the nitrating houses and brick walls rise.

Work begins on "Staff Village," twenty houses for "key" personnel along Otter Creek on the former Jennie Johnson Spoors property west of Highway 12.

William Joseph Cinealis dies of a fractured skull after being hit on the head when the truck he is riding on drives beneath a low-hanging pipe.

October

Construction workers employed at end of month: 10,974.

With help from excellent fall weather, fifty percent of construction is completed and thirteen percent of equipment installed.

Work is delayed because of shortages of stainless steel, structural steel, electrical switchboard panels and lead-covered cable.

Brickwork is completed on the Number One boiler in the power house, but it is a long way from supplying any heat. With winter approaching, a boiler to heat the administration building is installed and fired. Even with the new boiler running, it is chilly enough for women working in the Corps of Engineers office to ask for and receive official approval to wear slacks. "Good news for the girls!" says the *Badger Ordnance News*.

During the final week of the month, the number of pieces of construction equipment operating peaks at 1,473, including 434 trucks, 209 bulldozers, cranes, tractors and scrapers, and since work goes on night and day, 110 portable lighting plants.

Joseph Clarence Fjelstad dies of pneumonia after he is hospitalized with nine cracked ribs resulting from his collision with a heavy timber protruding from the side of a construction truck.

November

Construction workers employed at end of month: 9,040.

Midst delays, postponements and the onset of winter weather, rumormongers declare that, "it doesn't matter whether this powder plant is completed or not because it will never produce any powder."

Work is delayed by real shortages of plumbing and steam valves. Numerous leaks have been discovered and repaired in the fifty-eight-inch water main, but the river pump house begins pumping water into the lines. The filtration plant is running and the four-million-gallon reservoir is filled with treated water.

The 66-volt power line from Prairie du Sac is energized.

December

Construction workers employed at end of month: 7,005.

The month begins with below-zero weather and, just in time. Boiler Number One begins operating. By mid-month steam heat is circulating through part of the plant's piping. By the end of the month, a second boiler goes on line and steam is delivered to all the outside lines and to the buildings of Smokeless Powder Line C.

The 42-volt power line from Portage is energized.

Work is completed on the "Administration Barracks."

Work begins on a 500-unit housing development just south of the Administration Barracks. Called Badger, it will house workers and their families as well as a community center and school, a drug store, supermarket and other shops.

Over twenty-four miles of standard gauge railroad track have been laid on the site and marshalling yards have been built for each of the two railroads serving the property.

The Norton Company's concrete plant is dismantled after mixing 187,300 cubic yards of concrete, the equivalent of a block of concrete three feet high, three feet wide and 318 miles long.

A kerosene torch explodes and the main building of the neutralizing plant on one of the nitrocotton lines is destroyed by fire.

Frank O. Smith, a steamfitter, dies of internal injuries suffered after the wall of a hole he is working in collapses and drives a steel pipe into his abdomen.

January, 1943

Construction workers employed at end of month: 6,014.

By the end of the month 14,735 rail carloads of materials and equipment have been delivered to Badger. Approximately another 5,000 carloads will arrive by the time construction winds down in October.

On January 6, 1943, operations begin, with the first production of Oleum. On January 18, the ether house of Smokeless Powder Line C begins production and, on January 23, Line C produces the first powder made at Badger Ordnance Works.

Construction work continues throughout the rest of 1943, slowing down considerably in the coldest winter months, but picking up again in March, when there are still over 5,400 M&H contracted workers on the site, plus Hercules contractors and employees. By the end of the month, 320 buildings and other structures are completed. Over the spring, summer and fall, the power house, with a fifth boiler, is completed. The pulp and cotton lines, four smokeless lines, Badger Village and complementary infrastructure are also under roof and ready to run. The ethyl centralite (E.C.) Line is completed and starts making powder in early July. In November 1943, the conclusion of construction is symbolically recognized when the Corps of Engineers transfers command of the site to the Ordnance Department, which will contract with Hercules to complete the project. In early January 1944, for the first time in two years, Mason and Hanger has no subcontractors working at Badger Ordnance.

This first phase of construction, which created what the Army called Badger Job M-1 in 1942 and 1943, saw the completion of 911 construction "units," which included all built structures, from the cannon testing tunnel to the magazines, the flagpole to the river pump house. More than seventy-two acres were put under roof, an area almost one-quarter mile wide by one-half mile long. The buildings were spread over an area of about five miles by two miles yet linked by nearly thirty miles of steam pipes, eighty-eight miles of water lines, forty-five miles of sewers and 140 miles of electrical cables. Plumbing for production consisted of nine miles of pipes to carry acids, seventeen miles for alcohol and ether, almost four miles for hydraulic fluid, five miles for inert gas and over six miles for pressurized air. Mileage also racked up for standard gauge rail lines reached twenty-four miles, for tramways to move materials inside the plant, twenty-five miles, and for roads, seventy-six miles. All of it was surrounded by over fifteen miles of lighted fencing.

"The surprise of your life comes when you reach the edge of Sauk Hill and look down over what might be the suburbs of Chicago or Milwaukee or any other large city in the United States," wrote Wisconsin Dells news editor William Drumb after he visited Badger in early September 1942.

"Even on a clear day there were buildings as far as the eye could see. There are buildings of every conceivable shape, both large and small, mostly of wood, but some others were quite substantial in their makeup, apparently of modern brick construction.

"The buildings are mostly white and stood out in the bright sunshine like no other place that the writer has ever seen. When we say wonderful and astonishing, that does not half tell the story."[59]

It was "wonderful and astonishing," an exemplary marshalling of resources, people and management skill. For sixty years rumors have persisted of deliberate wastefulness, gouging of taxpayer money and thievery of equipment and materials during the wild days of the first construction blitz. Rumor says there is a bulldozer buried beneath the power plant, a portable generator in the sand somewhere, a home down the road built entirely of pilfered lumber and hardware. Rumor says construction workers sliced up lumber carelessly, poured twelve inches of concrete slab when six was specified, thinned the concrete mix by using one bag of cement when three were specified, buried precious copper and steel hardware and dumped usable materials on burn piles that glowed all night. Rumor says all these things, and more — none of which can be substantiated. When pressed to locate the equipment buried in the sand or name the house built with stolen goods, the rumormongers fall silent. Human nature being what it is, waste and thievery almost certainly occurred while building Badger, but deliberate and large-scale misuse of time or materials has yet to be documented and probably never will be. "The general disorders were waste of time, drunkenness on the job or maltreatment of government property," reported David Mayer, historian for the Corps of Engineers at Badger. "Cases as cited above were few. The records prove that the employees on the project, in general, exemplified a fine character, initiative and enthusiasm in their work." They had sufficient initiative and enthusiasm to assemble a sizable and complex industrial enterprise where none had stood before. Badger shipped its first trainload of finished product, 60,000 pounds of .30 caliber smokeless powder, to the Twin Cities Ordnance Plant on May 10, 1943, where it was loaded in M-1 rifle cartridges that were later fired anywhere from New Guinea to Normandy. By the less-than-realistic original schedule, the shipment was five months late. By the dictates of reality, it was on time for a war in which America's largest battles had yet to be fought.[60]

Badger M-2

In February 1944, with two smokeless lines running smoothly, the construction blitz started anew. The War Department ordered Hercules to build what it called Badger Job M-2, initially announced as a $24 million addition to produce three million pounds per month of "cruciform grain stick powder," i.e., rocket propellant. Once again, Mason and Hanger was the contractor, with a new set of subcontractors hired from several midwestern states, few of whom had worked on Badger M-1. Once again, plans changed early and often. By October 1944, the Ordnance Department, hustling to keep up with the accelerated demand for ammunition prompted by unexpectedly heavy use in Europe after D-Day and by major offensives in the Pacific Islands, expanded the plant's capacity to 5,525,000 pounds of powder per month and its cost to $38.5 million. Along with a $50 million addition to the Indiana Ordnance Works, and an expansion at Radford Ordnance, Badger was part of a $300 million expansion of American ammunition manufactories, with an emphasis on rocket powder. "So vital is the production of rocket powder, the Army has placed it on the critical list, ahead of planes, aviation gasoline, tanks and ships," said the *Badger Ordnance News* when it announced the addition.[61]

To turn out 5.5 million pounds of rocket powder a month meant that the Army and its contractors had to build over 500 buildings, or the equivalent of the production areas in the just-completed Badger M-1 all over again. They would have to extend the existing road, rail, water, sewer and electric grids, expand the security, safety, fire and administration infrastructure, and assemble another power plant. In a labor-starved market, they had to find an estimated 8,000 workers for another building blitz and then come up with over 4,500 people to make the powder.

Rocket propellant was "double-base" powder, a mixture of the nitrocellulose that was the "single-base" of smokeless powder doubled up with the high explosive nitroglycerin. In order to manufacture "rocket," as it came to be called, Badger had to increase its output of nitrocellulose and start fabricating nitroglycerin. It meant building another chemical works to mix and finish nitroglycerin, as well as a new production line where the nitrocellulose and the nitroglycerin would be processed into "rocket."

Survey and layout work for the N-Line, as the rocket assembly line was called, began late in the winter of 1944, but "actual construction of temporary buildings and grading of railroads" did not start until the end of April. The war in Europe would be over in a year and two weeks, and in the Pacific in about fifteen months, but no one knew it then. What the War Department and everyone else did know was that the Marines were still slugging their way island by island against the Japanese in the south and central Pacific, while the U.S. Army's Italian campaign was stuck on the beach at Anzio and in the hills at Monte Cassino, with no breakthroughs in sight. When the breakthroughs came, starting with the Normandy invasion in June 1944, the demand for powder only accelerated.[62]

The layout of Badger M-2 was based on the familiar practice of using space for safety according to the table of explosive distances. The nitroglycerin plant, which would manufacture the least stable, most powerful, chemical explosive made at Badger, was tucked away across the north-south railroad line in the hills 2,000 feet east of the smokeless lines and 2,000 feet north of the rocket line as well. Also east of the tracks and starting a few hundred feet from the southwest corner of the smokeless area, the N-Line ran east into the moraine for more than a mile. The ground was a bit rougher here, more rolling than on the flat prairie, and would require more grading and levelling than the smokeless area.

Work started in June on the nitrating and separating houses and by July a total of 1,992 workers were making good progress on the maze of plumbing at the nitroglycerin plant. Other areas lagged, chiefly because of the labor shortage. In this matter, Badger was its own worst enemy. When Badger M-1 was finished at the end of 1943, nearly all the construction workers moved to other jobs. Less than 1,000 of them were on the job in September 1943, and about 100 in December. Men who might have stayed to build Badger M-2 were long gone when work started on it. Badger M-1 also had more than 3,000 workers employed making E.C. and smokeless in April 1944, 4,000 in July and 6,000 in December. They were not available for construction of Badger M-2 even if they wanted to work there. From the spring of 1944 until the summer of 1945, over forty percent of Badger M-1 employees were female, and females were not hired for construction work in Sauk County, Wisconsin, in those years and for many more to come. So even if a woman wanted to hire on to build Badger M-2, she would not be welcome. Males employed at Badger M-1 were not welcome either. They were warned that if they left their jobs in operations they would be blacklisted for construction at Badger M-2. To further aggravate the situation, in April 1944, just when construction work was about to shift into high gear, the Ordnance Department sent orders to activate nitrocellulose Lines B and E, which had been on standby since their completion in the autumn of 1943. Their output would be needed for the new rocket line and to meet increased demand for smokeless powder. So, just when thousands of new construction workers were in demand, so were several thousand more production workers. In response to these demands in the spring of 1944, Hercules launched an intense recruitment program targeted at women. In the culture of the time, females weren't allowed to drive bulldozers or swing framing hammers, but they could dip nitrocellulose and muscle powder cans onto conveyors, all of which retarded progress at Badger M-2. The recruiters were successful enough to bring Smokeless Line B into production in May 1944, but E-Line remained Badger's "Ghost Line," vacant and idle until workers to run it could be found in September.

Had all the construction workers needed to build Badger M-2 on schedule been recruited and had they all appeared for work, where would they have spent the night afterwards? "Housing of workers has been a problem," reported Mason and Hanger to the Corps of Engineers in January 1945. With room for only about 1,600 construction workers reserved in government housing, space was tight. Production workers had gobbled up nearly all the livable space in the area, and even using the generous definition of the times, minimally tolerable places were occupied as well. The twin drawbacks to locating Badger Ordnance in Sauk County that H.N. Wendle had cited in November 1941—shortages of labor and housing—had not been overcome. Even with the trailer park at Prairie du Sac, the Administration barracks, the new

Gate 10 barracks, the apartments at Badger Village, the rental units in homes, farms and tourist camps, the shared houses and "hot beds" where workers slept in shifts, the local housing stock could not absorb another flood of construction laborers. As the tide of combat turned in favor of the United States in the fall and winter of 1944-45, and the threatening situation of early 1942 faded from memory, finding workers willing to camp out for Badger became increasingly difficult.[63]

"The combat GI Joes are not going to like the facts in this article," wrote *Milwaukee Journal* reporter Lewis French in early December 1944. "The Badger Ordnance Works in Sauk County desperately needs 1,200 more construction men...."But there is only a trickle responding to the employment appeals." French went on to decry the high rate of absenteeism as well. "...on November 25, the day the state deer season opened, there were 788 men of the 3,842 on the [construction] payroll who did not come to work." Two weeks earlier "it was raining, not a downpour but an unpleasant drizzle. That morning 998 did not show up for work nearly one man out of every four..."[64]

The N-Line fell so far behind schedule that equipment and machinery shipped on time sat "under canvas" waiting for slabs to be poured and walls to be raised to house it. In August 1944, with 147 buildings under construction in what should have been the busiest month of the season, the Badger M-2 construction crew numbered 2,554, about half the force needed to complete the work on time. Deadlines were pushed back once again, with partial production of rocket powder rescheduled for January 1945—if another 2,000 laborers and tradesmen materialized at the hiring hall. When they didn't show up, the best the Army could do was send 100 furloughed soldiers from California who agreed to work at Badger in return for their Army pay plus construction wages. The start of production was postponed until March now, and this deadline was met. Actually, a small amount of rocket propellant using nitrocellulose "paste" shipped in from the Sunflower Ordnance Works in Kansas was processed on January 30, 1945, but the first rocket powder made entirely at Badger was not produced until March 1945. In April enough of Badger M-2 was operating to produce at the rate of 1.5 million pounds a month, about thirty percent of the rocket area's planned capacity. By the time Germany surrendered in May, the N-Line powerhouse, using equipment transferred from the phased-out TNT operations at the Pennsylvania Ordnance Works, was making steam. Its four boilers were each capable of generating 100,000 pounds of steam per hour, compared to the five 175,000-pound units at the No. 1 Powerhouse. With its powerhouse running, the rocket plant was eighty percent complete and production was rising.

Construction continued through May, June and July of 1945, with all hands believing that the United States would need every grain of powder Badger could produce for a year or two to defeat the Japanese. Military strategists projected that the necessary invasion of the "Home Islands" of Japan, scheduled for the spring of 1946, would dwarf Europe's D-Day in terms of ships, planes, troops and fire power. The bloody conquest of their homeland would wreak enormous damage and death among the Japanese and claim an estimated one million American casualties. In July 1945, with the smokeless lines shut down and experienced workers transferred there, the N-Line was producing at the rate of 4.5 million pounds of powder a month. The Ordnance Department, anticipating production at full capacity, sent orders to build more loading docks and other shipping facilities. At the end of the month, the Engineers announced that the rocket area would be entirely finished by November 1, 1945. On August 6, the atomic bomb was dropped on Hiroshima, and on August 8 on Nagasaki. Six days later the Japanese accepted terms of surrender and the war was over. The demand for powder evaporated about as quickly as the flash of nuclear lightning over Japan. On August 13, America's last full day at war, Badger M-2—nitro plant, N-Line, magazines—was declared ninety-five percent complete. A week later, it was shut down. In the four months it had operated, Badger M-2 had produced a total of 14.4 million pounds of rocket powder, about the same amount it would have produced in ten weeks had it ever operated at full capacity. The rocket area was shut down before it ever really started. It had little impact on World War II, but the rocket plant, new, complete and ready to run by war's end, would have permanent consequences at Badger Ordnance.

Powdermakers — Polly and Pete

It looms large in memory and impact but the World War II period was deceptively brief for Americans. The United States was at war with Germany for three years and six months, and with Japan for an additional three months, less time than it takes to earn a high school diploma. A student living in the Town of Sumpter and starting high school in Prairie du Sac in September 1941 would travel to class on a Highway 12 surrounded by farms in peacetime. During the second semester of freshman year and most of sophomore year as well, the student would drive past a wartime construction site frantic with activity. In March of 1943, with the end of sophomore year on the horizon, the student would start to see four plumes of gray smoke pouring out of a new powerhouse, Smokeless Powder C-Line starting to run, followed by the D-Line in May and the E.C. Line in July. Not until the student's junior year was nearly over in May 1944 would Smokeless B-Line come on, while the E-Line would not start until September of the student's senior year. The N-Line would sputter on in the student's final high school spring, but all smokeless production would shut down by graduation day in late May. Two months after graduation, all of Badger Ordnance would be idle. The lines that operated the longest, Nitrocellulose C and D, ran for twenty-six months. B-Line made nitrocellulose for ten months, and E-Line for eleven. The E.C. Line made its specialty powder for twenty-four months, but the N-Line limped along for only about six months and produced only a fraction of the rocket powder it was designed to make.

Other acts in the Badger drama also had short runs. Badger Village was occupied for about twenty-two months and the school ran for only four semesters. The Prairie du Sac trailer park was inhabited for two periods, each lasting less than one year, about as long as the Gate 10 Barracks were occupied in 1944-'45. Plant guards kept an uninterrupted watch at the gates from March of 1942 but occupied the fence-line towers for no more than six months that year. Men from Jamaica and Barbados worked in production for about ten months, about as long as soldiers on furlough helped build the rocket line. Over 2,000 of the 10,500 acres of farmland taken for Badger in 1942 was leased back to farmers in 1943, after spending only one growing season unworked. Strange the Badger trip may have been during World War II, but it certainly was not long.

From the commanding officers down to the line workers, employee turnover was high. Badger had four commanding officers during World War II, who moved on as their careers dictated. Lt. Col. George F. Griffith, who had been executive officer at Radford, left Badger to command the Keystone Ordnance Works in Pennsylvania in October 1942. He was succeeded by Major Richard P. Staeffler, who stayed until July 1943. Major Victor L. Keldsen, who lived in Baraboo with his wife and two sons, was in command until June 1944, when he moved to the top spot at the Alabama Ordnance Works. For the final year of the war, in which rocket powder production was paramount, Major Charles Pruden moved to Badger after commanding at the Sunflower Ordnance Works, the largest of the rocket propellant works operated by Hercules Powder.

Major Wayne Hauck, the West Point graduate and career soldier, was the first of three Area Engineers responsible for building Badger M-1. He left the plant to supervise an expansion at Fort McCoy in June 1942. He was succeeded by Major Albert E. Stoltz, who served until December, and Captain Alfonso Tammaro, who turned the plant over to the Ordnance Department in January 1943.

Civilian workers regularly left Badger for military service, family demands or other jobs, including the four managers who ran the plant for Hercules: Henry Chase, Charles H. Gant, F.L. Boddicker and George Ripley. The "Badger Honor Roll" listed 2,356 men and women in uniform by June 1943 and would grow much larger by the end of the war. Knowing they were about to be drafted, many men working at Badger took several weeks or a month off from work for a final fling as civilians before entering the military. Bearing the brunt of family responsibilities, the women who made up a large minority of Badger workers tended to be absent from work more than men. Working mothers required more time off than other workers, whether the fathers of their children were in the military or at home. In November 1944, for example, an outbreak of flu kept about one out of every twenty Badger workers home, with three woman absent for every two men, but not because they were more likely to be sick themselves.

The state of war created its own demands. Many women powder workers took an extended leave to spend a final few weeks with a sweetheart bound for the military, took off for a month when a husband or boyfriend was furloughed home or left the area to live with a soldier stationed elsewhere in the United States. If any man took time off to be with a woman leaving for or about to serve in the military, no record of it was found. Some workers were strictly seasonal, including teenage boys too young for the draft and schoolteachers who hired on to make powder during their summer breaks. Also seasonal were men who farmed and who worked at Badger in between fall harvest and spring plowing. Such comings and goings would have been impermissible in the job-starved, prewar years but were acceptable during the war and tolerated by employers who needed workers.

Hercules Powder hired a total of 16,705 people to operate — not build — the plant during World War II. Men numbered 10,416, women, 6,289, to make a sixty-two percent to thirty-eight percent split between the Petes and Pollys. Hercules quarterly employment reports list five or six African-American "Negroes" at work prior to September 1944, when the first of over 200 men from Jamaica and Barbados, who were also classified as "Negroes," began at the plant. In July 1945, after all the Caribbean men had left, Hercules listed seventeen African-American men and one woman on the payroll. Except for the Jamaicans and Barbadians, only a few foreign nationals worked at Badger. They were usually prewar visitors trapped in the United States by wartime limitations on travel. Since they lived just about as close to Badger as most other commuters, Ho Chunk people signed on, and so did Ojibwas, Menominees and Oneidas from much farther away in northern and eastern Wisconsin. Indian, African-American, Hispanic and Asian workers were about as scarce at Badger as they were anywhere in 1940s Wisconsin, where cultural diversity meant people had ancestors from everywhere as long as everywhere was in Europe.[65]

The largest number of non-construction workers employed at the plant was 6,679 in June 1945, when four nitrocellulose areas were humping to feed the rocket powder lines. The trend over time was to hire more women, starting with about twenty-five percent in 1942 and rising to forty-eight percent in mid-1945. In 1942, a construction-only year, the 435 women on the payroll were employed in traditional female jobs as clerks, receptionists, typists, nurses, cafeteria workers and cleaning staff. Among the inevitable exceptions was Mary Brown, Portage, who signed on as Hercules' first female chauffeur in November 1942. In January of 1943, the first female guards were hired, nineteen women who were "in good physical condition, average height and weight" and possessed of "a high school education and a pleasing personality." Women entered the production force in March of 1943, when nine "WOWS," or Women Ordnance Workers, went to work in the nitrocellulose cutting room and ten in the "Purification" house. They were the first group of women who could have been called Polly the Powdermaker — if only the War Department publicist who invented Rosie the Riveter had been present at Badger. By the summer of 1945, 1,728 of the 3,148 workers in the smokeless powder department were women. Had the war continued for a few more months and the rocket area reached full capacity, the Pollys would have outnumbered the Petes and made up a majority of the workforce. One unconventional but revealing sign of the heightened presence of women at Badger was on the latrines. Built in 1942, Badger M-1 had thirty-one men's latrines spread throughout the production area, and fifteen for "ladies." In 1944, Badger M-2 was laid out with 46 "combination ladies and men's latrines."[66]

Women were productive employees and economical for the company, although not as economical as the pay scale might indicate, since on many of the more physically demanding jobs, two women were delegated to do the work of one man. The starting wage for male "helpers" in 1944 was $.82 to $.86 per hour with "opportunities for advancement to operators at $1.01 to $1.09 per hour." Women "helpers" started at $.66 per hour with "opportunities for advancement to class A helpers at $.75 to $.83 per hour [and] to operators at $.91 per hour." Both men and women received a $.03 per hour premium for working the evening and night shifts, and time-and-a-half for more than forty hours a week. Since Badger worked a six-day, forty-eight hour week — plus some seven-day shifts — everyone received at least eight hours of overtime pay. Top gross pay for a man, with shift premiums and eight hours of overtime, totalled $58.24 per week; for a woman it was $48.76. These wages are for production workers only, although other semi-skilled blue collar and pink collar workers — bus drivers, guards, janitors, power plant laborers — received comparable pay. Receptionists, typists and maids were paid less. Skilled tradesmen, supervisors, white collar professionals and managers were paid more.[67]

Female employees could take advantage of the services of the women's counselor, who was hired in February 1943, before women went on the production lines. As management acknowledged, there was "never any question about the need for the services of some person to serve as counselor to women employees at Badger. The most difficult problem to decide was whether or not the party selected should be a man or woman." Despite "no small amount of skepticism on the part of most operating heads who remembered the success Radford had experienced with a man in this position," Hercules hired Aimee White, a registered nurse who had attended Columbia University and worked as an "Industrial Hygiene Nurse" for the Wisconsin State Board of Health. "Many of the women who will be employed here will be working in an industry for the first time," White told the *Badger Ordnance News*, "and the transition from their previous occupation to their job here would probably be difficult, particularly in some cases." She did not say that the women coming to Badger had not worked at all; nor could she, since many, if not most of them, were farm women. Saying farm women had not worked, even farm women fresh out of high school, would be as inaccurate as saying they had not eaten, walked or breathed. White saw her role as "assisting the girls in getting started on the job properly" and "making herself available to help them adjust themselves to their work." Since Badger ran and women worked around the clock, Hercules hired a second women's counselor—Kathleen McKillip—in July 1943 to "alternate on the two night shifts." By the fall of 1944, the female workforce had grown enough to require the services of a third counselor, Elizabeth Sansun. All five of the women who held the job during World War II were college graduates and, except for White, had spent time on a powder line or in the lab at Badger. They were expected to "patrol the lines as often as possible, talking to girls and foremen [to] create harmonious working conditions and iron out individual problems."[68]

After one year, all employees, male or female, were entitled to one week's paid vacation and a Christmas bonus roughly equivalent to two weeks' pay. Medical care for on-the-job accidents and illnesses was provided at the plant hospital, while off-site medical and hospital coverage was available with costs shared by employees and Hercules. When the "wonder drug" penicillin came into general use in the summer of 1944, the Hercules health plan covered it. The company also paid for prescription safety glasses. It supplied fire-resistant coveralls for all production workers, head-covering turbans or caps for women and special metal-free "powder" shoes that would not spark and cause an explosion—all charged to employees at "less than cost." Gross pay was further reduced by social security withholding, special "victory" taxes and war bond purchases. As early as February 1943, Badger workers were deducting almost eleven percent of their pay to buy war bonds, a practice they maintained for the duration of the war. Union dues were not part of the equation for Hercules wage earners until the final few months of operation. Although its employees voted in favor of AF of L representation in October 1944, Hercules did not sign a union contract until March 1945.[69]

Just as the war stimulated the American economy and ended the great national Depression, a job at Badger Ordnance cured the depression afflicting many an individual pocketbook. First, Badger eliminated unemployment and created a greater demand for workers than Sauk and its neighboring counties had ever seen. Virtually anyone fit to work could now find a steady job at a rate of pay only dreamed of in the 1930s. In 1940, the average gainfully employed person in Wisconsin was earning $1,400 a year. A man clocking top pay on the production line at Badger four years later was making more than twice as much as $1,400 and a woman was earning about $1,000 above the 1940 average for all workers, male or female. Top Badger wages were also higher than the national war-inflated average of $46.13 in 1944. A man making $58 a week at Badger was taking home about the same amount as the Sumpter farmer on whose land the plant was built netted in his last year there. The difference is that there were one hundred workers making that pay for every displaced farmer. Had H.H. Thies, principal of the Sauk County Normal School in 1943, quit his $2,650-a-year job for the smokeless line at Badger, he'd have come out ahead at the bank, so would the county highway commissioner, who made $2,400 a year. For women in particular, a job at Badger meant a giant step up in pay. A woman working in a Baraboo print shop, in a trade traditionally friendly to women, was paid ten-fifteen cents an hour. A starting general duty nurse at St. Mary's Ringling Hospital received eight dollars for a twelve-hour shift, while the Sauk County Nurse, an experienced registered nurse, was paid a salary of $2,000, or about one dollar an hour for a forty-hour week. Country schoolteachers, most of whom were unmarried women whose wages topped out at $100 a month for a nine month term, could more than double their pay at Badger.[70]

The experience of young William Sisson was typical for men or women. A farm boy from Greenwood, Wisconsin, just out of high school, who came to work at Badger late in 1944, Sisson was used to working for a dollar a day, twenty-five cents an hour tops, yet he started at eighty-five cents an hour for shift work at Badger, and in three months was making over a dollar an hour. With that one dollar he could purchase one pound of Hills Brothers coffee, one pint of Borden's condensed milk, one pound of Kellogg's Corn Flakes, plus one quart of canned peaches, and still get change. He could also buy five gallons of gas, which he didn't need since he didn't have a car and couldn't buy one anyway. Sisson could walk to work since he lived in the Administration Barracks where the monthly rent for a single room was $18 - $21, including a fresh towel daily, change of bed linen weekly and maid service. Had he been a family man who needed extra space and if he could find a way to commute to and from, Sisson could have spent a little more than a week's pay to rent a seven-room house in Lake Delton for $60 a month. Since he was single, he would share the house with three or four other Badger workers and cut his share of the rent accordingly. Had he been so disposed, Sisson could have purchased a house in the new Victory Heights subdivision off of South Boulevard in Baraboo for $4,400-$4,950, and easily made the monthly payments of $31, had all the houses not already been sold before he came to work.[71]

The good pay for unskilled, inexperienced labor made it possible for Hercules to fill all the slots it had open to run two smokeless powder lines from May of 1943 until April of 1944. It was able to hire a cadre of men and women and dispatch them to Radford Ordnance for several months of training while construction was underway at Badger and before the jaws of the employment vise started to close. Once production expanded, acute shortages of workers for specific jobs occurred, especially for wringermen in the nitrating houses. The wringerman shortage caused delays and reduced output, but also expedited the hiring of women for production jobs. Overall, Badger ran smoothly for a year at fifty percent capacity, just as ordered. In June 1943, the C- and D-Lines produced 4.9 million pounds of smokeless and achieved a year end total of 55.3 million pounds. Badger filled the War Department's original 1942 order for 105 million pounds of smokeless powder eighteen months after start-up. In fact, had the plant been only half its size, with two nitrocellulose lines feeding two smokeless powder lines and one E.C. powder line, it could have utilized local resources of labor and housing with little stress and strain. In the fall of 1943, the employment picture was rosy enough for the Hercules employment office to lay off some of its people. By January 1944, the housing shortage had eased sufficiently that all but two of the ten Administration Barracks were unoccupied, while federal housing authorities had started removing trailers from the Prairie du Sac camp. Not until late spring of 1944, when all four nitrocellulose lines were ordered to switch on, and when construction work started on the rocket area, followed by production, would Badger swamp the resources of labor and housing as it had in the mad days of summer 1942.

Scrambling for workers, the Army, Hercules Powder, and Mason and Hanger launched intense labor recruitment campaigns in the summer of 1944. The powder company expanded its long-distance recruiting to nine midwestern states and the building contractor received permission from the War Manpower Commission to look for workers in nearly every state east of the Rocky Mountains. Still, most of the workers were expected to come from Wisconsin. Major Victor Keldsen, Plant Commander for Army Ordnance, attended the state convention of the American Legion in August 1944 and asked that "assistance in the recruitment of personnel for Badger...be the No. 1 objective of the Legion in this state." Legion posts received a packet of information on Badger and were urged to go to work. "If the exact day on which the European war will be over were known, a reasonable estimate could be made of ammunition requirements...but we cannot afford to gamble when the price is American soldiers' lives...If we fail to provide the ammunition, the powder, campaigns will be delayed and the loss of life will be greater."[72]

Hundreds of news releases and ads were placed in state newspapers. Eighty billboards touting the plant were erected on state highways and a new, brightly painted, lighted sign went up at the main gate. Exhibits were prepared for county fairs, ads and response cards hung in Madison city buses, and radio spots aired on Madison stations WIBA and WIBU. When the 1944 college football season started, Badger Ordnance sponsored University of Wisconsin broadcasts with a pitch for more employees. Advertising copy identified Badger as "Wisconsin's No. 1 War Plant," sketched out employment schedules and benefits, and declared that "Cannon and Rocket Powder is on the Army-Navy MUST HAVE RIGHT NOW! list." The *Badger Ordnance News*, which had suspended publication in August 1943 to conserve newsprint, was revived in May 1944 and focused on bringing current employees into the recruitment drive. "Tell your friends about work at BOW," wrote the editor in June. "Each employee gets another employee — NOW," was the slogan in September. Employees had already found "courtesy cards" in their pay envelopes which they were to give to prospective workers to return to the employment office for follow up. Considering the number of families with two, three or more members working at Badger, those cards must have circulated. The Hewitt family of Baraboo was typical. In 1944-45, Elizabeth Hewitt, who worked in the B-Line Dehydration House, had three daughters at Badger: Dorothy Schubring, E-Line Mix House; Gladys Steinke, B-Line Poacher House; Violet Voll, D-Line Press House. Frances Byers, of Wisconsin Dells, was the connecting link in an extended family of ten people who worked at Badger, including her husband William, daughter Virginia, sons Bernard, John and Virgil, sister Susanna Fosdahl, brothers Michael and Leo Moran, and niece Catherine Murphy. For many people, working at Badger was a family affair.[73]

In the summer of 1944, the "Badger Ordnance Works traveling employment office" hit the road. Consisting of a sedan with a set of trumpet speakers mounted on the roof and pulling a trailer painted red, white and blue, the mobile office was dispatched "to recruit men and women in neighboring and outlying areas where they would not ordinarily come to the plant." Guy Krumm of Hercules Personnel manned the office and could hire workers right off the street. In Madison, Krumm hired ten people a day for a week, then went on to Stevens Point and Marshfield, before circling back to Iowa, Grant and Lafayette counties. At the rate of ten people a day, and those coming from the largest city in the area, years would pass before the traveling employment office would drum up the 8,000 workers Badger would need when the rocket area was scheduled to start in January 1945 — but workers were hard to find in central Wisconsin. Krumm would have been elated to meet more women or men like Jean Toman and Dora Hein of Necedah, fifty miles northwest of Badger in Juneau County. Living beyond the range of the

Badger bus, which ran no farther than New Lisbon, the two women recruited Majorie Meyers, Ann Muha, June Toman, Ann Christenson, Bernadine Miner and Amelia Hedman. The seven women pooled their resources and hired a Necedah district school bus to pick them up at 4:45 AM for the day shift and get them to New Lisbon in time to meet the Badger bus. The Necedah bus then started its regular school route. In the afternoon, the driver took the schoolchildren home, then drove to New Lisbon to pick up the Badger workers and bring them back to Necedah by 6:00 PM. It was a long day but, "We just decided to go to work to show that we could do our part," said Toman, who had a five-year-old son. Hein and other women also had children and all relied on their extended families for child care.[74]

In addition to reaching farther away from Badger for workers in the fall of 1944, recruiters also targeted farmers, who made up the largest prospective pool of able-bodied men in the vicinity. "How about arranging your affairs for the fall and winter so you can make this all-important powder," asked the copy of an ad that appeared in Wisconsin newspapers in September. The appeal worked well enough to bring a young man like William Sisson down from Greenwood, but also attracted older men. After a Badger radio ad, John Kading, a comfortable, semi-retired farmer from Lowell in Columbia County came to work in smokeless. "I find the work quite interesting and an experience that would do any farmer good to see how big plants operate. There are many farmers here, which goes to show that without farmers we can't win this war."[75]

With the vegetable harvest season of 1944 drawing to a close, Hercules also sought out foreign-born cannery workers under contract to the War Food Administration, many of whom were already at work in Wisconsin. The WFA imported workers from English-speaking British colonies in the Caribbean region, and when their work was completed at a cannery in Illinois, a group of thirty-three men from the island of Barbados moved north to Badger. A few days later 128 men recruited directly from Jamaica arrived, and by October 1944 a total of 223 Caribbeans were at work in the plant. Initially contracted for ninety days, about one-third of them opted to return home at the end of December, but the majority stayed until the powder plant slowed down in the summer of 1945.[76]

Jim Crow dominated America in the 1940s. In its description of change houses and bathrooms at Badger, for example, the Army noted that "white and colored persons are not separated in this locality," and therefore segregated facilities were not necessary. In fact, though, the Caribbean men were segregated in housing and on the job in canneries and at Badger Ordnance. They lived in the remodeled CCC barracks at Devils Lake, and, if not prohibited, were not encouraged to make their way to Baraboo except for special shopping trips via the Hercules bus. Lacking other motorized transport and experiencing unfamiliar winter weather unlikely to encourage their hiking from the south end of Devils Lake to the courthouse square, the Caribbeans stayed in their barracks. Strict sequestering of dark-skinned Caribbean workers was not unusual in Wisconsin, but it was not absolute either. Although they lived in camps, Caribbean cannery workers in other communities were visible downtown and at church on Sunday, where they enlivened the musical segments of many a small-town Presbyterian, Methodist or Episcopal service. At Badger, the Caribbean men stayed at Devils Lake, played cards, shot pool and watched movies depicting the world outside the camp gates.[77]

Segregation was more difficult to enforce at work even though Hercules policy, like that of other employers, endeavored to keep the foreigners working together in a single group. At Badger, the safety rules decreed that production work be performed by relatively small teams of workers, and newcomers joined a work crew one or two at a time. Keeping the untrained, inexperienced Caribbean men in their own group was impossibly dangerous. Hercules policy makers also wanted to keep the Caribbean men away from the Wisconsin women and supervisors at Badger dutifully reported to Wilmington that they were toeing the color line here: "It was found best to concentrate the Negro workers on E-Line in the Nitrating Area and to eliminate female operators in this area, to avoid possible racial conflicts and ill-feeling," reads the report of the Nitrocellulose Department. As recalled by Hercules employees, however, Jamaicans, Barbadians and Wisconsinites—male and female—worked side by side as the job required. William Sisson said, "I worked with the Jamaicans. That was a pleasant experience. They sort of passed through...about a dozen worked in our area. Very efficient husky guys." Norma Clavadatscher remembered "Leslie and Leonard, from Jamaica. Such nice guys, they would run to help me pull the powder out..." Rita Slezak Lachmund worked with a Jamaican named Norvel Johnson. "He said he was saving his money to buy a taxi cab in Jamaica and after the war if I would go to Jamaica he would take me all over the island free." She never made that trip and didn't find out if Johnson was able to turn his American earnings into a Kingston cab. However, in the winter of 1944-45, small-town Wisconsin people worked together with Caribbean men, integrated by necessity of war.[78]

One group of foreign workers present practically within sight of the powder plant but not available for work were German prisoners of war. International treaties prevented prisoners from working in defense plants, but they could milk cows and chop silage in the countryside. Several hundred POWs, all German, worked on farms in Sauk County and a few of them liked it well enough to return and settle in after the war. POWs were also allowed to work in food-processing plants and in the summer of 1945 over 130 Germans were housed in a camp at the foot of Webb Avenue in Reedsburg. They worked in canneries at Reedsburg, Baraboo and Loganville, as well as on area farms. German POWs bused in from a camp in Dane

County also processed peas and beans at the Sauk City cannery. Bused around as they were from camp to canneries and farms — maybe even on Highway 12 past Badger Ordnance Works—the enemy prisoners of war saw more of Sauk County than the Caribbean allies sequestered at Devils Lake.[79]

As the demand for more workers intensified, and women already working at Badger proved their capabilities, recruitment of women intensified. According to the ad copy, by making powder at Badger a woman could bring home the "boys" — whether the boys in question were sons, husbands, fathers, sweethearts or brothers. Profiles in the *Badger Ordnance News*, which also appeared in community newspapers, focused on women working to hasten the return of their men. "You feel you have to do something when you have somebody over there to do something for," said Mrs. Kenneth Feight, Union Center, who worked on the Smokeless C-Line. She had a son and six nephews in the military. Elsie Westlund of Mazomanie had three sons in service. "I felt it was my duty to come to work here...Of course I would like to stay home and do fancy work, but I think I should work to help get this war over with..." Della Bliss of Hillsboro, a mix house operator whose photo was part of the newspaper ad campaign in the spring of 1945 said, "I like it fine at Badger Ordnance Works and would hate to quit before the end of the war because my two sons, Melvin and Elmer, who are in the armed services, cannot quit until the job is done."[80]

Ads recruiting female workers in 1944 and 1945 reflected a grudging change in attitude towards women working on the powder lines. Their presence in production was — like the integration of Caribbean men — a wartime necessity. Hercules policy was based on need, not desire. "Because of the shortage of men, it was recognized that women would be needed from the start..." When first hired in March of 1943, women were restricted to the "purification" work that took place after the cellulose was nitrated. By the summer of '43, after it became clear that Badger would be running only Smokeless Lines C and D, and that the supply of men was adequate, "it was decided to replace women operators in purification, with men on a replacement basis to as great a degree as possible." Women were not to be fired, but whenever a woman left her job, she would be replaced by a man. "Reasons for this decision were that women, while satisfactory operators in the main, presented a number of problems in supervision not usually encountered with men and provided a much less flexible crew, owing to physical limitations and the natural reluctance of women to change from job to job."[81]

Less than one year later, with all four nitrocellulose lines operating and the rocket area projected to start, necessity dictated a change in policy. "With the expansion it became at once apparent that a much larger percentage of women must be accepted and it would be necessary to use women in jobs never before considered suitable for them, especially on the dipping floor of the nitrating house." By November of 1944 the nitrating house of B-Line was an all-women operation, with twelve women "Dipperettes" on each of the three shifts. C- and D-Lines had both men and women in their nitrating houses, while the E-Line nitrating house continued to be an all-male operation, largely because the Caribbeans were assigned there. Women had been working throughout the plant since its start-up, and had long outnumbered men in traditional administrative, health care and housekeeping jobs, and also in the critical ballistic and "supersonic" testing labs. Yet it was not until they went to work in the nitrating houses that women were involved in every step in the actual manufacturing of smokeless, E.C. and rocket powder. They did not make the nitrating chemicals, but they did mix the chemicals with the cellulose and do all the refining work that followed. As for their performance: "Although the girls did present many problems and did not permit the flexible organization that an all-man area does, they did become first class dipper operators and, as usually, excelled men in detail work and cleanliness." Despite their first-class work, the women remained subjects of the prevailing view that female workers were auxiliaries of the males. When the war ended in Europe in May 1945 and Hercules started to curtail operations, "a greater percentage of males was retained" in the nitrocellulose department. For women ordnance workers, it was last-hired, first-fired, even if their work was required to bring "the boys" home.[82]

Getting To Work

For men or women, a job at Badger began with a filled-out application form. Construction workers not already employed by a subcontractor were hired out of the union halls in Baraboo or Madison. Production workers filled out their papers at the Hercules Powder Company employment office, which moved from Baraboo to the Administration Building in May 1942 and to its own building a few weeks later. Despite the shortage of workers and the compelling need to start making powder, no one was hired to work at Badger Ordnance without a background check. The Plant Protection Department had two divisions, the familiar, uniformed Guard Force and the more discreet Investigations Department, whose duty it was "to thoroughly investigate every applicant for employment and to conduct a further investigation of every employee if some act of his could be construed as detrimental to the war effort."[83]

The initial background check consisted of identifying applicants, checking the draft status of the men, taking fingerprints and forwarding copies to the

Federal Bureau of Investigation by way of the Sauk County Sheriff's Department. The FBI had yet to become the grand national storehouse of information on citizens, but it was working on it, and World War II gave the Bureau its first access to data on tens of millions of Americans, military and civilian. Between operating and construction workers, Badger Ordnance added about 25,000 names to the federal data bank. During World War II, defense plant security sat up and took notice if an applicant had a record of criminal activity, was a union activist or possessed some link to either communist or pro-German organizations. Central Wisconsin was not particularly notorious for such goings-on, including the criminal, but Madison was noted for leftist leanings among staff at the university, while Sauk County, like all of Wisconsin, was Germanic enough to incite a flicker of interest in curious investigative minds. In October 1942, Paul Davies, Superintendent of Plant Protection, asked *Capital Times* editor William Evjue what he knew about an organization known as the "Loyal Americans of German Descent." Davies was concerned because "we very thoroughly investigate all persons working at Badger Ordnance Works and...we have many applicants of German descent..." Evjue told Davies that the group had been founded by a Dr. Friedrich H. Roetter, "a German refugee in Madison [and] a former Berlin attorney who succeeded in escaping from the Nazi clutches" and gave Roetter's address to Davies.[84]

At its largest in the fall of 1942, the Investigations Department at Badger employed thirteen investigators and thirty clerks. As the proportion of clerks to investigators indicates, an employment investigation mainly involved paperwork and phone calls. A worker absent without notice from the job was likely to receive a phone call from an Investigation Department clerk. If the problem persisted, or if the worker did not have a phone, an investigator initiated "field work" both in the plant and out. Absenteeism impeded production and hurt the war effort.[85]

Infractions of the "Loose Lips, Sink Ships" variety also merited attention. Each investigator compiled "a contact file of informants" inside the plant which became "quite extensive" and "invaluable in the ensuing three years" of the war. This view of Badger, as a place where snoops checked up on workers who dogged it for a few days or who mouthed off in a patriotically incorrect manner, is the dark side of the pious "Praise the Lord and Pass the Ammunition" perspective on World War II defense work. Yet, by its own records the Investigations Department of Badger Ordnance Works conducted 24,581 investigations of employees in 1942, and racked up a total of 55,356 investigations by the time the office closed in October 1945. There was a war on. Badger Ordnance was a defense plant making munitions. Sabotage was an ultra-remote possibility, but the potential cost in human lives at the plant and on the battlefield was too great to gamble with, so the snoops went to work.

After clearing the background check, an employee's first chore was getting into the plant. The people who fought World War II, mostly born in the years since 1915, made up the first generation of Americans to know and love the automobile from their earliest days of life. Their parents were equally enamored of it, perhaps even more so, since they could remember the sluggish pace of horse travel and the inconvenience of the railroad. For all but the final year of the war, despite gas rationing, tire shortages and many an official message discouraging it, the large majority of Badger Ordnance workers — whether twenty-five or fifty years of age — traveled to and from work in a car. They were among the first exemplars of a phenomenon that would transform the American countryside in the decades after World War II. They lived in the country, but no longer had to live off it. They used modern autos and highways to sever their economic link to the land and brought about as profound a change to the rural landscape as Badger Ordnance did to the Sauk Prairie — but that was in the future.

During World War II Badger employees just wanted to get to work as quickly and conveniently as possible, which meant traveling by car. Sharing rides and gasoline ration points were standard practices, with three or four people coming to work in every vehicle. Workers who needed rides and drivers who needed riders were obliged to find each other through the plant Rationing Department, which kept track of travel needs and issued "Basic A Ration Books" for gas and tires. The Rationing Department staff, which included Sauk County Ration Board member Arthur MacArthur in its start-up days, also issued supplemental gas rations to drivers carrying passengers to and from work. Indeed, recruiters used the prospect of extra gas in the tank as an incentive for workers. Gas truly was rationed, but tires for civilian autos were so restricted that to say there were tires enough to ration is an overstatement. To keep rubber on the road drivers resorted to patching, recapping, driving slowly and staying home. Except for the plant managers living about a mile away at Staff Village, driving to work alone six days a week was not only an extravagance, it was unpatriotic — and well nigh impossible for anyone with only legitimate access to gasoline and tires.[86]

During the Badger M-1 construction period, the Army chartered special daily commuter trains to and from Madison with both the Chicago and North Western and the Milwaukee Road. The North Western commuter ran from April 1942 until the end of January 1943. The Milwaukee Road commuter ran for about two months during the peak of the construction season from June to August 1942. The fare was twenty-five cents one way. It was less than the cost of gas to and from Madison and certainly less than the wear and tear on an auto, yet, even though a substantial number of construction employees lived in Madison or on the train lines to and from Badger and the capital city, the commuter trains died for lack of patrons. Building contractors still had 6,000-plus workers on site. Equipment installers and Hercules had approximately 2,000 more, yet still the trains could not attract enough passengers to justify their claim to the track.[87]

Bus transport was more popular than rail travel, but still a losing proposition in a rural area with a thin population spread over seven counties. Bus service survived because it was the only practical way to bring all the workers needed to the plant and because the final operators, Mason and Hanger and Hercules Powder, were able to pass their losses on to the Army. In January 1942, one Wisconsin common carrier asked the Public Service Commission to approve a bus system for Badger similar to that later used by the consolidated school districts, whereby buses would run on local routes and stop at every farm where a Badger worker lived. Recognizing the impracticality of this notion, the Public Service Commission did not authorize the plan. In March, the Commission did license the Baraboo Range Transportation Company to carry all passengers, not just Badger employees, from stations in Madison, Prairie du Sac, Sauk City, Baraboo, Portage and Wisconsin Dells to and from the plant. The Company asked for permission to charge a realistic two dollars for a round-trip ticket from Madison, and $1.20 from Baraboo, but was forced to accept a lesser rate of two cents a mile one way, or seventy cents from Madison and twenty cents from Baraboo. Round trip, weekly and monthly tickets cost even less. Despite the lower fares, so few workers used the bus that, in less than two months, Colonel Griffith issued an alert to all employees that "it may be discontinued."[88]

Instead, the Baraboo Range Transportation Company went out of business and sold its fleet of thirteen buses to the Army, which then hired Mason and Hanger as operator and pledged to subsidize its losses. The fare was set well below cost at twenty-five cents per round-trip, with the first routes drawn to Baraboo, Reedsburg-LaValle and Portage. As construction of Badger M-1 picked up, the fleet expanded to twenty-five buses, each capable of carrying twenty-nine passengers. In July 1942, Leroy Jenkins of Spring Green organized the Diamond Transportation System as a common carrier to run buses to Badger from Dodgeville via Mount Horeb. Jenkins was just in time for the peak employment period and in August he saw his buses and those of Mason and Hanger packed with as many as fifty-five people. In the interest of safety, the buses were limited to forty-five passengers, only sixteen more than they were designed to carry. Demand often outweighed regulations and the buses were packed from windshield to taillight.

As winter approached, Mason and Hanger increased its fleet to thirty-seven and added more drivers. Pushing to complete the plant before winter set in, the builders had two ten-hour shifts working, with the last one ending at 4:00 AM, which meant bus drivers started work earlier. "Operations during this period were carried on under serious difficulties...the roads into the area were almost impassable because of mud. Buses were not properly equipped or heated, and as winter set in passengers and drivers suffered..." Added to the winter cold and the dark, breakdowns were frequent. No wonder people preferred to drive.[89]

After production began early in 1943, Hercules Powder took over the bus service. Facing protests from the drivers, the company agreed to raise their wages to eighty-five cents an hour and pay the same three-cent bonus production workers earned for shift work. Fares increased from the twenty-five-cent flat rate to a scale based on distance traveled: fifteen cents one way for a ten-mile trip; twenty cents for twenty miles; twenty-five cents for more than twenty miles. By spring of 1943, with nearly 4,000 people in operations and 3,000 still in construction, the buses carried only about 1,300 people a day. As operations stabilized and construction wound down, the Diamond Company went out of business and Hercules bus usage plateaued at less than 2,000 passengers a day.

In the spring of 1944, with all the smokeless lines running and construction of the rocket area underway, the buses filled again. Mason and Hanger reentered the bus business to carry construction workers and Hercules expanded its fleet by acquiring nineteen buses from the Keystone Ordnance Works in Pennsylvania. Since the construction crews worked two shifts and production people worked three, coordinating schedules was difficult, and sometimes one fleet of buses deadheaded back to the plant only a few hours before another group pulled out to pick up a different set of workers. At their peak in the spring of 1945, Badger buses started their runs at forty-five different places, passed through 111 cities, villages and towns in seven counties and clocked over 425,000 miles a month.

The routes ran on Highway 12 south to Middleton and Madison; on Highway 78 to Mazomanie, Black Earth and Spring Green; on Highway 60 to Spring Green, up Highway 23 to Plain and on Highway 14 to Gotham and Richland Center; on Highway 12 north to Baraboo, then to Reedsburg on Highways 33 and 136, then on to Wonewoc, Hillsboro and Elroy; also from Baraboo on Highway 33 to Portage and Wyocena and on Highway 12 to Wisconsin Dells, Mauston and New Lisbon; from the Dells on Highway 13 to Adams, on Highway 135 to Oxford and Montello, and on Highway 23 to Briggsville, Endeavor and Westfield. The number of buses rolling in and out of the lot reached 100 and daily ridership hovered around the 7,000 mark from February until July 1945. In this M-2 period with both production and construction in full swing, the bus system enjoyed its greatest use. Travel by auto was more difficult than in 1942, with gasoline still rationed and tires that were serviceable two years earlier now worn thin and threadbare. That in this final year of World War II operation Badger employed its largest number of women also increased bus ridership. In the 1940s, while a majority of working families owned a car, few owned two cars. If the family car was driven to work, the man drove it, not the wife or daughter, who may not have been licensed to drive anyway and who worked many miles away at Badger

Ordnance. The typical Badger bus at the height of ridership in the first half of 1945 was a bus full of working women.[90]

Traveling by bus was slower than by car and added to the hours allotted to the workday. The Necedah women who met at 4:45 AM to catch the Badger bus at New Lisbon put in a very long day, and so did Badger workers from Adams, Hillsboro, Montello, Muscoda, Wyocena, Richland Center and Madison — all of whom added two hours of travel time to their regular shifts each day. Traveling in winter, on roads maintained by resource-strapped county highway crews running undersized, primitive snow removal equipment, was often hazardous and always more time-consuming than in the warmer seasons. This type of long-distance travel for work on a regular basis was unheard of in rural Wisconsin until Badger Ordnance arrived and not even possible until the late 1930s when the vehicles and the improved highways they rode on were built. Erhart Mueller once conjectured that the War Department had laid plans to build a powder plant on the Sauk Prairie during World War I. Mueller never documented the statement, but, had a plant of any size approximating Badger Ordnance been built in 1917, workers would not have been able to commute to it. An entire company town, like Mason City at Grand Coulee Dam, would have to have been built no farther away than a person could walk or a trolley could run. In the 1940s, workers could live anywhere within driving distance and work at Badger Ordnance.

On the plus side, bus riders were not driving and could sleep, read, write to the sweetheart overseas, play cards, kibitz with friends and, as all workers do, sound off about work. That an investigations department informant might also have been on the bus casts a sinister shadow on the ride. Buses were safer than cars, with only two fatal accidents recorded for the millions of miles driven during the war. In March 1943, on an icy Highway 12 just south of the plant, bus driver Paul Leach was killed and four passengers injured in a collision with a truck. In April 1945, Mrs. Winslow White Eagle was killed when a car driven by her husband slid into a Badger bus near Lyndon Station.[91]

Badger buses were also used to carry schoolchildren from the Staff Village to Badger school and for shopping excursions from Badger Village and the barracks areas to Sauk City, Prairie du Sac and Baraboo. They also brought the Caribbean workers from Illinois and Indiana and carried them back and forth to work from their barracks at Devils Lake. When the plant guards marched in Madison's Independence Day Parade in 1944, they traveled there via Badger bus. Reaching even further afield, and hoping to decrease the amount of red ink in the bus transport ledger, Hercules set up a Greyhound station at the plant and in the course of eighteen months sold several hundred tickets for the long distance line at a ten percent mark up. With long distance auto travel all but impossible and with an unprecedented number of people traveling from one end of the country to the other, Greyhound bus service directly to and from Badger Ordnance was convenient and popular. One person who appreciated it was Rita Slezak Lachmund, powder cutter on the C-Line who, on the one weekend she got off every two months, was happy to pay one dollar for a Greyhound ticket home to Fond du Lac.[92]

Hundreds of Badger employees could get to work on foot. Residents of the barracks and Badger Village who lived directly across Highway 12 from the plant could walk into the plant. The shoulder of a high-speed highway has never been a safe place for pedestrians and several people were injured and one killed while hiking along Highway 12. In February 1945, Cyrenus VanWhy, one of the soldiers on furlough working on the construction of Badger M-2, died while walking along Highway 12 in a snowstorm, hit by a truck. The following spring, after discussing whether to install traffic lights, erect an overpass or dig an underpass, the state highway department authorized Hercules to build a concrete sidewalk parallel to Highway 12 and let pedestrians cross the road at their own risk.[93]

Inside the Gate

Once a Badger worker arrived for work by bus, car or on foot, he or she had to get past the Guard Force. In March 1942, the initial force of eighty-two guards on three shifts was assigned to twenty-four posts around the presumed perimeter and left to use their own cars as shelter from the cold. Mobility proved to be a virtue when guards parked on the wrong side of the plant boundary line and had to be redirected to the right spots. After the surveyors marked the property line, the first real guard shacks arrived. They were almost as portable as cars since they were built on skids and positioned at the gates by workers who slid them off the backs of trucks. The only creature comfort in each shack was a space heater fired by oil, coal or wood and which left the guard the option of alternately roasting or freezing depending on which side he turned to the heater.

Once the M-1 construction blitz was underway as many as 500 cars and trucks a day passed by the shacks and the guards acted more as guides than security men. Steering truckers to delivery points and laborers to work sites was more important than screening incoming traffic. In fact, keeping an eye on vehicles and people entering the plant mattered less than watching who and what was going out of the place. If a trunk full of tools or 2 x 4s left the construction site, it had to travel past a guard who was ordered to make spot checks, but not delay construction by inspecting every vehicle leaving the premises.

The Guard Force was organized by Lieutenant Otto Cronce, a veteran of the Green Bay police force. He was so highly regarded that Colonel Griffith went to Green Bay and asked the city council to release the Lieutenant from his obligations there so he could come to the powder plant. Even so, his stay at Badger was typically brief, ending in September 1942, when he was succeeded by Captain Louis Wilson, a former police Lieutenant from Richmond, Virginia, lately in command of the guards at Hercules' New River Ordnance Plant. From the initial corps of eighty-two men, the Guard Force grew to 207 men in the summer of 1942 and peaked at 409 men and twenty-two women in the spring of 1943. Male guards had to be physically fit, between the ages of thirty and fifty, and ineligible for the draft. Hercules reported having some difficulty finding men who could make the age cut and the fitness requirement. Among them was Frank Uphatel of Mazomanie, an artist whose woodcut prints had become the signature art for the works of August Derleth, H.P. Lovecraft and other authors in the late 1930s.

Female guards had to be between the ages of twenty-one and thirty-five, and preference was given to women "who were above average...with experience as social workers, nurses, teachers or in some type of supervisory work." These were hefty requirements, more than was asked of the men and more than the job required, since the first reason for hiring female guards was so female employees would not be searched by males. It was important work, but hardly required a college degree. The women guards were not assigned to the gates, and they were not armed, since they patrolled production areas on foot, making spot searches for matches and other potential fire starters. They also worked as clerks, issuing time cards and badges in the clock alleys.[94]

All the male guards received a minimum of two weeks' training in basic police work and were sworn in as auxiliary military police with authority to make arrests and hold suspects. They were issued a cadet grey uniform shirt, pants and jacket, with an over-the-shoulder Sam Browne belt where they could holster their Smith and Wesson .38 Special revolvers. The guns belonged to Hercules Powder and the guards carried them only when on duty and rarely in the production areas. There is no record of a guard ever having to fire his .38 except in practice on the shooting range. Early on, Hercules also supplied knee-high leather boots, popular among motorcycle police at the time, then reneged when shoe-leather rationing tightened. In the end, the company agreed to supply uniforms, but the guards had to supply their own footwear in the form of conservative black oxfords. The Guard Force was also equipped with a fleet of seven motorcycles, four sedans, six stationwagons and two trucks with passenger bodies to haul a crew of people to a trouble spot. As up to date as any county sheriff's department, they kept in contact with each other via two-way radios. In addition to their handguns, the guards had an arms locker containing a store of twelve-gauge shotguns and a quantity of tear gas canisters which, fortunately, they were not called upon to fire.

Security was surprisingly loose until production facilities were up and running in January 1943. In the summer and fall of 1942, for example, it was not unusual for construction workers to spend their off time on the plant site. Why men who had just completed a ten-hour shift digging, nailing or plumbing would return to the building site might be hard to understand until the alternatives are examined. A small room in a crowded house in Baraboo, a trailer parked on the asphalt at Prairie du Sac, a crowded bar and grill in Sauk City, were diversions whose attractions could quickly pale. By comparison, a summer's evening at the foot of the Bluffs, or on the river at Weigand's Bay or in a grove of trees outside an old farmstead, while hanging out with the guys from the work crew, must have been pleasant. The work site itself was its own source of entertainment and gave even veteran construction workers something to watch and jaw over in the manner of working people in any place and time. The image of a group of men, weary from the day, slouch caps cocked low on their heads, backs resting against the white pines that still stand in what was the yard of, say, the old Shimniok family farm, smoking cigarettes, talking quietly, sharing something alcoholic, while the vast building project of which they were a part was arrayed before them, is a rarely considered perspective on the Badger story.

Winter came, along with stricter security rules, and the picnics ended. In November, the sentry towers on the fence line were manned and a large search light was mounted on a tower at the pumping station to sweep over the dark waters of Lake Wisconsin. Late in January 1943, using nitrocotton from the Hercules Plant at Parlin, New Jersey, and acid from Du Pont's Kankakee Ordnance Works in Illinois, the C-Line initiated smokeless powder production at Badger. Security tightened immediately. "It was necessary to fence the entire operating line and to station guards at each of the sixteen openings to make match searches." The nemesis of security, hated by all safety-conscious Hercules staff because it was truly lethal, the humble match—in stick or paper form—created a lot of heat at Badger Ordnance. Fences went up, gates were mounted, shacks hastily hammered together, and the guards settled in for the cold winter nights. As the temperature dropped to minus-twenty-six degrees, "Every stove that could be found, regardless of kind or condition, was pressed into use...hauling fuel to these temporary posts required a two-man crew with a pickup for each shift."[95]

In May 1943 the D-Line, the Acid Area and the Ether Area were all separately fenced, adding dozens of "openings" that required more guards and shacks, but no heaters for a few months. In July, the E.C. Line, which was located just south of the Pioneer Cemetery, started operating and had to be enclosed behind its own fence, which meant more guards and shacks that had to be staffed round the clock. To fill them, Captain Wilson removed the guards from the perimeter towers and, although the crisis eased, never sent them back. Forbidding in appearance, the towers were hollow threats to intruders who never approached.

Finally, with the production area divided into a series of inconveniently fenced enclaves, difficult to secure and staff, the entire plant site was declared a restricted area. Construction workers were not allowed to wander around at will and their evening gatherings on the grounds halted. Matches and smoking were banned from the entire grounds in both construction and operations areas. All Hercules employees were required to enter and leave the property through the "Clock Alleys" at the rear of the parking lots. Here they punched in their time and picked up the identifying badges they were obliged to wear at all times on the premises. Hoping to avoid problems with badges lost or forgotten many miles away at home, Hercules did not allow its workers to take their badges off the plant site.

At the end of the Clock Alleys were the Search Houses where incoming workers were checked for cigarettes and makings, lighters and matches. Security was "very tight," as recalled by Rita Slezak Lachmund. "I had a purse two feet long. Put everything in it. Had one paper match, didn't know it was there. The guard found it. I got three days off with no pay."[96]

Restrictions and enforcement grew tighter as production increased. "Matches, lighters or other flame-producing devices are not allowed within the plant fence," was the rule. In 1944, possession of a single "safety" match meant a week off with no pay for the first offense, two weeks off for the second offense. Three strikes and you were out. Possession of a cigarette lighter meant two weeks off and no pay for the first offense, dismissal for the second. Possession of a "redhead" or "strike anywhere" match that could be lit on a wall or on the edge of a thumbnail meant immediate dismissal, as did possession of a firearm or ammunition. Smoking on the premises, no matter how the tobacco was lit, meant immediate dismissal. "They watched us very closely," said press operator Mardetta O'Brien. "Accidents could have meant our lives." Yet the rules were broken. William Sisson remembered a coworker on the nitrocellulose E-line who ducked out for a smoke regularly. "Four months after I left, they had an accident and I always wondered if it was him."[97]

Safety was a vital concern. The plant hospital was much more than a well-equipped first-aid station. At its peak in May 1943, the staff included seven doctors, eleven nurses and ten technicians, although a more typical roster had only three or four doctors and seven to nine nurses. In the event of an accident, the operating room and first-aid station, "could be put immediately into use...to treat casualties." The hospital's fifteen beds could be supplemented with twenty-two cots, while the guard room nearby was ready to be cleared "in the event of a disaster" and additional bedding set up.[98]

Yet despite constant exhortations to obey the rules and work safely, accidents happened. Ten workers died on the grounds at Badger during the World War II operating period, seven in explosions. "The major factors in production of smokeless powder by Hercules Powder Co. are safety first, quality second, quantity third," was the declared policy of the company. The safety lecture was delivered to new employees on their first day at work. They also witnessed "burning demonstrations" of the coveralls they would be wearing, with flame-proofed garments compared to untreated clothes, and saw what the acids used in the plant could do to clothing, building materials and, by extension, bare skin. Small bombs showing the explosive powder of nitrocotton were set off and a metal weight was dropped on a dollop of rocket paste to demonstrate its power to do harm when abused. They were also lectured on the dangers of toxic exposure to spray paint, carbon monoxide, mercury, lead, nitrous fumes and carbon tetrachloride. All applicants underwent a medical examination before starting at the plant and some received regular checkups afterwards. Food handlers were examined for sexually transmitted and other contagious diseases every three months. Lead burners had a blood count every six months. Most closely monitored were people exposed to dinitrotoluene, or DNT, an ingredient in the chemical cocktail used to solidify nitrocelluose prior to its being pressed, extruded and cut into grains of smokeless powder. People working in the diphenylamine mix house were regularly exposed to DNT and company policy dictated that "every two to four weeks" they have a blood count and be examined for weight loss and other symptoms. Workers who loaded smokeless powder and DNT into "sweetie barrels" were supposed to alternate weeks on another "outside" job. They were also ordered to pull their rubber boots up under the coveralls, keep their coveralls buttoned and their heads covered, to wash their hands and face "at frequent intervals" and eat a "well-rounded diet including fresh vegetables, milk and other such health-giving foods. While an explosion was the most obvious hazard at Badger Ordnance, more subtle dangers were also present.[99]

After going through the Clock Alley and the Search House, Badger workers then entered the Change Houses where they exchanged street clothes for specially designed, acid-resistant and flame-proofed garments. The smokeless area had twenty-nine change houses for men and eighteen for women, while the rocket area had ten for men and nine for women, plus one each for men and women at the rocket powder testing lab. In 1943, the first women working in the purification houses were "outfitted with special flannel slacks and shirts," while "Cutting House" operators wore heavy blue denim coveralls. A year later, everyone was wearing flame-proofed coveralls, "powder" blue for the women, blue-gray or rusty brown for men. Neither slacks, shirts nor coveralls had pockets, which could serve as potential repositories for matches or other combustibles, except for the tiny slit of a handkerchief pocket in the women's suits. Mens' coveralls had a rear pocket of webbing with openings too wide to hide a book of matches. However, in all but the warmest weather, the coveralls were worn over street clothes where forbidden items not discovered in the Search House could easily be hidden.

In the 1940s, shoes were made with leather soles and heels fastened by steel nails. As the leather wore, the nail heads were exposed and, when scraped over concrete or other hard surfaces, they sparked. At Badger, such sparks could be deadly, so workers were issued "powder shoes" with glued-on rubber soles and heels that did not spark nor conduct static electricity. Hard hats were not required for industrial or even construction jobs in the 1940s, but men customarily wore hats outdoors at all times and inside blue collar workshops, so they wore them at Badger. As in other industrial settings, women were required to wear headcoverings at Badger. Long hair and machinery had been recognized as a hazardous combination since the start of the Industrial Revolution and 19th-century female millworkers bound their hair tightly and covered it with kerchiefs. World War I-era factory workers wore a sort of floppy beret with a close-fitting brim and a looser top. Women starting at Badger in 1943 were issued tight-fitting turbans that were unpopular because they smashed hairstyles. When offered the option of wearing a soft cap into which hair could be loosely gathered, then enclosed by tightening a drawstring in the rim, Badger women took it. By the end of the war, more than three-quarters of them were wearing caps instead of turbans. The conversion to the cap was so pronounced that the headgear depicted is one way to date photos of women working at Badger. If the women are wearing turbans, it is almost certainly a 1943 photo. Women in caps date a photo as 1944-45, and probably later than the spring of 1944, when the rocket powder buildup started and the number of women employed increased dramatically. Seemingly superfluous, the question of headgear for females at Badger illustrates the double message American culture was sending to working women during the war. On the one hand, the press, radio, movies, even comic strips — heavily influenced if not controlled by the government — were telling them to be attractive for the "boys" in service, with short skirts and long, wavy hair in fashion. On the other hand, they were factory workers, expected to produce the materials those same "boys" needed to win the war. Tucking long hair into a loose cap instead of smashing it under a turban was one way working women responded to both messages.[100]

Hercules supplied laundry service for its workers' coveralls, but cleanliness was not its first concern. In order for the flame-proofing to adhere, the clothing had to be clean and the best way to insure cleanliness was for the company to do the washing. The laundry was in a sizable one-story building in the maintenance area, where high-powered washing machines and special detergents removed powder and chemicals from the clothing. Finding a "soap" that would remove nitroglycerin was a challenge, with the laundry eventually boosting the power of the cleanser it used to remove smokeless powder and chemicals by adding soda ash and harsh alkali soap. By January 1945, the Badger laundry was processing 25,000 coveralls for men alone. Five months later, at the peak of operations, nearly 43,000 coveralls were washed and flame-proofed. Missing buttons were replaced, rips and tears repaired, so loose clothing would not catch on machinery. After washing, flame-proofing and mending, women's coveralls were pressed, with creases in legs and sleeves, to present a neater appearance than was presumed necessary for the men.

Once safely dressed for the job, a Badger employee left the Change House and boarded a bus bound for his or her workstation. With such a sprawling operation spread over so much territory, the fastest way for a worker to get to his or her job station was to ride there. Busing also kept workers from wandering around the plant and, unintentionally or not, creating safety and security hazards. During World War II and in later periods of operation, few Badger employees ever saw any part of the plant other than their specific work area. Nitrating house workers went to the nitrating house and did not visit a press house or the magazines. Acid plant workers knew the acid plant only, ballistic lab workers knew no more about what a blending house looked like than they did about downtown Tokyo, nor did the Army or Hercules think they needed to know. Isolating workers from other parts of the plant was necessary for safety and security, but it was also in keeping with the philosophy of the assembly line which held that a worker needed to know his or her part of the operation and nothing more. Understanding how the work of an individual fit into the larger scheme of things was for management only, and only upper management at that. The assembly line approach was not restricted to Badger, nor to defense plants, but was the standard of American industry, whether it was at the Nash auto company in Kenosha, the Nekoosa-Edwards paper mill in Nekoosa or the Oshkosh B'Gosh overall shop in Oshkosh.

Badger had two intra-plant bus lines during World War II. Hercules ran one for operations and, since security was tighter when it was building Badger M-2 in 1944-45 than on M-1 in 1942, Mason and Hanger also ran buses to rocket area construction sites. With motor vehicles of any kind in short supply, the buses used inside the plant were about as basic as a bus could be and still have seats and wheels. Some weren't even buses at all, merely bus bodies mounted on semi-trailers, hitched to and hauled around by trucks. Nonetheless, the Hercules buses alone racked up over 10,000 miles as they wound through the byways of the powder plant delivering people to work stations.

Making Smokeless

The Badger Ordnance Works was a multi-faceted operation. Its support facilities — laundry, security, utilities, hospital, transportation, maintenance — required a staff of approximately 2,000 "non-productive" people in the summer of 1944, before the rocket line started to run. Yet the core of the operation, the reason the plant existed, was to make smokeless, blankfire and rocket powder. All the other facets of the operation, although necessary, were secondary to the manufacture of powder. After the initial mixing of nitric acid and cellulose, the process was one of purifying and refining until each grain of powder was exactly the same as the other. Soldiers and sailors firing the shells containing the propellant had to rely on load after load to fire exactly the same. In describing combat in Europe after D-Day, historian Stephen Ambrose tells of German soldiers who so feared the accuracy of American artillery that as soon as shells started to hit their position, they knew they had to move. One after another the shells would rain down right on target again and again. It was a victorious combination of finely crafted guns, well-trained gunners and flawless ammunition.

"Over four decades of interviewing former GIs, I've been struck by how often they tell stories about duds...shells falling near their foxholes and failing to explode. There are no statistics available...nor is there any evidence on why, but I've never heard a German talk about American duds. The shells fired by the GIs were made by free American labor; the shells fired by the Wehrmacht were made by slave labor..."[101]

Ambrose stated the goal of production at Badger Ordnance. No duds, no misfires, no shells exploding in the gun barrel, or falling short, or wobbling off target because the propellant carrying them on their trajectory was less than perfect. It wasn't enough to simply praise the Lord and pass the ammunition, but instead to praise the Lord and pass the ammunition that fired perfectly every time.

The production of perfect powder at Badger began in two places, the acid mix and weigh house and the receiving docks of the rail lines. In the acid mix and weigh house, fresh nitric and recycled spent acid — the explosive ingredients — were mixed with moisture-absorbing sulfuric acid according to the powder's formula. At the railroad docks, cellulose in the form of bales of cotton linter or wood pulp was unloaded from boxcars and temporarily stored in a warehouse. Cotton linter was a waste material from textile manufacturing that resembles the lint at the bottom of pants pockets. Wood pulp is lumber from a tree, with the bark removed, washed and ground into coarse flakes ready to be processed into paper. Badger purchased its cotton from the Southern Chemical Company and other suppliers, and its pulp from papermakers such as Weyerhaeuser and International. Shipments of linter and pulp varied in composition and purity, which meant that the chemicals mixed with them had to be adjusted and they had to be inspected for sand, stones, bark and other foreign objects. When it was invented, smokeless powder was made only from nitrated cotton and its supply limited by the quantity of linter, which, coincidentally, made cotton a strategic military resource. In the early 1930s, a Hercules chemist successfully substituted wood pulp for cotton, thereby guaranteeing a virtually limitless supply of a raw material basic to powder making just as much of the world was rearming for war. Of Badger's four nitrocellulose processing lines, C-Line was originally designed for cotton only, B-Line for pulp, with D-Line and E-Line equipped to handle either one.

From the warehouse, the cellulose—cotton or pulp—went to a dry house where it was inspected and picked clean of stones, dirt and trash, then shredded, dried and blown by large fans up a duct to the third floor of the nitrating house. There it was divided into high-grade charges weighing twenty-one pounds and low-grade charges weighing forty-four pounds each and stored in aluminum cans. Nitration began when a "dipperman" or "dipperette" poured a charge of cellulose into a dipping pot along with warm acid from the acid mix and weigh house, where they were agitated for up to twenty minutes. Dipping was an unpopular job, one that "people in other departments...conjured up a misconception and spread devastating rumors damming the nitrating area over several counties." Since workers holding other jobs were reluctant to transfer onto the dipping floor, "it was immediately obvious that only newly hired persons could be used." When operations were expanded in the second half of 1944, the "newly hired" meant women, who took over the B-Line dipping floor, and the Caribbeans, who monopolized the E-Line until the war was ending. As a concession to the "dipperettes," Hercules switched to lighter-weight cans for dipping charges and installed "steel slides" so the women did not have to "lift the 140-pounds of can and cotton." These were among the many alterations to equipment, buildings and procedures Hercules introduced to make lifting, hauling, dumping and other aspects of the production work less physically demanding on female workers. The men benefitted too and, as far as is known, none of them complained about being forced to lift lighter loads or having to slide cans and boxes on rollers and not lifting them at all.[102]

After nitration, the nitrocellulose was piped into a centrifugal wringer on the second floor of the nitrating house. Wringing, which resembled the spin cycle of a clothes washer, separated out the spent acid, which was then drained and piped back to the acid mix and weigh house for reuse. Nitrating-house wringer operators wore safety clothing from head to toe, including acid-resistant boots and gloves, plus flat-topped metal helmets that made them look like they were wearing water buckets on their heads. Each helmet had a small window on its front side and a bibb hanging down to protect the chest and back. Because of the agitation and friction of the wringing process, the wringer was perhaps the most fire-prone station on the production line. A wringer fire that flared into a small explosion shortly after operations began in March 1943 injured two workers. In April another wringer explosion resulted in the first death of a line operator at Badger. C-Line wringerman Harry Borland bent over to peer into the wringer basket just as the nitrated cotton exploded. Despite his safety helmet, he suffered a concussion, a collapsed lung and other injuries which brought about his death the following day. Borland's was one of six fatalities that made Badger's first year of operation its most deadly to its employees. It also contributed to the dark reputation that clung to the wringers throughout World War II. Wringermen were so difficult to retain in 1943 that production was threatened and the plant managers seem to have been relieved that they were not ordered to run beyond fifty percent capacity. In order to get maximum production out of the wringermen it had, Hercules ran the nitrocellulose department twenty-four hours, seven days a week from July 1943 until the end of 1944, when it allowed workers three days off for Christmas and two days for the New Year's holiday. Work resumed January 2 and there were no more days off for wringermen until the war ended. Minor fires continued to break out in the wringer houses, but, despite the grueling schedule, no more serious accidents or fatalities occurred there.[103]

After wringing, the nitrocellulose was inflammable and explosive, but still a long way up the purification pipeline from being smokeless powder. First it was drained out of the wringer and mixed with water in a drowning tank on the ground floor of the nitrating house. The resulting slurry was then piped to the boiling tub house, one of the largest production buildings in the nitrocellulose area, and which housed thirty-two closed tubs, each capable of holding 12,000 pounds of slurry. In the tubs, the nitrocellulose was heated to 203 degrees Fahrenheit and left to simmer for up to thirty hours. It was then boiled twice for five to eight hours, then rinsed in clean, cool water to remove any unmixed acid or cellulose. The slurry was then piped to the beater house where equipment similar to that used in papermaking reduced the coarse fibers of the nitrocellulose to a finer consistency and sodium carbonate was added to make the acid-heavy mix less corrosive to the plumbing.

The slurry was then piped into one of twenty-four pine tubs in the poaching/blending house where it went through a series of boilings, coolings, settlings and washings. It was passed through a fine screen to remove clumps and dirt, then piped into blending tubs where the high and low charges were mixed to achieve the desired content of nitric acid. Next, it was piped to the wringer house where about two-thirds of the moisture was removed.

Once out of the wringer the nitrocellulose could no longer be moved as slurry in pipes. It was now loaded into the steel cans of the Telfer monorail conveyor for easy transportation to the dehydration press house. The nitrocellulose now packed sufficient explosive power so that work stations in the "dehy" house, and the mix and press houses next in line, were separated by concrete walls and covered with blow-out roofs. If one mixer or press exploded, it would not take out the whole building or its crew. In the dehy house, the nitrocellulose was packed into vertical presses and injected with ethyl alcohol, which was formulated in the ether still house and piped to the dehy. The pressing forced out moisture, replaced it with the alcohol and formed the nitrocellulose into blocks. Using non-sparking brass forks, workers then broke up the blocks and loaded the material, which now resembled wood fiber attic insulation, into fiber drums before wheeling them on a covered walkway to the mix house. Here the nitrocellulose was mixed with a cocktail of ether, ethyl alcohol, dinitrotoluene and other chemicals that jelled the loose fibers into a solid but still plastic "colloid."

As a mix house operator in the spring of 1945, William Sisson, "had to load nitrocotton and mix it with a pail of ether, about the size of a milking pail...it was cold, if you put your hand in it would freeze." After mixing, Sisson unloaded the colloid, "it looked like cookie dough," and transferred it to a macerator where it was coarsely chopped "into globs," which were then dumped into "six little barrels" that he placed on a dolly and pushed to the "cheese house." Both men and women worked in the mix house, with the labor divided by gender. A foreman, two male operators and seven women ran three mixers and four macerators. Sisson loaded and unloaded the mixers and wheeled the colloid to the "cheese house." While he was gone the women cleaned the mixers and made ready for the next batch. The pace was brisk, with 50,000 pounds a day the norm. "Everything was timed and if you fell behind sometimes, [the foreman] would close the vents and you'd get drunk on the ether." In this state, the pace would pick up, time lost would be recovered and "we'd sing, jump around, dance." Sisson was a vigorous nineteen year old who had worked on the family farm his whole life, yet on the day after working on an ether high, "boy was I sore, we worked so hard." Closing the vents designed to remove harmful fumes was not in the Hercules manual of operating procedures, but the mix house was only one stop on the assembly line. If it fell behind, the pace of the whole line was affected and no foreman wanted to be responsible for a slowdown. It was easier to close the vents.[104]

Although mix house operators added several chemicals to the nitrocellulose, it was the ether they remembered. As William Sisson recalled, "Once you worked with ether you didn't smell too much else." The fumes were pervasive and persistent and remained with the powder and the people for several more stops down the production line. "I used to make my own sandwiches [for lunch], baloney and stuff. I liked honey and bread sandwiches. One day the ether absorbed into that honey and I got sicker than a pup. To this day I can't eat honey."[105]

As a teenager Lola O'Brien Huber shared a bed with her older sister

Mardetta, who ran a powder cutting machine. "Sleeping with her was like hanging out at the hospital on surgery day. The smell of ether never left her. It was on her clothes, in her hair and even on her breath." Mardetta remembered that "sometimes the ether smell would get to us and we would get a little lightheaded. Our supervisors seemed to notice it more than we did. We just felt a little silly. They would have us take a break and go outside...When the weather was nice, it wasn't so bad, but sometimes, in the middle of winter and the snow was deep...we still had to take a walk."[106]

What William Sisson called the "cheese house" was officially known as the block press house and it was the next stop on the line after the mix house. Here the nitrocellulose colloid was compacted into cylinders that were about one foot wide and two feet high, similar to a wheel of cheese. The cheese was then taken to the finishing press/cutting house where for the first time it could be accurately called smokeless powder. First it went through another set of screens designed to remove lumps and unmixed cellulose. Then it was forced through the finishing press, which could also be called the pasta maker because it extruded the colloid in long threads that were firm and supple enough to wind around a fork. The diameter of the thread was determined by the caliber of the powder: spaghetti for .30 mm rifle powder, elbow macaroni for 75 mm artillery grains, finger-thick rigatoni for 8-inch mortar sticks. The threads came out of the press and were coiled into canisters about two feet high, open on top and screened on the bottom so the alcohol and ether could vent off. The canisters were then wheeled to a cutting press where a dozen or more threads were fed past cutters that sliced them to the desired length of anywhere from one-quarter to one-and-one-half inch, and perforated them end to end. The perforations could be one simple round hole, or six of them. They could be slotted, grooved, star-shaped, or a combination thereof, all precisely formulated to control the rate at which the powder burned down to the thousandth of a second.

Badger made about 251 million pounds of smokeless powder during World War II. Almost 147 million pounds was in the form of cannon powder for use in a wide array of artillery pieces: 40 mm for naval anti-aircraft "pom-pom" guns; 75 mm, 90 mm, and 105 mm grains for field guns and howitzers in regular and "supercharged" loads; 155 mm for the versatile "Long Tom" field and anti-aircraft gun; powerful 3-inch cannon and 8-inch howitzer loads. Badger also made over 102 million pounds of powder for small arms: .30 mm for rifles and light machine guns; .50 mm for conventional heavy machine gun loads, plus incendiary, armor-piercing and armor-piercing incendiary machine gun loads for tank warfare. Completing the total was the 1.6 million pounds of E.C. or blankfire smokeless powder Badger made for use in signal guns, tear gas canisters and hand grenades. By September 1944, all of Badger's four smokeless powder lines were running and making a range of propellants. On the B-Line it was 155 mm artillery and 8-inch howitzer powder. The C-Line produced 3-inch gun, plus 8-inch and 105 mm howitzer loads. The D-Line made .50 cal machine gun powder and on the E-Line, 105 mm howitzer loads. Operating at full capacity, twenty-fours hours a day, seven days a week for thirty-one days in January 1945, Badger set a record for monthly production of smokeless powder — 15,910,253 pounds, more than one-half million pounds a day.

Once out of the cutting machine, all those millions of pounds of powder were no longer "green" but still far from finished. The individual grains were returned to their ventilated canisters and loaded into wood-framed, copper-lined "coffin" buggies, so named because they were long, narrow, covered and resembled a casket on a mortician's cart. The buggies were parked in the solvent recovery house where warm air circulated through the powder, removing the ether, alcohol and other solvents added in the mix house. Were the chemicals left in the grains, they would eventually evaporate, alter the burning rate and spoil the powder. Instead, they were vented off and recycled. The finishing process continued in the water dry house, where the grains were stewed in a warm water bath. After the water bath, small arms powder was dumped into a round, enclosed "sweetie barrel" where it was dusted with a bright yellow coating of powdered DNT. Then it was bathed again and shaken through a screen where excess DNT was rinsed away. Next, the grains went back into the buggies, which were loaded on trailers and pulled by tractors to the air dry house where they were deposited into 10,000-pound tanks to be bathed in heated air for three to twelve hours. Small arms grains were then transferred to glazing barrels where they were coated with black graphite.

Next stop was the pre-blending house where batches containing 5,025 pounds of powder were tumbled in a blending barrel to distribute any remaining moisture. The powder was then placed in bags, eighty-plus pounds to the bag, stored in a rest house and labelled. Even at rest smokeless was dangerous, so the windows of rest houses were painted to block out direct sunlight that might overheat the powder. While in the rest house, samples of the powder were taken to the lab where they were tested for uniformity and identified for final blending. Tolerances were minute, but in the next stop on the line, the final blending house, powder from a bag on one side of the explosive scale was mixed with powder on the other to achieve a final mix within the extremely tight parameters of acceptability.

Powder was tested at Badger and at other sites by the Army and Navy, but not always accepted, which meant Hercules was not paid for it. In addition to all the valid patriotic reasons for producing top-quality propellant, Hercules had the most basic commercial justification, a paycheck. Rejected powder, both from Badger and from other powder plants, including some that had been shipped overseas before its flaws were detected, was returned to the line, remixed with new material and reblended to meet standards. In

July of 1944, for example, 350,000 pounds of unsatisfactory .50 caliber machine gun ammunition was broken down and the propellant reblended at Badger to form .50 caliber armor-piercing incendiary loads.

Although it was near the end of the line, blending was no less dangerous than other stops on the way. Following instructions from the lab, workers moved bags from the rest house to the final blend house, and poured the grains into a copper barrel that rotated them for approximately ten minutes. In October 1943, Arnold Swansby and Alexander Libke were blending 1,600 pounds of E.C. powder when an explosion occurred. The force was sufficient to rattle windows in Baraboo and damage buildings at Badger in the vicinity of the blast. The damage to the plant would have been much greater had Badger not been designed with such accidents in mind and buildings separated by generous amounts of open space. The E.C. blending house itself and the barricades around it were completely destroyed. Swansby, age 30, was from Baraboo. He and his wife, Bessie, were the parents of four children. Libke, forty-five, was from Dalton in Columbia County. He and his wife, Anice, were the parents of two children. Both men died instantly in the blast, which was so concentrated by the barricades that traces of their bodies were hard to find in the debris. Other accidents had occurred at Badger, small fires and powder flashes were common, but what Hercules called a "shot" mean enough to demolish a barricade was a devastating reminder of the lethal power of the material fashioned at the plant. It was only propellant for blank cartridges, but it still killed two men.

"The operating personnel group was badly shaken," read the company report, and a few people transferred out of the E.C. line to what they thought were safer jobs. Changes were made in procedures and equipment. All metal scales, funnels, dipping buckets, scrapers and shovels were replaced with wooden or fiber equipment. Leather padding around packing machinery, which might conduct sparks, was replaced by conductive rubber and grounded. No direct link can be proven, but the fatal blast of October 1943 might have "shaken" Badger employees into a heightened awareness of safety. The plant produced one-quarter billion tons of powder after the E.C. line blast and ran to within one month of the final peace before another explosion snuffed out workers' lives.[107]

The final step in the finishing of smokeless powder took place in the can pack house. Fiber drums carrying the blended grains were loaded onto tram cars and delivered to the top floor of the can pack. They were emptied into a hopper, passed through a screen for one last check and loaded into either copper or stainless steel-lined plywood boxes that were about the size of a tall kitchen waste basket. As they sat under the hopper with the powder trickling down, the cans were gently shaken to settle the grains. When filled, the cans were closed with a gasket-lined lid and sealed airtight like mason jars packed with tomatoes. An identifying number was stenciled on the side and then it was off to either the magazine or the rail shipping yard.

The magazine area, where powder was stored prior to shipment, stretched over 3,300 acres in the southern part of the plant. Buildings were spread out to take advantage of space and the shelter of the rolling terrain. A total of eighty frame buildings held smokeless powder, with fifty-six small arms magazines tucked behind barricades and another thirty-four cannon powder houses without barricades. DNT was held in yet another eight barricaded frame storehouses, and twenty-two structures with concrete walls and earth berms were reserved for rocket powder. Already packed in shipping cans or boxes, the powder was carefully loaded and unloaded by hand and carried to and from the magazine by a fleet of thirty-four powder trucks.

In the final months of 1943 and early 1944, the magazine was filled to capacity and, "emergency measures were necessary several times in order to obtain shipping orders on accepted lots...in order to provide space for current production." The demand for smokeless powder had so fallen and the magazine grown so full that in April 1944, Badger shipped out 13,000 tons of wood pulp that it could not process into powder. Time and the course of the war relieved the congestion. After the Normandy invasion, Badger couldn't produce smokeless fast enough and the magazines became little more than waiting rooms in an oversized freight depot.[108]

The situation certainly brought happiness to "Bud" Williams and other railroad executives. Badger Ordnance and the war had revived the struggling Baraboo line of the Chicago and North Western and pumped new life into the moribund Sauk City line of the Milwaukee Road. In addition to the thousands of carloads of construction material hauled into the plant during each of the two construction blitzs, the railroads also delivered machinery, coal, chemicals, cotton, wood pulp and flawed powder to be reblended. Depending on the weather and production needs, the power plant burned anywhere from fifteen to fifty million pounds of coal a month, all delivered by rail. On the way out, the trains hauled 200 to 500 carloads of powder and chemicals nearly every month from February 1943 until July 1945.

Badger also had its own railroad system, starting with a pair of steam locomotives in 1942. As soon as operations began, the steam locomotives, which often threw sparks and hot cinders out of their stacks, were replaced by three diesel-electric locomotives. They were used to move coal hopper cars, acid tank cars and other standard gauge freight cars to and from the North Western yards at the north end of the plant and the Milwaukee yards at the south end. "A great deal of switching and clerical work could have been eliminated had there been but one railroad serving the plant," lamented the Hercules reporter, forgetting that when the plant was being sited, service from two railroads was touted as an asset.[109]

Easier to manage was Badger's own narrow gauge railroad which moved powder through the plant during processing. The three-foot-wide lines radiated out from the nitrocellulose areas and crisscrossed the smokeless finishing area, but did not reach the Oleum and acid plants. When the rocket area was built, a single narrow gauge line ran into the paste area and forked onto the "paste house roads" in both the east and west rocket roll area, but did not extend to the magazines. Even if it did not run through the entire plant, the narrow gauge did stretch for over twenty-five miles and made 360 turnouts as it linked one production house to another and another and another. With bulldozers knocking them out of line and curious laborers fiddling with the switches, standard gauge trains derailed several times during the construction period, but the narrow gauge cars stayed on the track, which is why they were used in facilities like Badger. The single tramway derailment on record might have been disastrous. In July 1943, the narrow gauge was pulling a flatcar carrying 5,500 pounds of .30 caliber powder in buggies on its way to the Number 2 Glaze House. Thirty feet away from the house, the train derailed, one of the buggies tipped, spilling some powder, which burst into flame. The rest of the powder caught fire but did not explode and burned itself out. No one was injured and no damage was done, except to the tram car.[110]

Making Rocket

The production of rocket powder was similar to that of smokeless in that it started with the mixing of basic ingredients followed by purifying and refining to exact specifications. At Badger rocket powder was made by the "solventless" or dry extrusion method, which was later heralded as one of the most significant of the ballistic technologies perfected during World War II. Considering the extensive use of rockets in warfare during the second half of the 20th century, it must be so. In the United States, the solventless method was developed in 1941 by the Army's National Defense Research Committee, the California Institute of Technology and Hercules Powder. When it brought the technique to Badger in 1944, Hercules was already producing quantities of dry extrusion powder at Radford and at the Sunflower Ordnance Works in De Soto, Kansas. Hercules' largest plant, Sunflower turned out eighty millions pounds of "rocket" by the end of the war. Much larger than smokeless, the grains of rocket powder were about the size of a billy club or Little League baseball bat. In the wet extrusion process used in the United States prior to 1941, rocket grains tended to warp out of shape as the solvents evaporated, causing them to misfire. In the solventless process, green powder or "paste" was dried before it was extruded, thereby holding its shape and firing according to specifications.[111]

The first ingredient of rocket powder made at Badger Ordnance was unfinished nitrocellulose from the original plant. Second was highly explosive nitroglycerin, a mixture of glycerol, nitric acid and sulfuric acid formulated in the new nitro works at Badger. Production began when a tank full of nitroglycerin was wheeled into an earth-bermed storehouse to be mixed with the chemical diethylphthalate. After mixing, the nitro and the diethyl were carted to the premix house where they were introduced to the nitrocellulose at the ratio of about two parts nitro to three parts nitrocellulose. Workers delivered the chemicals to the premix house in tanks they pushed down from the storehouse in two-wheeled carts darkly identified as "angel buggies," to identify the choir the buggy driver would be singing with should he have an accident in transit. The "angel buggies" rode on what the company identified as "wheeling walkways" but the workers called "angel walks." They were covered wooden ramps raised on trestles built over the low spots in the moraine area where the nitro plant and rocket area were built. Nitroglycerin is notoriously unstable and volatile, but it can be poured, pumped, even jostled, without exploding, as it had to be to make rocket propellant. How much it can be bumped or jostled is the life and death question.

On July 19, 1945, less than one month before the war ended and only one week after the *Badger Ordnance News* had announced that the plant had operated for over 1.2 million "man-hours without a lost-time accident," the dangerous potential of nitroglycerin was demonstrated. At 7:30 in the evening, in storehouse B of the west rocket area, Erwin Pugh, William Denny, Ellsworth Goff and Mark Shearer were preparing to wheel a mix of diethyl and nitroglycerin down the angel walk to the premix house when "an explosion of unknown origin" occurred. The "shot" unleashed the power of approximately 7,500 pounds of nitroglycerin, reducing the storehouse to matchsticks and the surrounding barricades to twigs and sand piles. Windows rattled in Baraboo, Prairie du Sac and Badger Village, and the mushroom cloud rising over the rocket area was visible in front of the sun from across Lake Wisconsin in Columbia County. The storehouse crew was a typical Badger mix of experienced and new workers. Foreman Erwin Pugh, 52, had been in the acid department since May 1943. Ellsworth Goff, 31, had been on the smokeless line since March of that year. When smokeless production wound down in the spring of 1945, they transferred to rocket. William Denny, 45, an Oneida from the Green Bay area, had been at Badger for about a month. Mark Shearer, 27, from Douglas County, had started at Badger on July 4, 1945. All of the men were married, all but Denny had children. So ferocious was the shot that the body parts that could be found were gathered and buried in a common grave at the Kingston Cemetery, with Reverend Charles Lee of Baraboo offering the final blessing.[112]

Except for the loss of Storehouse B, damage to the plant was superficial and repaired by company carpenters. Production resumed almost immediately, and, so none of the rocket finishing area would be idle, Hercules ordered 310,000 pounds of rocket paste from Radford and Sunflower to be delivered for August. That eight children lost their fathers, four widows were made and four men died, was terrible. Consolation can be found in the fact that it was not forty or 400, that the design and operating procedure of the plant were such that since fatal accidents could not be entirely prevented, they could be minimized. Just as the overall advance of an army is not halted by the loss of four soldiers, powder production was not seriously hampered—as intended. It is the callous logic of war, where death is not restricted to combat.

Production continued as it had before the explosion. Once in the premix house, the nitro and diethyl were combined with the nitrocellulose and heated water to form a slurry, agitated briefly and piped to the final mix house. Here the slurry was mixed with another chemical cocktail, one of whose ingredients was candelilla wax as a solidifier, then piped to a wringer, which spun about eighty percent of the moisture out of it. The resulting cake, which was called "paste," was then carefully scraped out of the wringer, weighed into sixty, forty-pound lots and bagged. Despite its name, rocket "paste" was not smooth or creamy, but coarse and flaky, like attic insulation. Loaded into tramway boxcars, the bags were carried to the pre-dry house where they were laid out on wooden racks to dry. Depending on the weather, the bags sat in the pre-dry for three to seven days while circulating air pulled moisture out of the paste inside.

The bags of paste then took another ride on the tram to the paste breaker/blender house where twelve bags of paste, each from a separate mix, were poured into a blending barrel for cross-blending. Once the paste was in the blending barrel the workers turned on the switch to start the barrel rotating, then left the building. If they were following the rules, they waited on the safe side of a concrete wall, since the 720 pounds of rocket paste in the blending barrel could detonate into quite an explosion. After twenty minutes of blending, the paste was loaded into fiber drums and traveled in buggies to the paste rest house. Badger started processing its first rocket powder on January 30, 1945, using paste finished to the rest house stage at the Sunflower Ordnance Works. In order to jump-start production at Badger, and confident that it could obtain paste from Sunflower, Hercules had directed the contractors building Badger M-2 to complete the paste finishing houses before completing the paste fabricating houses. As soon as the finishing houses in the West Roll area were ready to run, in February 1945, Hercules ordered 170,000 pounds of paste from Sunflower. Since Badger's rocket dry houses were still under construction, the Sunflower paste was dried in the E.C. dry house before entering the west rocket roll house and moving down the line. Not until mid-March 1945 were Badger's pre- and final rocket mix houses, as well as the nitroglycerin plant, up and running. By the end of the month, the first carload of Badger rocket powder was shipped to the Naval Ammunition Depot in Hawthorne, Nevada. In April, enough of the rocket area was completed to produce the first million pounds of powder made entirely at Badger Ordnance.[113]

Whether it started life in Kansas or Wisconsin, rocket paste came out of the rest house and entered the roll house. Here, it was fed into rolling machines that were the industrial version of an old-fashioned wringer clothes washer, with one roller above the other and the paste squeezed between them. With the top and bottom rollers operating at different speeds, the paste was flattened and congealed into a wafer-thin sheet, about a yard wide, and no more than five-hundredths of an inch thick. Rolling was warm work, and the paste generated its own heat, so the roll houses were air conditioned. Cool as they were, in less than six months of operation Badger's rocket powder roll houses reported over 200 fires, all small and quickly doused by the operators or the sprinklers. After cooling, the sheets were stacked four high and passed five times through another set of rollers which compressed them to a thickness of one-sixteenth to one-eighth of an inch. The sheets were then rolled up like ancient scrolls and slid into pigeon-holed carriers mounted on buggies for transport to the slitting/carpet roll house. Here the rolls were slit into four-inch-wide strips, inspected for tears, bumps and gross impurities, then tightly wound into fourteen-inch-diameter "carpet" rolls.[114]

After eight hours in a heated rest house, the carpet rolls were transferred to the final press house for the dry extrusion that was the hallmark of the whole process. Nitroglycerin could tolerate a bit of rough handling, but it could have a short fuse when exposed to the high pressure of dry extrusion, and the final press house was designed for that eventuality. At its center sat the vacuum-powered, horizontal press into whose breech four carpet rolls were inserted and out of whose muzzle an endless snake of rocket powder was extruded. The floor beneath the press was a concrete slab, the walls and roof around it a ten-foot-diameter concrete arch, thirty-six feet long and buried under five feet of earth. One end of the tunnel-like arch was closed with a concrete wall separating and protecting the operator from the press and its lethal load. There were no observation windows in the wall and the operator ran the press without seeing it, but by observing gauges. The other end of the arch opened into an all-concrete rectangular cutting room with exit doors and windows. Should a blast occur, the earth-bermed concrete arch would deflect shock waves, heat and flames away from the operator and other crew members, and out the openings and roof of the cutting room. All production buildings at Badger were designed with an explosive accident in mind, but nowhere does this eventually so readily come to mind than in the concrete bunker of the rocket line's final press house.

The work of the press house was to extrude and perforate a length of powder and feed it directly into a cutting machine that trimmed it into rocket grains designated Mark 18, Mark 19 and Mark 22. All three were shorter, wider and shaped differently than the most common rocket grains manufactured for the Vietnam war, which were smooth 27-inch cylinders, 2.75 inches in diameter, with a star-shaped perforation from end to end. World War II rocket powder was cruciform or X-shaped, with all four arms of equal length and width. Vietnam-era rocket powder was also milled to a perfect finish in a wood-doweling machine. Ends were sealed with ethyl cellulose washers and the sides wrapped with six layers of ethyl cellulose tape. Although milled into a perfect X, and trimmed precisely square on the ends, World War II powder was not taped.

After it was trimmed, the sticks went back into buggies for a trip to the annealing house, which was divided into bays holding fourteen buggies each, where it was gently heated to relieve the stress of extrusion. Once cooled, the sticks underwent a "Supersonic" test to detect holes, cracks and the presence of foreign matter. The sticks of powder were submerged in a tank of water where FM sound waves were bounced off them. Readings were recorded by pens on a continuous sheet of graph paper which the scanner operators then read. The process and the equipment was new and "as may be expected, considerable difficulty was experienced at first." It was in a "constant state of flux" due to changes in equipment, personnel and the size of the grains tested. Time was also a factor, with barely six months passing from the day the supersonic equipment was set up until it was shut down — permanently. Had the war lasted longer, Badger rocket scanners would either have perfected the supersonic process or proved that it was unsatisfactory. In the Korean and Vietnam periods, the supersonic testing equipment was replaced by the fluoroscope. During World War II and later, rocket powder was also inspected by the means of X-rays and also at the ballistic lab, where in June 1945, over 900 rounds of rocket were fired.[115]

Through its use of the solventless method, supersonic scanners and X-ray cameras, the plant at Badger was on the cutting edge of rocket propellant technology in its days. With its finished capacity of 5.25 million pounds of powder a month and brief running time, it ranked among the largest, and least used, facilities of its kind in the world. On August 14, the government of Japan accepted the American terms of surrender and brought World War II to an end. The final weapon of victory turned out to be an explosive more powerful than any manufactured at Badger or any other conventional ordnance works. In the eyes of many observers in 1945, the atomic bomb had made war an unthinkable option. Arms manufactories, even non-nuclear facilities like Badger Ordnance, could dismantle their equipment, demolish their buildings, dispose of their powder stores. They would not be needed in what would truly be a post-war world.

At Badger, the victory celebration was subdued. The place was still crammed with enough powder to blast Sumpter Township into a lunar landscape and workers showed their respect for it. The plant fell silent for a two-day holiday, with only a minimum of workers on hand to watch over the power plant and other infrastructure, plus a guard force, of course. The bars and the churches were crowded in Baraboo, Prairie du Sac and Sauk City. Toasts were drunk to President Truman, the deceased FDR, Admiral "Bull" Halsey and General "Ike." Prayers were offered in thanks for the men and women who would be coming home and in sorrow for those who would not. Maybe somebody raised a glass to the ordnance workers or was grateful for the powder plant, but it did not make the headlines. It was easy to overlook that fact that World War II was a struggle between the world's modern industrialized nations, won by the most modern and industrialized. It was fought in Europe, Asia, Africa and the Pacific, but World War II was won in the Douglas Aviation plants in California, the General Motors tank factory in Detroit and on the smokeless line of Badger Ordnance Works in Sauk County, Wisconsin.

Boom, Bridge, Bypass and Bar

A rock thrown into a pool of still water generates concentric waves that gently roll to the far ends of the pond. In the still pool of Sauk County, Badger Ordnance hit the surface like a barrage of shrapnel. The shots peppered the surface from bluff to sandbar, kicking up random waves that lapped back on themselves and slapped unevenly over the county, the state and the nation.

One of the smaller shots hit the water in late December 1941 when Governor Julius Heil placed a telephone call from Madison to the home of Henry Ringling on Eighth Street in Baraboo. Ringling had just returned home from Washington, D.C., where he had obtained a commission in the U.S. Coast Guard for himself, and was napping when the phone rang. Not quite as "keen" as he would prefer to be for a call from the Governor of Wisconsin, Ringling listened as Heil asked him to welcome to Baraboo the Army officer who would be in charge of building the new powder plant. Yes, Ringling replied, his wife, Jean, would invite the Major and Mrs. Hauck to dinner. Yes, Ringling himself would inquire about housing for the family. Not until after he hung up the phone did Ringling think to ask just how important was Major Hauck that his welcome to Baraboo warranted the personal concern of the Governor?[116]

The Ringlings were the Ringlings, after all, and Heil wanted the people bearing Baraboo's most famous name to extend their hospitality to the

Haucks and, by extension, that of Sauk County and Wisconsin to the powder plant. The Governor also had other political fish to fry, or actually other roads to travel, and Henry Ringling was an important figure in the conservative wing of the Republican Party. Heil was up for re-election in November 1942 and had been around politics long enough to know that the race to the Governor's Mansion was more often than not won by the candidate who promised to pour the largest volume of concrete. Badger Ordnance would require that concrete be poured on more than a few state highway projects and Heil wanted Ringling's conservatives to credit them in his ledger.

Since it was presumed at this early date that the majority of Badger workers would come from points south and east of Sauk County, state highway upgrades were aimed at those points of the compass. Highway 12 was to be expanded to four lanes adjacent to the plant and connected directly to Highway 60 and the Sauk City bridge. The rerouting would eliminate the roundabout diversion of traffic through the business districts of Prairie du Sac and Sauk City. Highway 113, which zigzagged along the fencerows of Dane and Columbia counties, would be straightened and relocated to a more direct route along the North Western tracks from the east side of Madison to the river opposite Merrimac. Since it passed through the grounds of the new powder plant and might be a security risk, Highway 78 was to be cut off at the plant boundary and relocated across the river. The new Highways 78 and 133 would deliver travelers to the Wisconsin River opposite Merrimac, but they would not get them across the water, so Governor Heil proposed to realize the dream that had stirred the souls of Merrimac folk since the days of the pioneers. He approved the plan of the state highway commission to break the bottleneck caused by the ferry and build the long-sought-after Merrimac bridge. Because of the strategic shortage of structural iron and steel, it would have to be a "timber bridge," but it would still cost about $960,000.[117]

The Governor was all for the new roads and bridge, provided the federal government picked up the entire tab. He even went so far as to personally drive his Cadillac onto the ice of Lake Wisconsin to inspect bridge sites. Washington balked at paying the full amount and wanted a fifty fifty split with the state. Pressed by proponents of the bridge, Heil countered by saying, "So many people in the state...think the Merrimac bridge will win the war for America and that the Governor is the instrument causing the delay." He reluctantly agreed to a seventy-five-percent-federal, twenty-five-percent-state split but anything more from the state was out of the question. "Otherwise," he said, "they can build it themselves."[118]

Congressman Stevenson announced "tentative" federal approval in March 1942, but his eyes tended to be bigger than his stomach when it came to federal largesse. He also wanted the federal government to improve Highways 33 from Baraboo to Portage and Highway 14 from Spring Green to

Mazomanie in case saboteurs somehow shut down Highway 12 from Madison. Stevenson notwithstanding, the federal government withdrew support for the Merrimac bridge and for all the roadwork except the Highway 12 expansion to four lanes and its extension to Sauk City. Governor Heil went down to defeat at the hands of Progressive Orland Loomis, who swept Sauk and Dane counties, and who had built his reputation not for the concrete he poured but for the electric power lines his state agency had run throughout rural Wisconsin.

While state and federal officials supported the extension of Highway 12 directly to Sauk City, local officials in Prairie du Sac opposed it. When the state highway commission held hearings on the project in Baraboo, local Attorney Harold M. Langer and Assembly Representative George F. Woerth, of Sauk City, spoke against it. "Wisconsin has few drives of greater scenic beauty than the stretch from the business district of Sauk City to and beyond the business district of Prairie du Sac," Langer told the commissioners. "Prairie du Sac wants to live and let live. Its residents do not want to be taken off the state highway." Woerth then pointed out that the state was spending $100,000 to attract tourists so, "It would be folly for the highway department to shoot people through the state at the greatest possible speed."[119]

Once work on Badger Ordnance began, Prairie du Sac and Sauk City would have all the living they could handle whether or not tourists stopped for coffee downtown. The impact of what was not yet identified as one of the state's first bypass routes was not evident until after World War II. Twenty-five years earlier, when the federally funded state road system was developed, Main Street merchants had been very successful at keeping the improved highways running past their doors, even if it meant congested streets, roundabout routes and traffic delays. Old Highway 12, later identified as "Business 12" in Sauk City, Prairie du Sac and Baraboo, too, was a fine example of a federal and state highway diverted for local business interests. After Philipp's Boulevard in Sauk City and the Prairie du Sac town road known as "Yankee Street" were widened and improved into the new Highway 12, the strategy of shooting people at the greatest possible speed out and around village business districts — and the suburban sprawl it engendered — was set in concrete as the pattern for the future.

One man who would have been happy to shoot quickly through Prairie du Sac or onto a bridge at Merrimac was Lieutenant Thomas H. Blount, U.S. Army Ordnance. About the same time as Governor Heil was urging Henry Ringling to chat up Major and Mrs. Hauck, Blount was at the Picatinny Army Arsenal in New Jersey, taking training in "Contract Administration." A newly varnished second lieutenant, Blount completed his course work and, in late January, received orders to report to the Badger Ordnance Works at a place he had never heard of called Merrimac, Wisconsin. He set out by car and, as he recalled, "After four nights on the road and overnighting in Madison, I took the recommended route [Highway 113] for Merrimac." It was snowing and, "the road suddenly disappeared under a sheet of ice that looked as though it might be covering a lake. After asking 'how to get to Merrimac' I was told that one could drive over the ice, but I declined."[120]

He did make his way to Merrimac via the Prairie du Sac bridge, but found no sign of an ordnance works. "I stopped at the Merrimac post office. The Postmaster [Harry Johnson] told me that he didn't know where any offices or personnel were, but that he had about twenty sacks of mail addressed to the Commanding Officer...that he would like to deliver or have someone pick up, as he needed the room they were taking up...Maybe the people at the county seat over at Baraboo might know something."

Blount returned to his car and set out for Baraboo, following a county snowplow as it opened one lane of the road. In Baraboo, he found no one on the streets but a policeman who could not tell him where the office of Badger Ordnance was located. The officer then sent Blount to the telephone exchange where an operator said she didn't know the location, "but there were a lot of strange calls being made from the National Guard Armory... upstairs over the Chevrolet Garage." There Blount found the offices of Mason and Hanger and Army Ordnance, unoccupied except for a lone female secretary. "They were expecting me," he said, tongue in cheek.

The following day Blount and Captain S.G. Hilliard, second in command to Colonel George Griffith, drove back to Merrimac and stuffed the twenty mailbags into their car. Upon returning to Baraboo, they made arrangements with postmaster R.J. Osborne to receive mail at his station. Badger Ordnance had acquired its permanent address and was no longer identified as being "in the vicinity of Merrimac."

The flood of mail that had engulfed Johnson's store was now directed to Baraboo, where it would increase the volume of mail processed at the station to an amount comparable to that of a city the size of Madison. In October 1942, Postmaster Osborne announced expanded hours of service. For the first time in its history the Baraboo station would, except for Sundays, operate all night. Mail deposited in the lobby by 1:00 AM would be dispatched on the first mail train of the day, Monday through Saturday. On Sundays mail deposited as late as 8:45 PM would make the late train. The service windows would be open 8:00 AM to 6:00 PM, Monday to Saturday, and the money order window would remain open until 5:00 PM on Saturdays, the day after payday at the plant. It has been said that Badger Ordnance forced Baraboo to mature as a city, to take on the trappings of a much larger community. One place it happened was at the Post Office.[121]

Lieutenant Blount was only one of the 10,000 men whose expected arrival in Sauk County prompted state officials to launch what the Madison newspapers luridly announced as a "War On Prostitution in Powder Plant Area." Leading the campaign against "vice" were Dr. Charles A. Harper, state health officer, Hugh A. Jones, of the motor vehicle bureau, and John W. Roach, director of beverage and tobacco tax collection. "When a large number of workers are assigned to a community on a defense project such as the Badger Ordnance Works," said Dr. Harper, "it has been definitely in evidence that certain undesirable individuals are liable to come into the community and prey upon the people." The top state health administrator was concerned because Wisconsin men called up for the military had been found to have the lowest rate of sexually transmitted disease of any in the nation and Harper wanted to keep it that way. Sharpening the doctor's reasonable statement into a hook for a series of circulation-boosting news articles, *Capital Times* reporter Robert Fleming declared that "the threat of a terrible disease will face every resident of Dane, Sauk, Columbia and the nearby counties."[122]

The traditional seat of vice as it was known in rural Wisconsin was the village saloon, where a back room or upstairs flat might be used for prostitution. Local law enforcement looked the other way as long as the operation remained limited, discreet and publicly non-violent. In the 1920s the combination of the automobile and Prohibition enlarged opportunities for illicit activities by supplying the means and the motive to move out of town. Roadhouses and dance halls opened in rural locations where they sold illegal liquor, and sometimes more, beyond the easy reach of town constable or county sheriff. The motor courts and "motels" which also began to crop up along the highways in the 1920s, and which allowed guests direct, unscrutinized access to their rooms, further expanded possibilities. Even family-oriented vacation areas, such as Devils Lake, Lake Delton and Wisconsin Dells, had their "no-tell motels." When motorists started hitching living quarters to the back bumpers of their cars, they only improved on the combination of mobility and anonymity created by the motel. If the motel was not quite respectable, the trailer camp verged on the downright damnable.

Prohibition ended in 1933 and the Depression hurt tourism, but neither the roadhouses, the no-tells nor the trailer camps closed. To protect themselves some towns attempted to regulate the sale of alcohol within their limits. Sumpter restricted the sale of beer and liquor to the strip along Highway 12 at the top of the bluff, while the town of Prairie du Sac permitted only the sale of beer within its confines. The drinking public had to detour to the Sauk villages, Merrimac or Baraboo, where alcohol in all its popular formulations could be imbibed. Baraboo had fifteen saloons in 1941, Prairie du Sac and Sauk City each had a half-dozen. If any of these establishments also had back rooms devoted to sex for pay, it was well hidden.

The campaign on prostitution triggered by the arrival of Badger Ordnance was the second such war the government had waged in the area. In the summer of 1940, the United States Army brought 60,000 National Guard and regular Army units to Camps McCoy and Williams for the largest peacetime training maneuvers it had yet held. In preparation, Federal Bureau of Investigation and state beverage tax officers swept down the byways of west and central Wisconsin, including Sauk County, warning tavernkeepers that, despite the presence of tens of thousands of apparently distractable male soldiers, vice would not be tolerated. They also visited county sheriffs and local police to remind them that prostitution existed in their communities because law enforcement winked at it. While it was not one of the counties where the war games were held, Sauk was included in the sweep. Highway 12 was on the route of march for Guardsmen from Illinois and southern Wisconsin and units would bivouac overnight on the vacant ground between Sauk City and Prairie du Sac.

Having rehearsed the drill in 1940, state beverage tax personnel — minus the FBI — were ready for the campaign of 1942. In February they called a meeting at the American Legion Hall in Baraboo attended by 112 tavern operators, plus a larger number of town, village, city and state officials. Roach opened the meeting by asking every local government unit to adopt a 1:00 AM closing time for taverns. Dr. Harper took the floor next and, invoking his powers as state health chief, urged the local officials to adopt strict zoning and "sanitary" codes for trailers. He then made a constitutionally questionable threat to "quarantine" taverns, motels, trailer courts and houses suspected of harboring prostitution. He also announced that the state had extended its special wartime trailer camp code to all areas within twenty-five miles of Badger Ordnance. The measure empowered state health officers to inspect trailers and regulate the number of people living in each unit as well as the entire camp. Roach then came back and told the tavernkeepers to keep their establishments "spotless," and to revert to the pre-1920s custom of not welcoming women on the premises. The tavernkeepers were also warned to keep drunken patrons from driving, and, with the war on, to "be alert for spying and sabotage." Roach, whose inspectors monitored tobacco sales, legal slot machines, illegal gambling and the sale of alcohol, now added vice and sabotage to the list of activities on their watch list. In March, April and May they visited the taverns and village halls of Sauk and neighboring counties. Again they reminded the proprietors, police chiefs and sheriffs that the vice laws were to be enforced. By the end of June, just as the number of construction workers at Badger was about to peak, John Roach announced that, "To the best of our knowledge, there is no prostitution in Sauk, Juneau and Adams County near the army camps and defense projects." Having announced that the war was won, the vice police returned to Madison, and the locals discreetly returned to business as usual.[123]

Drinking establishments did not have to offer any stimulation other than the alcoholic to do very well during World War II. More people were making more money than ever, with fewer distractions to spend it on, so many a dollar passed over the bar tops. Clifford Schweppe remembers living about ten miles west of the plant in Denzer during the war. His parents, Ed and Iva, had moved to a place there after losing their Prairie farm to the powder plant. Their new house had plenty of room so the Schweppes took in as many as fourteen male boarders. They were men away from home with no responsibilities other than to show up for work at Badger. After work, they "joined the crowd with other plant workers for fun and refreshments at Goerk's Tavern," the only place in Denzer with lights on after dark.[124]

The scene at Goerk's in Denzer was duplicated at every tavern within driving range of the powder plant: on Ash Street in Baraboo, Broadway in Wisconsin Dells, River Street in Portage and on Highway 12 from Sauk City to Springfield, Middleton and Madison. Despite warnings from the state's vice warriors, local officials were unable — and not all that willing — to enforce a uniform and early closing time near a place with thousands of shift workers who were awake at all hours of the day and night. With the second shift at Badger ending at midnight, neither tavernkeepers nor customers wanted the doors to close at 1:00 AM. The notorious "Bombshelter" tavern in the basement of Baraboo's Warren Hotel is remembered for never closing at all, while other watering holes locked their front doors, shuttered the windows and let customers sneak in the back door. Tavernkeeping was a growth industry during World War II. Liquor tax revenues, based on sales, returned to local governments in Sauk County increased nearly eighty percent between 1941 and 1945. While a portion of the gain was due to price inflation and higher tax rates, increased consumption at the Wigwam in Baraboo, the Du Drop in Sauk City, the Twin Arches in Prairie du Sac and — don't forget — Goerk's in Denzer, accounted for much of it. Already well established in the county, tavern culture was stimulated by the war and flourished without a setback until the 1950s, when television and off-premise sales gave tavern patrons a reason to stay home at night.[125]

Restaurants also did very well during the Badger boom. They faced the very pleasant situation of having more people hungry at more times, and with money to dine out on. "It seemed as if someone was always eating breakfast when dinner should be served," recalled Badger school principal Clarence Krumm. Nolden's Cafe and Bar, which occupied an enviable location adjacent to the Up-To-Date Auto garage on the bridge corner in Sauk City, was typical. Newly built in 1932, and with a side door opening directly onto the parking lot of the Up-To-Date, the red brick cafe and bar catered to travelers obliged to drive by on Highway 12. Adolf and Natalie Nolden, counterman and cook, had kept late hours on summer weekends. Now they ran round the clock. Natalie's famous, in Sauk City, chicken dinners and cherry pie, as well as breakfast, were on the menu at all hours, as were customers coming through the door of the cafe.

While plant workers lived in houses where they had kitchen privileges or where they ate at the family table, they also relied on restaurants to prepare meals they took to work or to simply put variety in their diets. Rita Slezak

Lachmund, who lived in a Baraboo boarding house, was fond of the Bar B Que Restaurant. "Rabbit meat didn't need a [ration] coupon, so we ate a lot of rabbit [at the boarding house]. It was baked, fried, boiled and every other way. One day I sat down and looked at my plate of rabbit. All I could see was a little bunny sitting on my plate. No more rabbit. It made a vegetarian out of me." Not quite, for Lachmund also got away from rabbit dinners at the Barb B Que, where she was served all the coleslaw, french fries and ground beef barbecue she could handle for thirty-five cents. "It came on a tray. There was so much I couldn't eat it all."[126]

A simple thirty-five cent meal at a diner in Baraboo, and a young woman's ability to pay for it, does a better job of illustrating the economic impact of Badger Ordnance in Sauk County and World War II on the United States than an encyclopedia full of bar graphs, pie charts and macroeconomic elucidation. Ten years before she came to Badger Ordnance, in the early 1930s, if she could have found a job, Lachmund would have been paid about ten cents an hour. A thirty-five cent meal would have cost nearly a half day of labor and she probably would not have bought many of them. As a line worker at Badger in the early 1940s, Lachmund earned about ninety cents an hour. A thirty-five-cent, all-you-can-eat barbecue cost about twenty-five minutes of work and she could afford it just about any time she wanted a night off from rabbit meat. Multiplied by the trillions of other pennies circulating through the wartime economy, Lachmund's thirty-five cents ended the Great Depression and laid the foundation for the rise in the American standard of living that took place in the post-war years.

Badger workers spent a part of their paychecks on beer and barbecue, but they also made deposits in savings accounts. They were aided by the wartime moratorium on the manufacture of major consumer products. In the years after the war they put money earned during the war into big ticket items — homes, cars, appliances — they could not buy until peace returned. Farmers invested in new equipment, livestock and materials that brought about the greatest change American agriculture had seen since the Civil War. No survey has been made to count the number of tractors, station wagons or bungalows purchased with money salted away by Badger employees during World War II, but anecdotal evidence is abundant. In April 1945, for example, the *Badger Ordnance News* reported that Donald C. Bayles, who worked in the instrument shop, and his wife, Kathleen, were the first recipients "in this area" of a home loan under the provisions of the GI Bill. As a Marine veteran who came to Badger after he was injured in service, Bayles was eligible for a low-interest, federally guaranteed loan from Baraboo Federal Savings and Loan. The home the Bayles family purchased was in Kathleen's home town of Adams, which further illustrates the scope of Badger's economic impact. They made the down payment with money that both Donald and Kathleen, who had worked at Badger until her son Michael was born in late 1944, had earned at the powder plant.[12]

Badger Ordnance is remembered for what it took away from the farm families, the scenery and the environmental health of the Sauk Prairie. What it gave in economic opportunity is often forgotten, even though the list of men and women who sought and found relief from city unemployment or rural poverty at Badger Ordnance is long enough to fill all the telephone directories in Sauk County.

The immediate impact of Badger Ordnance was, of course, the impact of people — hundreds, occasionally thousands, of people in places where not so many people had been before. They crowded restaurants, filled movie theaters, swamped medical facilities, bumped up school enrollments and raised the level of activity in every community in the area. All the villages were small enough so jokes were commonly made about the sidewalks being rolled up in every one of them at night. Now the sidewalks were occupied late and the formerly funereal quiet of the streets was broken at 11:00 PM and 7:00 AM by the blasts of automobile horns sounded by drivers summoning their car-pooling passengers to get up, out and to work.

Lola O'Brien Huber was in high school when her family moved from St. Croix County so her father could take a job at the plant. The nine O'Briens crowded into a run-down old house near Glenville, on the eastern edge of Baraboo. Huber remembers hiking all the way into the Square to see a movie, getting into a long line to buy a seat and learning the tickets were all sold out just when she reached the sales booth. If she went to the ice cream parlor next door, she had to wait for a seat and for a cone with two scoops. On V-J Day a Baraboo reporter stopped people on the street and asked how they felt about the end of the war. One man answered, "the best thing is that I won't have to wait in line anymore."[127]

Despite all the talk about wild times during the Badger boom, church attendance increased. People with loved ones away from home, or away from home themselves, had more than sufficient reason to spend time on their knees. Church and other volunteer organizations acquired more members and became more active in their traditional work and in the home-front war effort. Just as more men occupied bar stools, more men also sat on hardback chairs at the First Methodist Sunday school, the Lutheran League meeting and the Catholic Holy Name Society. Rita Slezak Lachmund remembers that when she came off the night shift on Sunday morning she went directly to a packed St. Joseph's Church in Baraboo for the 9:00 AM Mass. Unable to change out of the casual clothes they wore to work, and with the church filled anyway, Lachmund and her friend Lorraine Leigel went up to the choir loft. "One day one of the nuns said, 'the Blessed Virgin would cry' if she saw those clothes we had on. My friend Lorraine said, 'the Blessed Virgin didn't have to stay up all night making ammunition.'" Nor did the nun, who was overlooking the testimony to their faith the young woman were giving by their coming to church after work when they might easily have gone home to bed or out to breakfast with a beau.

Another place where powder plant workers and Catholic nuns came together was at St. Mary's-Ringling Hospital in Baraboo. While the hospital, the largest in the county, was not as primitive as Mayor Law of Madison had charged in 1941, it did not have the resources to handle the influx of patients who came with the plant. The doctors, nurses and nuns struggled through the first Badger construction blitz with the aid of dozens of women who volunteered for Red Cross training and worked as nursing aides. During the second blitz in the summer of 1944, Badger school principal Clarence Krumm brought his daughter to St. Mary's for an emergency appendectomy. The procedure was completed without a hitch, but Krumm was upset by the "chaos of inadequate space, insufficient services and inefficiency." He called on Colonel Keldsen, Army commander at Badger, who agreed to seek federal funding for a hospital addition. At the same time, Sister Mary Marcella, superior at the hospital, embarked on a community fund-raising program for a $260,000 addition.[128]

With Keldsen's help, the hospital received $129,000 in federal funding, plus a commitment on his part and that of the civilian contractors to contribute to the "community" share of $65,000 to be raised for the facility. Hercules employees were regular contributors until the end of the war. Completed in December 1945, the hospital addition came too late to serve the Badger workers who helped pay for it, but remained a lasting reminder of their positive presence.

Presence of a different sort came with John and Clara O'Brien after they moved to their house outside Baraboo in 1942. John caught the bus and went to work as a carpenter at Badger Ordnance. Clara loaded five of their children in the family car and presented them for enrollment at the Glenville school and at Baraboo High. The scene was duplicated over and again throughout Sauk and neighboring counties. New people moved in. One or two parents went to work at Badger. The children went to school. In September 1942, enrollment in the Sauk City schools rose from 309 to 440 students. Parents living at the Prairie du Sac trailer camp sent 170 additional students to the village grade school in 1942, doubling the number of pupils. The old brick school, built in 1891, did not have space for the newcomers but erecting a new school was out of the question during the war, so classes were held in church basements, the village park pavilion and the Knights of Pythias hall. In Baraboo, the increase held steady at 180 students in 1942 and in 1943. Closer to the plant, near Devils Lake, the Park View one-room school, which had been closed for fifteen years, reopened with forty pupils. The following year, Park View had seventy students attending classes on two shifts. The King's Corners school, which had run on and off for years, was about to reopen in 1942, based on the announcement that "Conway's Acres" trailer court would be built about a half-mile to the west. The trailer court was not built and if King's Corners did reopen, it was only for a year. In 1943, it was replaced by the Badger School. Other rural schools in the vicinity of the plant also showed increases, but these too were eased when the Badger school opened.[129]

The list continues: sixty-one students at Merrimac State Graded in 1941, 142 in 1943, with two classrooms of the discontinued high school reopened for them; Delton State Graded, fifty-four students in 1941, 184 in 1943, attending classes in the basement; North Freedom, seventy-six pupils in 1941, 109 in 1943, requiring the hiring of an additional teacher. The exceptions were the Reedsburg school system, where enrollments declined during the war, and all the high schools, where young men left early to join the military, to work on farms or at Badger Ordnance.

Although, as County School Superintendent Kurt Schoenoff reported, "the location of the Badger Ordnance Plant and the impact of the war has had serious effects on the total education program in Sauk County," the problems were localized and brief. Some schools, like Park View, Prairie du Sac and Sauk City, reported big jumps in enrollment for one session. The 1942 session was the most trying, with enrollment throughout the county—not counting Baraboo and Reedsburg—rising from about 4,200 to 5,000 students. In 1943, however, the increase fell to about 600, including Baraboo and Reedsburg, and stayed at that level until the war ended. The overall increase for all public schools—city, village and rural—in the years between 1941 and 1945 was just over 600 students, or about ten percent. The schools experienced stress and strains, but not any greater than they would during the postwar baby boom, when all the city and village schools experienced crowded classrooms and split schedules at one time or another.

The pressure on the school systems was not as great as it might have been for a number of reasons. The construction workers who made up a large majority of the newcomers descending on the county were exclusively male. They were not likely to bring children with them and they did not stay very long even if they did have children. Many workers commuted, especially the production people. Their children were in schools in Madison, Richland Center, Westfield or any of the other communities within driving range of the plant. Finally, not every family that moved to the county was as prolific as the O'Briens. The 1930s had seen a baby bust, with the rate of growth in the American birthrate declining significantly, resulting in fewer children in the 1940s. The baby bust stirred into the baby boom during the war, but those children were a few years away from school age. Even at the Badger School the number of preschoolers was at least as large, if not larger, than the number of school-aged children. The Badger School itself, which opened in September 1943, made a difference by keeping grade school students out of the other county schools and by sending relatively few high school students to them. Consequently, the presence of Badger Ordnance in the county certainly had an impact on the county schools, but it did not cause a crisis.

The additional children in the Badger Ordnance defense area prompted the federal government to fund big-city-style summer recreation programs in Baraboo and the Sauk villages. Under the direction of Robert Gollmar, Chairman of the Sauk County Council of Defense, each community organized a recreation council and hired a recreation director. The council agreed to raise money to pay for equipment, the villages supplied space in the parks and schools, and the federal government paid the wages of a Recreation Director. In Sauk City in 1942, for example, Director Steve La Due organized "manual training" programs for boys, crafts for girls, and softball and playground games for all the youngsters. Adults citizens were invited to pitch horseshoes or sit in at card parties.[130]

In Prairie du Sac, the recreation committee announced an ambitious program that included playground activities, the swimming pool, musical performances and home economics for people of all ages. "The committee wishes it understood that the recreation facilities are open not only to the residents of our community who were here a year ago, but to all newcomers who are now here or who may come later. A cordial invitation is extended to our new neighbors..."[131]

The Prairie du Sac officials were cordial even though the powder plant prompted a great deal of anxiety in the area. Rumors abounded, rose and spread like steam from the powerplant stacks. It was said that fumes from Badger's acid works would kill vegetation on and near the grounds and eat the leaves off the trees on the surrounding hills. On the other hand, as many powder plant officials said at public gatherings, Badger Ordnance was sited on the fertile prairie so crops and trees could be planted on its grounds. Ground cover was important because dust blowing off bare soil could contaminate the powder, or set off explosions when abrasive dust scraped against and sparked on metal posts, pipes and safety chutes. Plantings were also required for camouflage because Badger was a prime target for aerial bombardment, which was also why a military airfield was going to be built on the Emma and Glen Kindschi farm. The threat from the air was real enough for some innocents to believe that the trailers at the Prairie du Sac trailer camp were parked staggered in their rows not to increase privacy in the crowded lot, but to minimize damage inflicted by an enemy fighter plane strafing down the line. This fear did not apply to Badger itself or to Badger Village, where buildings were aligned in rows as straight as a gun barrel. However, unless the carrots and cabbages in the ten acres of Victory Garden on Highway 12 count, nothing was ever planted for camouflage at Badger during World War II.

Most of the rumors involved the people moving to the area, starting with the number of men, women and children who would descend on the county.

Estimates ranged wildly and reached as high as 20,000-plus, but the actual number of flesh and blood humans never rose that high, even in August 1942, when approximately 14,000 people were at work on the plant. While the volume of new people was real and significant, it was never as large as rumored and the peaks were always temporary. From its high in August 1942, the number of construction workers fell to zero by the end of 1943. While nearly 6,000 people were on hand to operate the plant in the spring and early summer of 1945, the number dropped like a rock out of an airplane as soon as peace returned in August. One of the oft-expressed concerns about the plant—that when the war ended laid off-workers would remain in the county, clog the welfare roles and bankrupt the taxpayers—never materialized.

The type of people moving in, especially the construction workers, caused the greatest worry. In the worst case scenario, the carpenters, plumbers and 'dozer operators were rootless bums and drunken brawlers, all accompanied by slatternly female camp followers. A few fit the description, but most of them were simple working Joes whose calling required that they travel from job to job. Many of them were veterans of New Deal public works projects—like the Grand Coulee Dam—and were permanant transients. They planned to finish up at Badger, then move on to the next big war plant, air base or highway project, just as they had been doing for years. Early in 1942, Colonel "Barney" Oldfield told a Prairie du Sac audience that it was their patriotic duty to open their homes to plant workers.

"What if I don't like the person who comes to live with me?" one person asked.

"Then ask him to leave and find someone else," Oldfield responded, with the emphasis on the finding. In a time of crisis, the national interest overrode personal discomfort.

Of course, some newcomers had their reservations about Sauk County as well. After getting her car stuck in a snowbank, a woman from one of the southern states started to complain about the winter weather. Said the policeman who was pushing her out of the snow, "Lady, if you think this is bad you should have been here the year before the war."

To which she responded, "Do you mean to tell me that people live in this country in peacetime?"[132]

The limits of hospitality, or at least tolerance, of the people of Sauk County were not seriously tested by the great and ugly divider of Americans, race. The building contractors and Hercules Powder recruited workers in southern and other states with large populations of African-Americans, especially for the rocket area blitz. African-Americans were themselves taking part in their great migration from the cotton fields to the defense plants of Chicago, Detroit, Milwaukee and other industrial cities of the north. Yet few African-Americans came to Sauk County either to build or operate the powder plant.

It is likely that they were not recruited, and were, in fact, turned away by Hercules and the building contractors.

A few anecdotes and records exist. In describing housing built for Badger, the Army routinely pointed out that separate quarters were not necessary for blacks and whites "on this project," as it was in the South. Separate housing was not necessary because Badger would not employ enough African-Americans to warrant it. When the Caribbean workers hired on, they were housed in separate quarters at Devils Lake instead of the barracks on Highway 12. A story is told in Baraboo that when Major George Griffith came to town in 1942 his Army car was driven by an African-American. On one occasion, Griffith stopped at a diner and went inside with his driver, who was probably in uniform. The black soldier was told he would not be served, whereupon both he and Griffith left. Neither the black soldier nor the white officer was ready to break the color line in Baraboo. On the other hand, there is the letter from Oliver Jackson, who identified himself as a "Negro worker for Mason & Hangar, Prairie du Sac." Jackson wrote to the *Wisconsin State Journal* in the summer of 1945 and the letter was reprinted in the *Sauk County News*. "Speaking for the Negro group as a whole, since our stay here in your county and state, we as a group of people, have been treated swell...

"You have been good to us. You met us with a smile and we hope to leave you with a smile. We have been treated better here than in any state we have been in and in the near future we hope all of the states and counties will look on you and take a pattern from your state and county.

"We feel welcome, we walk welcome, we sleep welcome, and we eat welcome."[133]

No other mention of Jackson and the "Negro group" has been found. They probably lived in the trailer camp on the edge of Prairie du Sac, which was federally run and supposed to be color blind. The people were cordial. Perhaps some progress had been made.

No Room in the Chicken Coop

Probably nowhere else outside of the plant was the impact of Badger and the people it brought to the area greater than in the realm of housing. In 1942, federal housing authorities surveyed the housing stock and projected that Baraboo and Portage, without any significant additions, could absorb 1,500 new residents, while Reedsburg, Sauk City and Prairie du Sac could take 1,000 each. The Wisconsin Dells/Lake Delton area could handle as many as all the other communities combined if the hotels and resort owners went out of the tourist business. The customary week or two-week rental units could be leased for longer terms. Summer cottages could be improved for year-round living, which, in most cases, meant installing insulation, heating and plumbing.

The federal authorities did not mention the need for plumbing in Lake Delton nor did they offer advice on how the other communities would handle the population increase, only that the communities should and could welcome the newcomers. H.N. Wendle, in his prewar report to Hercules on the plant site, said that new housing would have to be built for at least half of the 10,000 construction workers he estimated would be hired to build the plant. Once construction was over, operating personnel could occupy the housing. The engineer's superiors must have skimmed this section of his report, for the initial plans for Badger called for the construction of housing for about ten percent of 5,000. Called the Administration Barracks or "North Badger," the housing complex started with five single-story, H-shaped, wooden-framed buildings, each capable of housing 104 individuals in single, double and four-person rooms. Built in back of Lyman Kindschi's homestead, which was still standing west of Highway 12, the complex also had a mess hall with seating for 200. Since it stayed open twenty-four hours a day and welcomed visitors, the mess hall became a popular spot for workers starting or coming off their shifts at the plant, even those who did not live nearby.

The rooms were heated and furnished and came with maid and linen service. The bathrooms were communal, as was each building's sole telephone. Monthly rent for a single room was $18, or $21 for a corner room with two windows. Doubles rented for $15 per occupant, and the four-person units went for $12 per person. Since even the lowest-paid laborer at Badger was bringing home over $100 a month, barracks living was a good deal. More workers might have taken advantage of it, but the first building did not open until November 1942, when construction was slowing and the workforce was already twenty-five percent down from its summer peak. When Barracks No. 1, which was actually the fourth unit to be completed, opened in December 1942, it was reserved for women, none of whom were construction workers, but who would start making powder in the spring of 1943.[134]

Despite the housing shortage, the barracks were not fully occupied in 1943 and a good part of 1944. They fit Badger's wartime pattern of offering too little, too late. By January 1944, only one men's and one women's barracks were open, with no more than 120 total occupants. In April of 1944, occupancy fell to 92. The trend reversed when work began on the rocket area the following month and construction and operating workers filled the barracks by Labor Day. Anticipating a demand for more housing once the rocket area started production early in 1945, the Army hired contractors to construct five more barracks. In design they were all but identical to the original five, except for Barracks No. 10, which was divided into two-room apartments for married couples without children. Among the first tenants of a single-room unit in the new Barracks was young William Sisson, who moved in late in 1944 and remembered that, "when I moved in there was still sawdust on the floor." For a nineteen-year-old away from home for the first time, barracks living was "nice." All the basics of life were within walking distance — work, food and a measure of recreation. He could eat hot meals in the mess hall, or walk to the A & P at Badger Village and buy his own groceries for sandwiches and snacks. Communal bathrooms with hot running water were no hardship for someone who had grown up with a hand pump and an outhouse.[135]

The apartments for married couples were less popular. More like hotel suites than apartments, they did not have their own bathrooms or kitchens. Tenants had to use the communal bathrooms and eat at the mess hall. Although Badger was hiring new workers by the thousand when the apartments opened early in 1945, many units went unoccupied. For the summer of 1945, they were opened to female teachers recruited to work on the rocket line, and who doubled up to rent the family units.

Two additional H-shaped buildings were trucked onto the barracks site and assembled in the fall of 1944. Longer and lower than the other barracks and painted dark green, these buildings would have housed approximately 250 people, but the war was finished before they were. As soon as the war ended in August, work halted on all housing projects. Rooms were no longer rented and the barracks emptied as fast as workers were laid off or quit. In October 1945, the last tenants moved out and the barracks were closed.

Also to meet a demand that did not fully materialize, the Army constructed four barracks, each capable of housing 200 men, just inside Gate 10 and west of the rocket line. The interiors were Army-style open rooms with double-deck bunks for fifty men, plus toilets and showers. They opened in February 1945 and closed in May. At peak occupancy in April, the Gate 10 Barracks, capacity 800, housed no more than 290 men.

Another attempt to create new housing quickly at the start of the Badger project took place at Prairie du Sac. In May 1942, the Farm Security Administration announced plans to purchase 105 acres of the Raymond Heintz farm, half of which was in the village limits, and build a 550-lot trailer camp known as the Merrimac Defense Housing Project. Heintz, who drove a milk route in Dane and Sauk counties, kept the family home and six surrounding acres. On fifty-one of the acres sold to the government, contractor Edward Kraemer and Sons began grading and graveling roadways and parking pads for 250 trailers that would be owned by the FSA and rented to Badger workers. Another 300 lots were laid out and improved as spaces for workers to park their own trailers. The contract awarded to Kraemer for all construction work totalled $124,500.[136]

In January 1942, Prairie du Sac had, like other cities and villages in the area, enacted a trailer camp ordinance so strict that it all but banned the facilities within corporate limits. Local lawmakers were concerned about villagers who would rent out vacant lots or their backyards to trailer owners and turn their communities into shantytown vice pits. By the time construction began, the state had extended its own strict trailer ordinance to the "Badger defense zone" and the federal camp became legal in Prairie du Sac.

Government owned and managed, the Prairie du Sac camp was intended to set the standard for any new privately owned trailer parks. The roads and parking pads were blacktopped or gravelled so the camp would not turn into a mud hole. Concrete sidewalks were poured along all the streets, which were made "extra wide" so cars would be parked on the pavement instead of helter-skelter throughout the camp. A concrete walkway ran from the sidewalk to the doorway of each trailer. The camp was lighted, had mail delivery and garbage collection. It was connected to the Prairie du Sac water and sewer lines, at federal government expense. Sixteen combination shower and toilet buildings—heated—were placed on the site, spaced so that no trailer was more than 125 feet away. Six laundry buildings were also strategically placed around the camp and equipped with washing machines, laundry tubs, ironing boards and "plenty of outdoor clothesline space." Since it housed working people, the camp would have a "nursery school" for preschool-age children of plant workers.

An olive-drab "expandable" government trailer with space for parents and children rented for seven dollars a week. Trailer owners could rent a parking pad and use the services of the camp for $3.50 a week. For construction workers with families who had been living in places as remote as the Grand Coulee Dam, an improved trailer camp adjacent to a homey village in Wisconsin was as nice a place to live as any they had known. The winter could be challenging, but schools, stores, churches, parks, a movie theatre and other amenities were within walking distance. Even the radio reception for the big shows from Chicago, New York or Nashville was clear compared to the static-clogged airwaves in the western wilds.[137]

Yet, despite its amenities and the demand for them, it is not clear if the Prairie du Sac camp was fully occupied until the final year of the war. Prior to the camp's opening on October 1, 1942, a total of 345 individuals and families registered either to rent a government trailer or a parking pad. However, the Federal Housing Administration, which ran the camp, was unable to obtain a full complement of trailers to rent, nor had a full 300 trailer-owning individuals pulled in. With the rocket area building blitz underway in June 1944, the FHA asked for, but did not receive, 350 new trailers for Prairie du Sac. Trailers enough were on the site by January 1945 to house 275 construction workers with families, plus another 450 men sharing trailers dormitory style. If each one of the families had one child, then almost 1,300 people were living in the Prairie du Sac camp in the first seven months of

1945, nearly doubling the already swollen population of the village.[138]

The influx and sheer number of new people strained the social fabric of a community which had settled comfortably into its position as a farm market center where the only strange faces belonged to summer tourists passing through on their way to someplace else. Now residents complained about unattended children "too much on their own" and about unsafe streets where "a woman or young girl was subjected to the nastiest remarks and offers going down the street, not only in the evening, but during the day time." The village board ordered saloons to lock up at 10:00 PM on weekdays, 11:00 PM on Saturday, and to stay closed all day on Sunday. Instead of following the customary practice of jailing drunks until they sobered up and then releasing them, Police Chief Miller began hauling them before the local justice of the peace, who levied stiff fines as punishment. When a "bunch of women...who were clever at picking up the loose change of indifferent and hazy drinkers" started frequenting the bars, Chief Miller rounded them up and chased them out of town.[139]

The situation climaxed in March 1945, when a 20-year-old sailor named Martin Scholl, home on leave, got drunk and obnoxious in the Twin Arches tavern. When city police tried to subdue him, Scholl resisted, claiming that since he was in the Navy, he could not be arrested by civilian lawmen, only by the Navy Shore Patrol, which of course did not have a brig in Prairie du Sac. The chief then summoned the other city cop, along with three sheriff's deputies, and the five men started to hustle Scholl out of the bar and up Water Street to jail. A crowd gathered, estimated at about 300, even though it was a Thursday night in March — in itself a fact that illustrates the state of affairs in Prairie du Sac. Scholl must have been a tough nut for the five officers to control, because Deputy Herbert Kraemer felt compelled to pull out his gun. A melee ensued, with Scholl still fighting, spectators — drunk themselves — pushing and shoving, yelling that a "boy" in service should not be arrested. Someone grabbed Kraemer's arm, his gun fired and sent a bullet into the right shoulder of Joseph E. Mullen, a Badger Ordnance worker who was part of the crowd. The slug passed through Mullen's torso and lodged in his spine. He was carried into Mueller's drugstore and, before a doctor could be summoned, died.

Martin Scholl, sobered by the shooting, spent the night in jail at Prairie du Sac. He was transferred to Baraboo the following day and held long enough so that the Navy would consider him AWOL — absent without leave. County Judge Henry Bohn dismissed all charges against him, stating that the punishment meted by the Navy to an AWOL sailor would suffice. County coroner Otto Pawlisch found that the death of Joseph Mullen was accidental and that there were no grounds to prosecute Deputy Kraemer. The body of Joseph Mullen was sent home for burial in Dubuque, Iowa.

In Prairie Du Sac, the mob scene and shooting death on Water Street prompted calls for a civic cleanup. With annual elections coming up on the first Tuesday in April, reformers quickly placed a referendum on the ballot to ban the sale of alcohol in the city. Soon members of the business community began to have second thoughts. The war was almost over in Europe and, even though Japan was holding out, the conflict would end and life would return to normal in Prairie du Sac. It would be a peaceful farm market town again and, as Mayor Bert Geigerich said, the farmers "have been used to coming to town for their family and farm buying and before going home the old man probably had a few beers and schnapps...This is distinctly not a dry town under normal conditions and most of us in business would hate to see the village go dry." The referendum vote, for which no record can be found, must have been defeated, because Prairie du Sac did not go dry. As soon as the war ended, the trailer camp emptied, the population balloon deflated and memories faded of mob violence and a shooting death on Water Street.[140]

Life was consistently peaceful and uncrowded at the government housing project known as Staff Village. Laid out on a winding lane near Otter Creek on the old Jennie Johnson Spoors place west of Highway 12, the village had fifteen two- and three-bedroom frame houses, five with garages, all reserved for plant managers. They were probably the only rural homes in Sauk County connected to a treated water system and sewerage disposal plant. Mail was delivered to curbside boxes on the village's single street, which no one ever got around to naming, but was lit at night and plowed of snow in the winter. Plant workers maintained the houses and the grounds, and in the spring plowed and disked a Victory Garden for villagers to share.

The houses were comfortable, but would not compare favorably with many of the farm homes demolished for the plant. They had yards and were luxuriously private and spacious compared to the cramped, thin-walled units at the trailer camp or Badger Village, but were modest by the upper-middle-class standards to which their occupants were accustomed. Managers capable of running a plant of Badger's size and complexity were in demand at defense plants throughout the country. They could serve the war effort just as patriotically in San Diego, Seattle or many another location that was more urban and less wintry. Some accommodation had to be made, even in a national crisis, and Staff Village was it, but only for the fifteen families who resided there. Most Hercules managers scrambled for housing in Madison, Baraboo or the Sauk villages just like everyone else. Although its name hints that Staff Village was reserved for the military brass, it was instead set aside exclusively for civilian Hercules executives. Neither the top Corps of Engineers nor Ordnance Department commanders lived there. No special accommodations were made for them since they had to follow orders no matter where they were sent.

For years, the rumor has persisted that the staff houses were supposed to be built of brick instead of wood. As reported by Erhart Mueller and others, the contractors tripped on a snafu. They wasted good brick on housing at some unspecified place in the balmy southern United States when they should have used it to make snug houses in frigid Wisconsin. The wooden houses that should have been built in the South went up at Badger. However, a house made of wood is not necessarily colder than one made of brick and Wisconsin has plenty of cozy wood frame houses. Also, brick construction is commonly found in the southern states, not because it makes a warmer house, but because brick is immune to the wood-eating insects that abound there. So maybe the bricks went south to foil the termites. Also, relatively little use was made of bricks at Badger. Wood frame construction was faster, cheaper and suitable for buildings designed for demolition in five years. If wood was the right material for poaching, press and pre-blending houses, why not for people houses too?

Jennie Spoors' house, also wood frame, was located on the southern edge of the Staff Village parcel, but not on the same street as the staff houses. It was also used for worker housing, but not until after it had served as headquarters for a special unit of the Wisconsin State Police assigned to the area for traffic control and emergencies at Badger. By the summer of 1944, the troopers were gone and the house was remodeled into a seven-room dormitory for twenty-eight male workers hired for the rocket blitz. The Spoors house was one of fourteen farmsteads taken for the plant but outside the fence, all or parts of which were still standing when the war ended. Six of them, along with a dozen summer cottages at Weigand's Bay, were leased for housing during the war.

In addition to the government-owned and -operated housing built for Badger, there was a development that can be identified as a public-private partnership. Early in 1942, Baraboo real estate man George Martiny assembled all the pieces for the aptly named Victory Heights. Located on the north side of Baraboo's South Boulevard, just west of the city athletic field and swimming pool, the residential subdivision consisted of 111 city lots set out on winding roads named Rachael, after Martiny's daughter, and Russell, after Russell Case, the previous owner of the land, with a third called Badger Drive, for obvious reasons. Myron Park, in the southwest corner of the subdivision was named after Martiny's son.

Putting together a residential subdivision during World War II was no mean feat. Building materials were rationed, new appliances unavailable, and heating and plumbing fixtures — furnaces, sinks and water closets — scarce. Martiny is remembered as not the kind of man who let the grass grow under his feet. He had to be active in order to hunt down the building materials and fixtures he needed for his houses. He also had to be enough of a salesman to persuade the War Production Board to designate Victory Heights as "Private Defense Housing," which meant preference — along with federally guaranteed financing — was given to those well-paid Badger Ordnance Workers. As long as the plant continued to run and a home purchaser stayed on the job, no mortgage payments would be missed. Prices ranged from $4,400 to $4,950 for single-story two- or three-bedroom frame houses.

After an initial burst of construction, homes went up at a slow but steady pace, with the war ending before all the lots were occupied. Nonetheless, Victory Heights was the only residential development connected to Badger Ordnance that was intended to become a permanent part of its community. Not only was it present, Victory Heights was expanded when the plant was reactivated for the Korean war. Conditions dictated that he be a scrounger and Martiny was accused of using scrap lumber, but he did build real houses instead of thin-walled prefabs. Victory Heights also represented the only significant increase in housing that Baraboo had seen since World War I. Individual houses had gone up in the 1920s and '30s, but many years had passed since a local developer was ambitious enough to plat out 100 lots with for-sale signs on them.

Other citizens who attempted to ease the housing crunch did it one house or room at a time. Not too many years had passed since the days when city and village dwellers kept dairy cows, horses and poultry in barns and sheds on their lots. Many of these places were refitted for rental housing, with carriage house lofts among the most coveted spots. Vacation cabins were remodeled into year-round houses, which is one reason why the Park View school, just outside Devils Lake State Park, reopened and filled to overflowing. Cabins on both sides of the Wisconsin River from the Prairie du Sac dam to past Wisconsin Dells were turned into housing for workers.

In private homes unoccupied bedrooms were rented out, basements improved, porches enclosed. The Dresen Brothers, who had been the most important builders in Sauk City and Prairie du Sac for generations, advertised plans and materials for homeowners to build an "Extra Bedroom," which could "be done easily and inexpensively." In the tightest periods workers would share a room and a "hot" bed, but not at the same time. When one worked, the other slept—as long as they were not assigned to the same shift at the plant. It was possible for three workers to coordinate their shifts and share the same bedroom, but if such a hotter bed existed, it has yet to be discovered. The rules of use varied from house to house. Some renters had full kitchen privileges to cook for themselves, others were allotted space for snacks in the refrigerator and cupboards, while others relied solely on the landlord to put food on the table.

With housing in such great demand, rents quickly soared to twice their pre-powder-plant rates. To halt further increases, in August 1942, the federal Office of Price Administration froze all rents in Sauk, Dane and Columbia counties at their March 1 level. The March 1 rates were at least double those of the pre-powder-plant days but not necessarily an unfair benchmark. Demand for, or even interest in, year-round rental housing in Sauk County prior to October 1941 was all but nonexistent. Were they locked into pre-powder-plant rates, property owners would be much less likely to rent their rooms and houses. Calls for patriotism and good citizenship were helpful, but a larger rent check made it easier to live with strangers in the living room and a hot bed upstairs.

While attention was drawn to housing in the cities and villages, rural areas also supplied places for newcomers to live. Many 19th-century farmhouses were built to house multi-generational families, with lots of kids and hired help too. By the 1940s, families were smaller and farmhouses had empty rooms to rent. Many estimates have been made of the population increase in Baraboo and the Sauk villages due to the powder plant. By comparison, no one has attempted to count how many people rented rooms on farms in the towns of Sumpter, Honey Creek, Freedom, Prairie du Sac and Baraboo, which were nearest to the plant, or in the Dane County towns along Highway 12, or even farther from the plant. Although they did not turn up on any of the official radar screens, farmers played a significant role in meeting the housing shortage.

Perhaps the most unusual farm building converted for housing was Harold Weidenkopf's "Silver Dome" poultry house. Located off Highway 12 on the south edge of Baraboo, the Dome was a three-story structure, 48 x 72 feet, with a barrel-vaulted roof rounded on one end. Viewed from the rounded end, it looked like dome. Inside the 8,000-square-foot building was a concrete slab at ground level, where up to 1,000 frying chickens were fed up to market weight. Above were two floors used for feed and storage. It was insulated and had indoor plumbing, heating and electricity. In the spring of 1942 Weidenkopf offered the building to "some contractor or anyone else..." who could make use of its spacious interior. He was willing to sacrifice his poultry for the cause. "Those chickens would do just as well 50 miles from here or for that matter in someone's skillet." The chickens were moved and the Dome was remodelled into housing for several hundred construction workers, who occupied it during the building booms of 1942 and 1945.[141]

For some farmers it was not necessary to rent rooms in the house or space in the chicken coop. In April 1942, Arthur Olson, his wife and their son Robert pulled a trailer into the yard of the Ken Premo farm west of the powder plant. Originally from Richland Center, Olson started as a construction worker, then moved into production after the plant opened. Using a farm pump for water and an outhouse for waste, the Olsons rented space from the Premos for nearly four years. Another son, Arthur Junior, was born while the Olsons lived in the trailer, and by the time the Olsons left the Premos in February 1946 young Robert was ready for school. Had the plant remained open, who knows how long they would have stayed?[142]

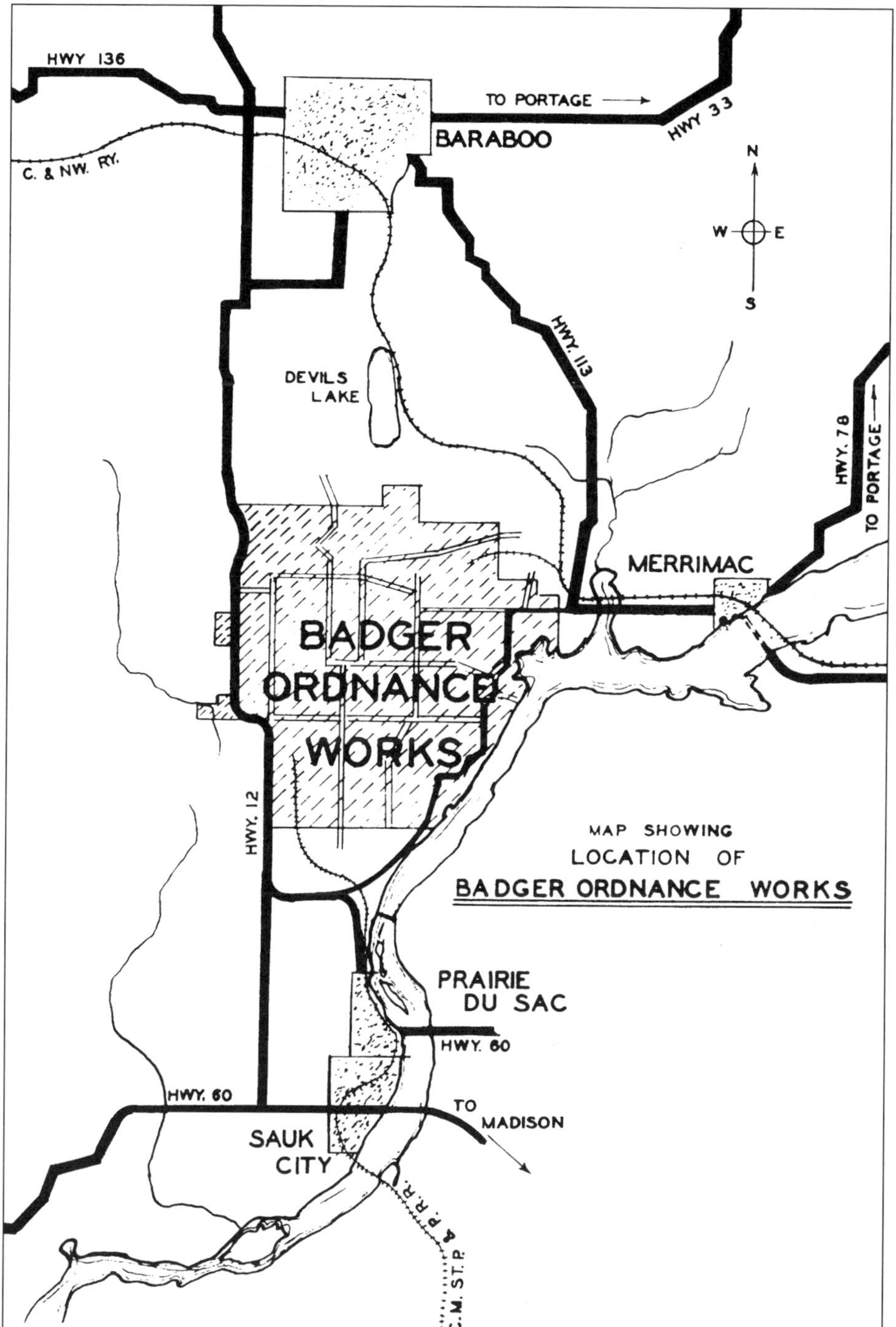

A 1942 Army map of the original 10,565 acres of the Badger Ordnance Works with county and town roads still crossing the grounds. (BAAP)

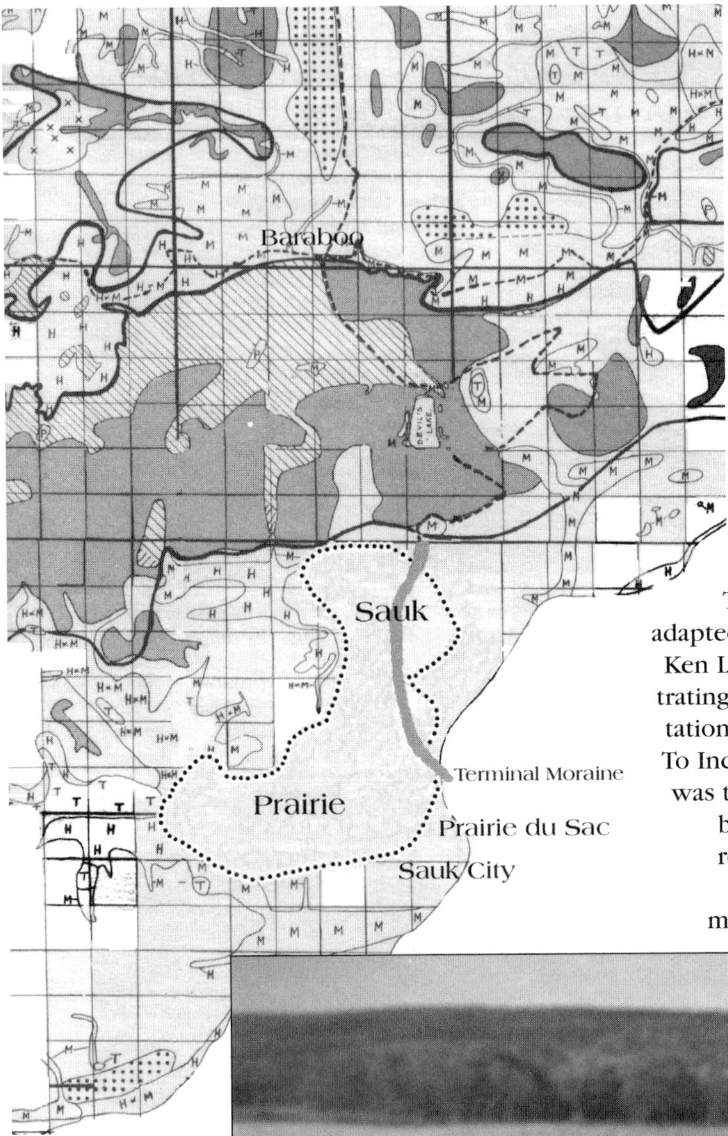

The Sauk Prairie, adapted from naturalist Ken Lange's map illustrating the native vegetation of Sauk County. To Increase Lapham, it was the "sack prairie" because its shape resembled that of a partially filled miller's grain sack.

Prairie grass meets the Baraboo Hills. (Paula Bunch)

Above. Regular fires set by humans maintained the prairie and "oak-opening" savanna grasslands of Sauk County for thousands of years. (Paula Bunch)

Below. Perhaps one thousand years old when they were plowed over by settlers, the Kingston Mounds were among the hundreds of ancient earthworks on and near the Sauk Prairie. (Sauk County Historical Society)

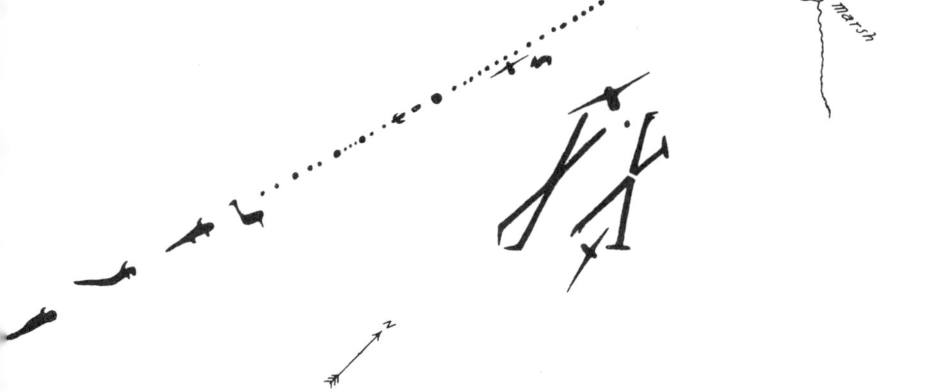

Yellow Thunder, the aged leader of the "Winnebago" Indians, refused to accept the 1837 treaty removing his people to reservations in Iowa. They would have been exiled from Sauk County by the United States had Yellow Thunder and his band not repeatedly returned to their homeland, where they remain today as the Ho Chunk. (Dells Country Historical Society)

Above. The ox-drawn breaking plow transformed the Sauk Prairie to farmland in about ten years. (Sauk Prairie Historical Society)

Below. The Shimniok farm house on the "Middle Sauk" Road where Prescott Brigham built an inn to shelter travelers crossing the prairie. (BAAP)

From left to right, Jacob Hatz, Robert Hatz and Monroe Nolden with a load of silage on the Hatz farm in the 1930s. (Wayne Hatz)

Above. Teddy the cow dog herding the Hatz family cows down a prairie road. If the cows strayed into a neighbors' field, "everyone" turned out to bring them home. (Wayne Hatz)

Below. A threshing machine on the Irvin Leutscher farm. The local threshing ring in which neighbors helped neighbors remains a symbol of the cooperative nature of farming prior to World War II. (BAAP)

The Louis and Amy Eschenbach farm on the eastern side of the powder plant. (BAAP)

One of two houses on the William and Amelia Burkhardt farm near the Lutheran Church, now part of the magazine area. (BAAP)

The Walter and Edith Kindschi farm on Highway 12 at the southwest corner of the plant. (BAAP)

The Julius and Agnes Steidtman house just north of Gruber's Bay, where the family had lived since the 1850s. (BAAP)

The yard of the Jacob and Emma Gasser farm which straddled the head of Gruber's Bay. (BAAP)

The Virgil and Ida Roick house near the center of the plant in the Ball Powder area. (BAAP)

The summer cottages of the Walter Rudolph and Dean Van Matre families at Weigand's Bay. (BAAP)

The Evangelical or German Methodist Church was organized by Henry Thilke in 1860. The site is in the magazine area. It was one of three churches, three schools and three cemeteries on the grounds. A small burial ground used by the Miller family was discovered on the Alex Steuber farm while Badger was under construction. (BAAP)

George Jenewein's one-pump gas station on Highway 78 and Weigand's Resort were the only non-farm commercial properties taken for the plant. (BAAP)

ᴧUKEE JOURNAL

Wednesday, October 29, 1941 | Daily .... 3 Cents | Sunday .... 10 Cents | 40 Pages—**Final Markets**

65,000,000 Powder Factory Is Given to Wisconsin Hamlet

Crimean s Pierced, mans Say

roops Barely .10 From Rostov-on- aim; Navy Base, Periled

Associated Press

oops were reported to
a break-through Wed-
the Crimea peninsula,
's Black sea naval base
l, while other Nazi col-
d at the gates of Ros-
n what appeared to be
hase of the four month
Ukraine.
atches said Red army
ere already planting
ines in Rostov and pre-
ve the city of 500,000 a

German attacking col-
d to be only 10 miles
er of the big Don river
uards the northern ap-
Caucasus oil fields. In
military commentators
e the city had fallen
d be open for swift ar-
ns to strike south to-
ve center of the oil re-
p, 175 miles below Ros-

er Is Admitted

n newspaper Pravda
"great danger" for
Donets basin, the Cri-
tov, but declared that
prediction of victory
had collapsed.
d the Nazi high com-
rown almost the entire
army, artillery and
nine-tenths of the air
Russian campaign.

You Can Put It Down Now, We'll Help Fire the Last One

Land Acquired at Merrimac; U. S. Approves

War Department Sanction Announced for Project of Hercules Co.; Plant Also to Make Acid

Construction of a $65,000,000 powder and acid works by the Hercules Powder Co. of Wilmington, Del., at Merrimac, on the Wisconsin river in Sauk county, has been given final approval by the war department, it was learned in Washington Wednesday

Representative William H. Stevenson La Crosse Republican, in whose district the new plant will be located, said at La Crosse Wednesday that construction was scheduled to start as soon as contracts could be let and that the Hercules concern had already acquired many of the thousands of acres of land necessary for the project.

It's News in Merrimac

News that Merrimac had been awarded a $65,000,000 plant caught the village entirely by surprise Wednesday. No word of the intended construction had reached the ears of the inhabitants. Villagers speculated on the probable site and, especially, upon the impact that the huge industrialization would have upon the community's quiet life.

Jobs for Thousands

Stevenson, whose Washington office was notified of the war department's final approval, said that from 5,000 to 10,000 men would be employed on the project.

The plant will be completed by Jan. 1, 1943 It will provide 300,000 pounds of smokeless powder a day, 3,000 pounds of diphenylamine, and sulphuric acid "for all purposes."

[Diphenylamine is a crystalline compound, of pleasant, aromatic odor, obtained by heating aniline with aniline hydrochloride, and by other methods. It is of importance in the manufacture of many dyes and in stabilizing explosives and affords a delicate test for nitric acid. Sulphuric acid is a heavy, corrosive, oily liquid which is of the greatest importance in many processes, such as the manufacture of chemicals.

Stevenson said the new project would be known as the Badger State Ordnance works and he revealed that the $65,000,000 figure represented an initial" outlay. He said he believed the cost eventually would be in excess of the $65,000,000 estimate.

Allotment of Funds

Of the money to be spent, according to Stevenson, $42,000,000 will be spent for buildings that will stretch for miles on the sandy terrain of Sauk county; $21,000,000 will be spent for equipment, $1,020,000 for land

The most expensive War Department project in Wisconsin history, Badger Ordnance Works was statewide news in October 1941. Except that it would be near Merrimac, the site of the plant was unknown at the time of the announcement.

Left. Leo T. Crowley, rags-to-riches Madison entrepreneur, chairman of the Federal Deposit Insurance Corporation and the most powerful Democratic politician in Wisconsin, was the "president's friend" who used his influence to site Badger Ordnance near the line of the Chicago and North Western Railroad. (Wisconsin Historical Society)

Below. Baraboo's Clark Wilkinson, left, and John Lange, right, were among the local business leaders who worked to bring the powder plant to the Sauk Prairie. (Sauk County Historical Society)

Above. Garth and Nora Premo, at left, and John Litscher. Along with Emma Kindschi, Premo and Litscher led the landowners who opposed the siting of the plant on the Prairie and who later challenged the federal government through the condemnation proceedings. (Jay Premo and Lorene Litscher Harrison)

Below. Five Kindschi brothers and their wives, four of whom lost farms to Badger Ordnance. Left to right, Walter and Edith Kindschi, Ralph and Celia Kindschi, Frieda and Ernest Koenig, Emma Kindschi, Roy Kindschi, Emma's husband Glen, Essie and Lyman Kindschi. (Sue Mueller)

Above. The plant where millions of yards of concrete were mixed for the nearly 1,000 buildings that went up in the first Badger building blitz of 1942. (BAAP)

Below. As the worker wheeling the concrete buggy for a foundation in the rocket area in 1944 could testify, construction was labor intensive when Badger was built. Over 12,000 people labored on the site in August 1942. (BAAP)

Above. The TNT area, July 31, 1942. Construction was simultaneous at Badger, with hundreds of individual projects underway at one time. Plans changed often. The day after this photo was taken the TNT facility was cancelled.

Below. Framing the boiling tub house on the nitrocellulose F-Line in January 1945. The masonry barrier wall was designed to contain explosions and limit damage. (BAAP)

Above. Power House #1 generated steam for production and heating throughout the plant. (BAAP)

Below. Prime examples of prewar industrial technology, a double row of compressors furnished pressurized air to operating machinery. (BAAP)

Above. The River Pump House at Weigand's Bay on Lake Wisconsin was the intake point for the millions of gallons of water necessary for propellant production. (BAAP)

Below. Ammonia storage tanks were part of the extensive acid production facility at Badger. (BAAP)

Above. The main administration building and the 14' x 26' American flag presented to Badger by the Baraboo American Legion Post in 1942. (Gordon Emery)

Below. Surrounded by piping, the Paul Shimniok farm was one of twelve vacated farmsteads still standing inside the fence at the end of World War II. (BAAP)

Above. Laid out with plenty of space in between, the "houses" of the B and C Nitrocellulose and Smokeless Powder Lines stretch east from Power House #1. The T-shaped building at front left is the plant hospital, with maintenance shops at front center. (BAAP)

Below. Badger Ordnance at the end of World War II, with the new Rocket Area spreading out towards Lake Wisconsin, and the H-Barracks and Badger Village in the foreground. (BAAP)

Above. Loading sulfur for burning to make sulfuric acid at the Oleum plant. (BAAP)

Below. Bales of cotton "linter" go into the breaker. Powder production started with the mixing of acids with linter or wood pulp to make nitrocellulose. (BAAP)

"Dippermen" made the first mix of nitric acid and cellulose on the top floor of the dipping house. When she took the job in 1944 Catherine Hangar, Baraboo, became Badger's first "Dipperette. The job required that she wear coveralls and a heavy woolen jacket impregnated with acid and fire retardants. (BAAP)

After dipping, excess acid was spun out of the nitrated cellulose in a wringer. Prone to burst into flash fires and small explosions, the operation required that wringermen wear metal helmets, rubber gloves and body armor. (BAAP)

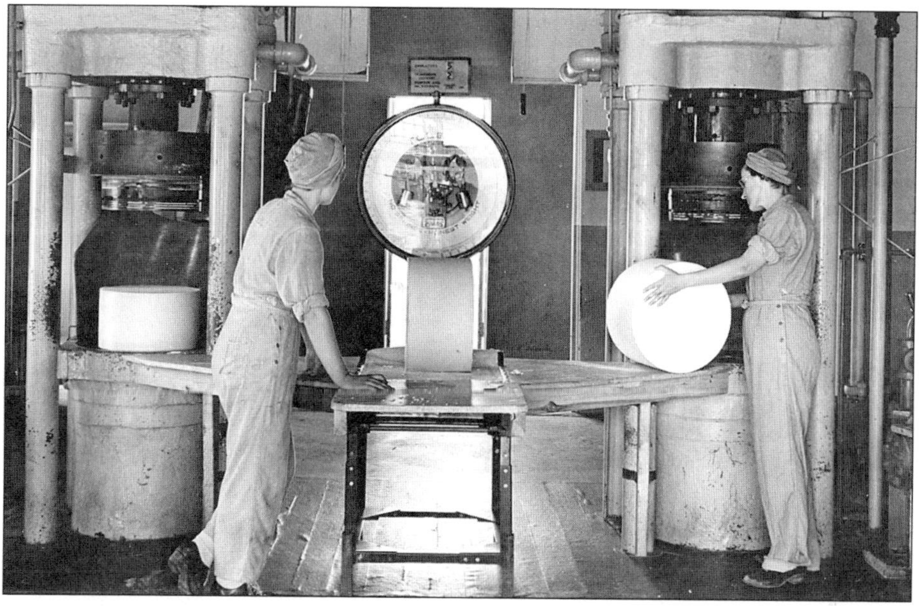

Above. After a number of refining processes, the nitrocellulose entered the "Dehy" house where it underwent the first of several pressings that shaped it into a cylinder resembling a wheel of cheese. (BAAP)

Right. In its final pressing, the "cheese" was forced through a set of dies that extruded it into strands as thin as spaghetti or thick as a pencil. (BAAP)

Right. The "spaghetti" was then cut to the desired length, which varied from a fraction of an inch to four inches long. Despite its name, individual grains of smokeless powder were actually small cylinders. (BAAP)

Left. Charles Warren, Adams, removing smokeless grains from the "sweetie barrel" where they were coated with dinitrotoluene. (BAAP)

Above. To make rocket powder, nitrocelluose was mixed with nitroglycerin, then pressed into "carpet rolls." (BAAP)

Below. Carpet rolls were then loaded into an extrusion press which shaped the powder into "sticks" of the desired thickness and shape. (BAAP)

Above. Badger made cruciform or cross-shaped rocket sticks for World War II and Korea. (BAAP)

Below. Vietnam-era rocket sticks were smooth cylinders with cores hollowed into a star shape. (BAAP)

Above. Finished products were test-fired in the ballistic lab for small arms powder and at the cannon range for artillery loads. (BAAP)

Below. Powder was held for shipping in barricaded or bermed magazine buildings. The "igloo" depicted here was specially constructed in the 1950s to store ammunition for the Truax Air Force base in Madison. (BAAP)

Above. A truck parked in front of an "angel" walk on which workers wheeled nitroglycerin delivered an apt message. (BAAP)

Below. The roof of a blending house on the E.C. Line peeps over the top of the barricades isolating it. Two men died in an E.C. blending house explosion in October 1943, but the barricades prevented further loss of life. (BAAP)

A three-story ether still house equipped with a safety chute down which workers were to flee at the first sign of a fire or explosion. (BAAP)

Above. A fire and explosion in 1966 destroyed Ball Powder Hardening House #2 and injured eleven workers. (BAAP)

Below. Vietnam-era workers receiving an award for safe operation. Although fatalities occurred, Badger operated accident-free for millions of man hours. (BAAP)

Top left. Lt. Col. George F. Griffith, Badger's first commanding officer.

Top center. Major Victor I. Keldsen, World War II.

Top right. Lt. Col. Oliver Williams, Korea.

Middle right. Lt. Col. Russell O. Enoch, Vietnam.

Bottom right. David C. Fordham, Commander's Representative, 1976-2002.

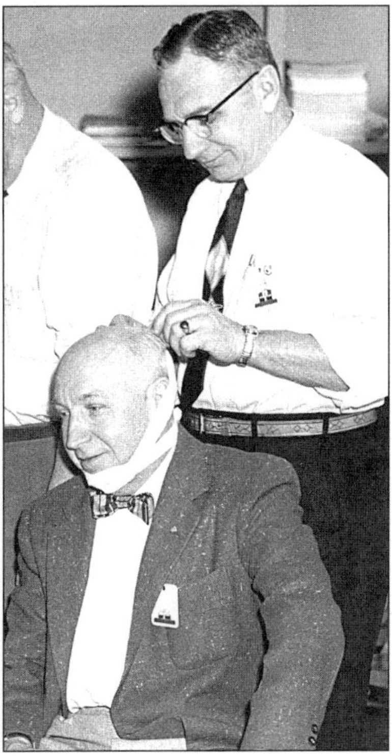

Left. John Troyer, Commander's Representative, 1959-66, bandaging the head of veteran manager Louis Reuter at a first aid workshop in the 1950s. (BAAP)

Worker recruitment brochure and billboard, 1944. The construction of the rocket area and increased demand for ammunition after D-Day forced Badger to expand its workforce to more than 6,000. (BAAP)

Above. Taking applications at the personnel office during the recruitment drive. By the end of World War II, almost one-half of Badger workers were female. (BAAP)

Below. The mobile employment office that toured Wisconsin in the summer of 1944. (BAAP)

Above. World War II artillerymen firing a shell propelled by powder made at Badger. The soldiers depicted visited the plant during the war and presented an oversize copy of the photo in gratitude to the powder workers. (Sauk County Historical Society)

Below. An American infantryman on patrol in Vietnam carrying a M-16 rifle loaded with Ball Powder ammunition made at Badger. (Wisconsin Historical Society)

Above. As a forward air controller in Vietnam Gary Dikkers, Madison, flew a single-engine O-2A armed with 2.75" rockets powered by Badger propellant. (Gary Dikkers)

Below. An Army UH "Huey" helicopter armed with 2.75" rocket pods. (Gary Baran, US Army TACOM)

Above. Children playing at the model government trailer camp built at Prairie du Sac in 1942. (Sauk County Historical Society)

Below. Aerial view of the Prairie du Sac camp, which offered trailers for rent and parking spaces for privately owned units. (BAAP)

Above. Housing was so scarce during the 1942 construction blitz that the "Silver Dome," a three-story chicken house on the Harold Weidenkopf farm near Baraboo, was remodeled to house workers. (James Pierce)

Below. Buses brought thousands of workers from eight counties to the plant. In World War II, they were among the first residents of rural America to commute to an industrial job. (BAAP)

Above. Baraboo's George Martiny acquired land, located scarce materials, obtained federal mortgage financing and built Victory Heights to house Badger workers. (Myron Martiny)

Below. Barracks #10, with two-room apartments, no kitchen and a communal bathroom, designed for married workers in 1942. (BAAP)

Above. George Martiny and Badger school principal Clarence Krumm. (Myron and Peggy Krumm Martiny)

Below. The multi-unit apartments of "South Badger" were erected in the mud of Lyman Kindschi's cornfield. (BAAP)

Above. The A & P grocery and the Walgreen's drug store made up downtown Badger Village in World War II. (Sauk County Historical Society)

Below. The community center, with wooden-floored gym, playground equipment and other facilities, was completed just as World War II ended in the summer of 1945. (BAAP)

Above. A two-bedroom family unit in Badger Village, with a coal-burning heater, electric hot plate and cozy dining table. (BAAP)

Below. A one-bedroom "demountable" house erected at Badger in 1945 and soon moved elsewhere to ease the postwar housing shortage. (BAAP)

Above. Class session in the Badger School, where teachers and children coped with the special circumstances created by World War II. (Sauk County Historical Soceity)

Below. Children and teacher in round-the-clock child care at Badger School. (Sauk County Historical Society)

Baby boom at Badger Village. During World War II and for the six years after, when it housed University of Wisconsin students, the village was filled with mothers and their children. (Sauk County Historical Society)

Above. The Bluffview Courts partners, Donna and Ron Nichols, Wayne and Myrtie Hatz, Bill and Enid Fuchs, Woody and Milly Zantow. (Wayne Hatz)

Below. After housing powder workers and university students, Badger Village was transformed into condominium units for senior citizens. (Wayne Hatz)

Designed in the 1970s to be the centerpiece of a completely modernized Badger, the new acid works is a state of the art facility that, except for test runs, has never operated. (BAAP)

Above. Anti-war demonstrators confront Sauk County Sheriff's deputies at the gate in 1966. *Inset.* Sheriff Mike Spencer, who defused a tense situation with fair treatment and good humor. (Mike Spencer)

Below. A Cessna 150 of the type used to drop homemade bombs on the plant in January 1970. The bombs were harmless, but a plane crash could have been disastrous. (EAA)

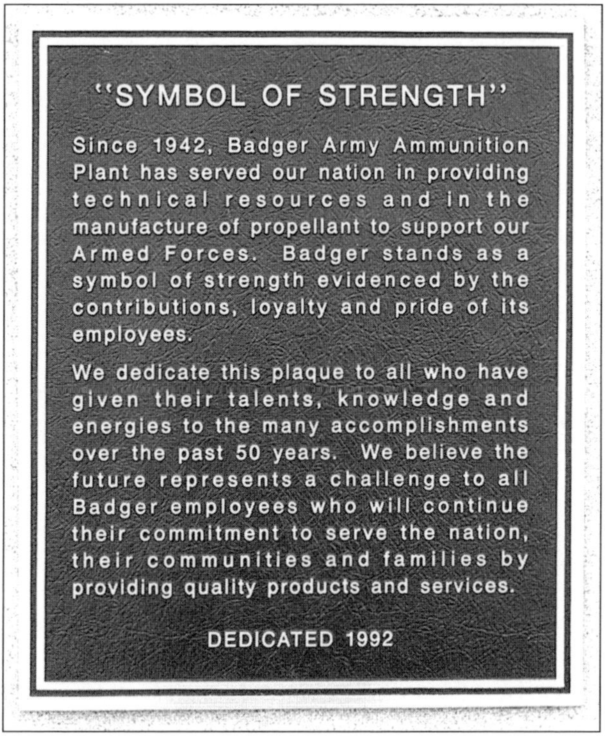

Right. The powdermakers' memorial at the plant. (NPP)

Below. Re-enacting an ancient practice, an environmentalist uses fire to restore the Sauk Prairie. (Paula Bunch)

An unidentified wringer operator from the Korean period of operations. The wringing process, in which excess acid was removed from nitrated cellulose, symbolizes powder production at Badger—a simple yet hazardous industrial process necessary for national security. (Wisconsin Historical Society)

Badger Village

If any place can stand as a metaphor not only for Badger Ordnance but for the entire United States during World War II, it is the housing complex built for workers and their families outside the plant gate on Highway 12. All the brevity and intensity, the ill-timing and wastefulness, the patriotism and sacrifice of Badger Ordnance in those years; all the broadening of personal horizons, the experimentation with new ways of working and living, the opportunities for change and growth, and the desire to create a better world for future generations that characterized the war years can also be found in the story of the wartime housing complex best known as Badger Village.

When the War Department announced its birth in October 1942, it was Badger Ordnance City. By the end of the war, the name was simply Badger, as officially authorized by the United States Post Office at its branch office on the site. Starting in 1946, when those parts of the development not already demolished or relocated began to fill with University of Wisconsin students, the term Badger Village came into common use. It is often heard to this day, even though the village was renamed Bluffview in 1960.

The Badger Village of World War II was divided into at least four sections — north, south, west and in-between. North Badger was the barracks area first built for construction workers in the summer and fall of 1942. Work on "South Badger" started early in 1943 on thirty-eight acres of Lyman Kindschi's old corn ground. The northeast corner of the plot was about one-quarter mile south of the main powder-plant gate, due west of Powder Line E. It was designed for, and residence was limited to, production workers and their families, with no space allotted to construction personnel no matter how badly they may have needed it.

South Badger had ninety buildings divided into 500, two-, three- and four-room apartment units. By War Department math, they could house 1,800 "Badger Ordnance Workers and their families." The village would also have an elementary school with extended day and summer programs, a twenty-four-hour child care center for preschoolers and students and, later, a "nursery" for children as young as fourteen months. This was the Badger which existed for not quite two years during World War II, with its post office, barbershop, beauty parlor, A & P grocery and Walgreen's drugstore. It was a place crowded with the children of young mothers turned into ordnance workers and single parents by the war, and where Principal Clarence Krumm and his staff created a "school" that was as much a comprehensive service center for children, parents and community as it was an institution of learning.[143]

South Badger had been running for about a year when, in September 1944, additional housing was erected on the in-between section of Badger. The National Housing Authority put up 250 individual "houses" on the forty acres in between South Badger and North Badger. Although the NHA identified the units as "houses," they were not much larger than the expandable trailers at Prairie du Sac, minus the wheels, and plus a connection to the powder plant's water and sewer lines. None of the houses was ready for occupancy until the spring of 1945, and less than half were inhabited before the war ended. In January 1945, the War Department decided to add a community center to the "in-between" section, with space for stores, offices, meetings and sports. It was completed in July, less than one month before the war ended, and hardly saw any use during the conflict.

Finally, in the spring of 1945, work started on "West Badger." Located directly west of "South Badger," the new section consisted of 350 single "houses" placed on concete pads laid out in a street grid and flanked by concrete walks. Approximately 300 of the units in "West Badger" were installed by the end of the war. None was occupied. Had the war and powder production continued and had these units been fully occupied, about 2,000 more men, women and children would have been living in Badger housing. Had all the barracks been fully occupied as well, including the Gate 10 units, 5,700 people could have called Badger home. In "West Badger," as in other parts of the development, by the time the housing was finished, it wasn't needed.

The housing best known as Badger ("South Badger") welcomed its first fifty tenants early in the summer of 1943. Among them was Mabel Gillingham, who had been riding the bus from the village of Gillingham in Richland County since starting work in the smokeless area in May. A single mother of two school-age children, Gillingham was the worker "South Badger" was built to house. She and her children took a one-bedroom unit that, "had one really long room and then you went into the bedroom." Mabel's daughter Loretta McCarthy recorded a different first impression, which she shared with many other Badger villagers who hailed from the country. "It had an inside bathroom which was really nice in those days."[144]

For reasons of speed and economy, the Federal Public Housing Administration, which built and managed Badger, purchased prefabricated multi-unit structures from the Structiform Company of Chicago. The ninety prefabs cost $1.2 million, about $150,000 less than units built on the site. They were trucked up Highway 12 in sections and set on concrete pilings poured before, during and after the site had been churned up by the installation of water and sewer lines. The mud left by these excavations figures prominently in memories of all Badger residents, children and adults.

"There are times when the clotheslines are accessible only in hip boots," as school principal Clarence Krumm wrote in 1945."[145]

He went on to describe the apartments as, "Those chicken coops! Those box cars!" In a two-bedroom unit designed for two parents and two children, "There was a living room...a snug fit for a 9x12 rug; a kitchen of approximately the same size; bathroom with no tub, just a small shower; no basement, and beyond the two clothes closets and two cupboards in the tiny hall, no storage space...Bedrooms?...approximately ten feet square."[146]

The apartments were trailer-size, butted end to end, and separated by thin wallboard partitions through which sound traveled as if no wall was there at all. "The noise of the heavy trucks on the highway, the broadcasting system at the plant, the conversation, children and radio in the next apartment," assisted by occasional blasts of cannon and rocket fire from the testing grounds at the bluffs, kept the atmosphere at Badger lively and loud. Privacy was a rare commodity in such tight quarters. If you knew nothing about anyone when you arrived, you knew everything about everyone by the time you left.[147]

Badger was a company town created to house workers for a powder plant operating around the clock. The company, Hercules Powder, acting as an agent for the United States, not only welcomed but recruited two-worker families and women whose spouses were in the military. Two workers living in the same apartment made the most efficient use of company housing. The wives of servicemen were valued employees because they needed jobs to support themselves and would work hard and long to end the war and bring their men home.

The two-worker families and the military spouses brought special needs to the company town, almost all of them involving children. Badger ran twenty-four hours a day, often — but not always — seven days a week. Shifts ran 8:00 AM to 4:00 PM, 4:00 PM to midnight, midnight to 8:00 AM. Workers changed shifts every two weeks, with a day off in between, so the working week usually did not begin on the traditional Monday, nor was Sunday necessarily a day of rest. Workers also had to sleep and, when they worked the evening or night shifts, they had to sleep during the day. All of these facts of working life in the company town of Badger led federal officials to authorize a system of care for the children of plant workers that ran like the plant — twenty-four hours a day, seven days a week. To the credit of unnamed federal administrators, child care was in the program right from the get-go. They did not just open a school at Badger, dismiss the students at 3:00 PM and — as could easily have happened — suddenly notice that the kids were on their own because Dad was in New Guinea with the 32nd Division and Mom was dipping nitrocellulose on the B-Line. Instead, they designed a child care facility of which the elementary school was the centerpiece of a table full of services for working families. "Our schools," said principal Clarence Krumm at the village dedication ceremonies, "are going to take care of the children of workers so that the workers do their best."[148]

Although federally funded, the Badger School was part of the Sauk County rural school system. It was regulated by the state and administered by the county as one of its multi-grade schools. As such, it was nominally under the supervision of the county superintendent of schools, Kurt Schoenoff. New to the job in 1941, Schoenoff did not play a prominent role at Badger, where the star spot was reserved for Clarence Krumm. He had resigned his job as principal of the Wonewoc school to seek a better-paying job as a "carpenter" at Badger. It says something about the pay scale for educators that, with his experience as a teacher and school administrator, with a master's degree in Adult Education, Krumm could do better as an unskilled carpenter at the powder plant. No matter, for he was not hired anyway. Instead he was called to Madison for an interview with the State Superintendent of Public Instruction, the formidable and forward-thinking John C. Callahan. The state superintendent, whose advocacy of rural school consolidation and busing would make him the most unpopular person in rural Wisconsin in the post-war years, was very interested in the new Badger school. In many ways, Badger proved to be the kind of school—multi-graded, located in a village and with a sufficent number of students to support a wide range of classroom and extracurriclar activities—that Callahan envisioned as the replacement for the venerable one-room school. Although he had no official authority in the matter, Callahan, noting Krumm's MA in a field other than elementary education, appointed the would-be carpenter as principal of Badger. Along with Maybelle Bush of the University of Wisconsin Extension, the superintendent opened doors to state and federal resources that Clarence Krumm used to good advantage. As Krumm said, "Mr. Callahan was very helpful in anything we had to do."[149]

In order to fit into the legal framework of the county and state system, the Badger school had to be authorized by the local school district. Since it was located within the boundaries of District 4, Town of Sumpter, in territory served by the one-room Waterbury School, the new school was officially listed as the Badger State Graded School, District 4, Town of Sumpter. After receiving Superintendent Callahan's firm endorsement, Krumm was "elected" principal by the District 4 Board: Herbert Yanke, Alfred Anderson and Ella Moely. They had seen their district lose about half of its territory and tax base to the powder plant, along with about one-quarter of its student body. For the 1942-'43 term the Waterbury School had twelve students in attendance and an operating budget of $600. In the summer of 1943, the District 4 board members discovered that they would become trustees of a new school with eight elementary grades that would cost $89,000 to build. Enrollment for 1943-44 was projected at 300. Badger would also be the only rural school in the county to have a kindergarten, a library and teachers for music, arts and physical education. Badger would also provide round-the-clock child care and operate extracurricular activities ranging from art contests for primary students to dances for the Teen Age Canteen social club.

At the annual meeting in July 1943, the twelve remaining tax-paying landowners in District 4 authorized nearly $47,000 for the elementary school operating budget. Included for the first time in District history were free textbooks. Excluded for the first time in memory was free ice cream for all students and parents at the annual school picnic. Next came the "nursery" budget of $72,000, which covered care for preschoolers, followed by the "extended school" budget of $42,000 for care of school-age children after and before classes. Of the $160,000 budgeted for Badger's first year, District 4 taxpayers were obliged to fund $967.52. The federal government would make up the shortfall, which meant that the Badger school budget required the signature of the President of the United States. Clarence Krumm recalled that he had to wait to purchase furniture and hire teachers in the summer of 1943 because President Roosevelt was too busy discussing strategy with Winston Churchill to initial the Badger budget.

While concrete finishers, carpenters and plumbers finished installing the residential prefabs and erecting the school and day care buildings, Clarence Krumm attempted to find books, desks and teachers in time for the school's scheduled opening on the day after Labor Day. He reviewed the applications of over 200 teachers, seeking people with both classroom and non-classroom child care experience. He hired three complete faculties of eight people each and lost them due to contractual problems before signing the staff that came to work on opening day. The building was finished on time but desks, books and other materials for all of the 147 students who showed up had yet to arrive. The grade schoolers sat on metal folding chairs hastily borrowed from the plant, but as kindergarten teacher Mary Wiegand Edwards remembered, "We sat on the floor." It was probably less of a hardship for the five-year-olds than the teacher. In time, equipment and material arrived and the school grew. Badger ended its first year in June 1943 with 253 students and opened the following September with 305. During the course of the 1944-45 school year, enrollment soared to over 400 students attending class in two split shifts.

Children entering the Badger school came from all over the United States, from rural, village and city schools, or from no schools at all. Children from the southern states had accents so foreign to Wisconsin ears, they were all but unintelligible—as were the Wisconsin accents to many of the Southerners. Many of the newcomers stayed only briefly before moving on, with some of the Southerners recruited in the fall of 1944 learning, in January 1945, that they did not really like winter in Wisconsin. "Nothing was ever settled," said kindergarten teacher Carol Bohn Hulterstrom. "[150]

"You might have twenty-five youngsters in a grade at the start and have twenty-five at the end. Only fourteen or fifteen would be originals who enrolled in September."[151]

The teaching staff was as transient as the student body. "We always seemed to have a list we could fill." Through federal government listings, teachers from as far as New York, Maryland, California and Alaska found jobs at the Badger school, then left almost as quickly as they came. At its peak enrollment in the spring of 1945, Badger employed twenty classroom teachers, five "special teachers" and nine substitutes. Some found the circumstances of the school and the village not to their liking. Others, when they learned the pay scale at the plant and compared it to their teacher's wages, took a job right across the road.[152]

Facing the dilemma of teaching a wildly fluctuating number of students from varied backgrounds, Krumm developed a special philosophy for education at Badger: "The Badger schools will begin with the pupil where he is, study his needs, and in meeting those needs provide for his further growth and development in line with his interests, possibilities and capacities."[153]

Since few of the incoming students brought along records or report cards, grade placement was less than scientific. "Each child was placed in a group corresponding to the grade in which the pupil said he or she came in...." then grouped by age and perceived ability. "From the very first, teachers, pupils and parents were indoctrinated with the idea of 'group' instead of a grade placement." Traditionally numbered grades were replaced by groups with names appropriate to the time and place. Kindergarten students were the "Coast Guards." Six-year-old students were the "Army"; seven year olds, the "Marines"; eight year olds, "Flying Tigers"; nines, the "Air Force"; tens, the "Rangers." What would be identified as the upper elementary grades at other schools were the "Seabees" when Badger opened. For the second year, the military jargon was extended to all eight groups. The military theme was reinforced by the school building, where the library became the USO. The Arts and Crafts room was the Camouflage Department, the dining room the Canteen and the bookstore the Quartermaster Department.[154]

At Badger, "Teachers followed the principle that the child and his development was more important than grade lines." The concept extended to grading students as well. They did not receive report cards with numerical or letter grades to indicate progress. Instead parents received a visit from the teacher who discussed their child's progress. When the idea was proposed at a Parent-Teacher Association meeting, a few parents expressed reservations, but a large majority voted in favor of it. The visits began a few weeks after the start of classes and, unless the mother was working at the plant, were unannounced. "The first visits...were in the nature of social calls where teacher and parent became acquainted and used the child as a common meeting ground." Later in the school year, parents also received a "general letter" outlining their child's progress. Parents were also invited to visit the school at anytime and did, even if it was in the middle of the night. All of these practices were consistent with Krumm's view of the school as a means to turn a settlement of strangers, whose common bonds consisted of a paycheck from the plant and a child in the local school, into a community.[155]

While the philosophy and methods of the elementary school extended beyond the traditional three Rs, Badger pushed farthest into new territory with its day care and social services. In 1943, the terms in vogue were "extended school care" for school-age children and "twenty-four-hour nursery" for preschoolers. Both were child care programs designed to answer the question posed for parents by Clarence Krumm. "How could working mothers who left home at 7:00 AM or returned home at 6:00 PM be assured that their children had proper care and supervision?"[156]

"We had to go to school on the shift when Mother went to work," remembered Lornetta McCarthy. "When she was on the night shift, we went to school and had to stay on the cot."[157]

Children ate, slept, learned and played in the school. "Pupils were allowed to check in early and remain late. Sometimes they ate three meals at school." Carol Bohn Hulterstrom remembers one five year old who spent so much time in extended care that she asked his mother to give him a Saturday night bath—at the school—so he would recognize her. Mary Edwards Weigand remembered being criticized by one mother for holding the woman's child longer than the mother felt a teacher should. "They were there so much," said Weigand. "I felt sorry for him."[158]

With children spending whole days and nights at the school, Krumm used an unorthodox tactic to learn how the children would like to fill all those hours. He asked them. A survey was sent to the elementary schools in the vicinity of Badger and a total of 2,500 students, with probably a little help from their teachers and parents, told "what they did before and after school each day, on Saturday and Sunday." In this last decade of the pre-television era, the youngsters reported that "besides chores at home," they liked music and rhythm, reading, outdoor play, indoor play, drama, arts and crafts. Krumm then used these findings to develop "a new curriculum" that was applied to the extended-care program during the school year, and also to the summer "extended home care" program, which of necessity began on the day after regular classes ended in June and continued until school resumed in September.

In addition to the extended-care program, Badger also had "the first nursery school in Wisconsin conducted on a twenty-four-hour program." The first students in the "nursery school" were between the ages of two and five years old, with a total of 438 youngsters enrolled during the war. Teachers were harder to find than students. "Nursery teachers were very difficult to obtain in 1943," and Krumm found that he was competing with defense contractors from all over the country for the limited supply of the university-trained preschool teachers he preferred to hire. A total of eighty-six teachers worked in the nursery program, with "the average tenure...three and one-half months." He also learned that university professionals were not necessarily the best candidates for the round-the-clock regimen, which required at least as much nurturing as it did teaching. "Occasionally the high school graduate proved more valuable than did the nursery-trained teacher..." One teacher whom

Krumm singled out for praise was "Grandma" Isabelle Stekl. " Past eighty years of age, she worked two hours in the morning and two hours in the evening...She was the one who took the place of their grandmother to tuck them in at night and be on hand when they awoke in the morning...her stories were splendid to drift off to dreams on." A woman who was a baby herself at the time of the Civil War was a comfort to children trapped in the circumstances of another conflict eight decades later.[159]

In August 1944, with the plant charging full steam into the rocket area expansion, Krumm was asked to accept children younger than two years of age into the nursery. "Many of the prospective employees at the present time are mothers with young children," wrote plant manager C. H. Gant to the school principal. "Most of the women with older children have already been recruited...Many of these young mothers are wives of servicemen...This group of women with young children, in addition to being comparatively large in number, is, because they are young and have an extra incentive to work in a war project, the most desirable type of female worker now available."

Accordingly, the official minimum age for admission to the twenty-four-hour nursery care program fell to fourteen months. However, if the need arose, younger children were accepted. "Putting it mildly," Krumm wrote, "we said children must be off the bottle and 'house broke.' Even at that we couldn't always be sure." Krumm applied for and received federal funding. Care for a total of fifty-one of these youngest of children was implemented and continued until the end of the war. How this pioneering program would have fared if the war had continued beyond August 1945 is guesswork. How much, as appears certain, it would have had to expand had all those hundreds of vacant family housing units filled with the young mothers who were "the most desirable type of female worker" is equally unknown.

What is known is the Badger school attracted a great deal of attention from professionals in the fields of education, public welfare and health, social work, university Extension and school administration. Alerted by numerous press reports about the experimental school and extensive child care services at Badger, Parent-Teacher Associations, clergymen and others with an interest in children and schools visited the place. Education students from the University of Wisconsin were at the school on a regular basis. Krumm reported that on an "average" weekday, forty visitors trooped through the halls of the school, with fifty curious people showing up on weekends. As part of the "curriculum," older students acted as guides for the tourists. Interest was so great that Krumm and the state superintendent's office organized a "training conference" which brought 185 educators to the school for three days in the fall of 1944. Housing was so tight in Sauk County that, although workshops were held at the school, the conferees ate and slept at the Hotel Raulf in Portage, the only hotel in the vicinity that could feed and house them.

The conference concluded that the "wartime child was also the peacetime child," with identical needs for love, security and family. Reporting on the experience of the twenty-four/seven child care, and inadvertently revealing the reality of life in a wartime housing complex, Krumm concluded that "it is possible to establish and maintain a school atmosphere of security, activity, happiness and good physical and emotional health in the midst of homes permeated by confusion, rush and insecurity."[160]

While children living in a crowded housing development in a strange place with one parent completely absent and the other away for many hours of the day and night would be expected to experience confusion and insecurity, Badger residents had another reason for anxiety. Clarence Krumm just happened to be looking out the window of his office towards the plant when an explosion occurred. It was either the blast that killed two men on the E.C. line in October 1943, or the July 1945 rocket line shot that killed four. "I saw this huge cloud...[the force] broke windows, took the back end of the window and just caved it in. Everything stopped. The first thing was to get the children in. We were confined to our rooms until given clearance."[161]

Loretta McCarthy was one of the children Krumm gathered inside. "I still remember the explosion...It went up as a huge mushroom. I still remember the fear...[afterwards] other people standing in the yard waiting if our parents were going to come home. I still remember when I saw my mother walking diagonally [across the grass to get home faster] how relieved we were."[162]

To cope with the confusion and insecurity its children faced, the Badger school took what was later identified as a holistic approach. Students received a physical exam upon entering the school. Teachers were instructed to keep a health record for each child and to monitor the children's physical and emotional status more closely than they would in a conventional school setting. The monitoring revealed that nearly three-quarters of the children exhibiting academic problems also had "physical handicaps" that had to be addressed.

Student health was also closely monitored because, in the crowded housing conditions, the outbreak of a contagious disease could be catastrophic. The school employed its own nurse and cooperated closely with the public health programs administered by the county nurse, including examinations for tuberculosis and other communicable diseases and immunization for whooping cough, diphtheria and smallpox. In addition to working in the school and being on call round the clock, school nurses made home visits to provide pre- and post-natal care to both children and mothers, as well as basic information on nutrition, health and child development. In its two years of operation the Badger school served a total of 1,051 children. "There were no major accidents, no siege of contagious diseases, and no juvenile delinquency among those enrolled."[163]

As did the entire Badger Ordnance experience, the school also provided a glimpse into the American future. It gave a preview of family life as it would develop in the last half of the 20th century, with a growing number of young mothers separated from family, juggling work and children, minus a husband and father.

The Badger school delivered its services in a surprisingly small physical plant. The original building was an ell-shaped structure with six classrooms, two kindergarten rooms, three playrooms, three nurseries, a library, arts and crafts room and an "auditorium." The auditorium was fifty-two by thirty-two feet with a nine-foot-deep stage along one end. A large "playground," unpaved until after the end of the first school year, introduced the children to the famous Sauk Prairie mud. The concrete went down by the summer of 1944, in time for the students to enjoy the spray of water from the sprinklers installed with it. Across another, smaller open space to the north of the school stood another ell-shaped structure housing the village administrative offices, drugstore, grocery, post office and storage. An addition with nine classrooms, four nursery rooms, a heating plant and store rooms was completed in March of 1945. It turned the ell-shaped school into a fish hook, but, as usual at Badger, was not ready until after demand for it peaked and only a few months before the war ended.

The school building was a busy place, with the auditorium the site of most of the extracurricular action. On Sunday it served as a church, with Christian Fellowship, Lutheran and Catholic services following one after another and a classroom down the hall set aside for Sunday school. During the week movies flickered on a portable screen. The Boy Scouts, Women's Club, Junior Hi Club and the Teen Age Canteen for "teens" as old as twenty-two years met there. Both the Junior Hi and the "TAC" sponsored dances for their members and decorated the hall for Halloween, Christmas and New Year's. The volunteer fire department, Parent-Teachers Association, Badger Tenant Association, Plant Recreation Association and the Town of Sumpter Board met there. Other groups, ranging from gin rummy players from the barracks, to the organizations that ran horseshoe, kid's baseball and women's softball leagues, also used the auditorium. This list "gives a good idea as to how the school was made the center of things."[164]

The center did not hold long after peace returned. As the plant wound abruptly down, the first workers to be laid off were women, many of whom lived at Badger. The housing units emptied and many were hauled off the site to other locations before the first snowfall. The elementary school opened in September 1945 as a multi-graded village school minus the extended services that had made it special during the war. After getting a taste of the new and the experimental, Clarence Krumm could not go back to running a conventional village school. Instead he entered another new field as a provider of audio-visual educational programs and materials designed to enliven the curriculum at many another school.The idea of a school-centered community did not die. In the year 2000 the United States Department of Education authorized a program entitled the "Twenty-First Century Community Learning Center." Schools are to welcome students and community residents of all ages and to be open, if not round-the-clock, at least well beyond the scheduled class time. A total of 903 school districts, ten in Wisconsin, are participating and are setting up "learning centers." The guidelines for the program could have been written by Clarence Krumm, borrowed from the practices he and his staff instituted at the Badger School. Once again, the impact of Badger Ordnance Works reaches far beyond the boundaries of its place and time.

The Condemnation Settlement

In January 1945, as the workers at Badger Ordnance began their brief fling into rocket powder, another procedure generating much heat and light was drawing to a close. The United States Attorney for the Western District of Wisconsin filed for the final decree in the federal government's taking of 10,565 acres of farms, fields and homes for the plant. He authorized a final payment of $2,500 to Wisconsin Power and Light for 319 mostly submerged acres and "flowage rights" on Lake Wisconsin. Confident that the power company would not appeal, the attorney concluded that the process of condemnation and taking was complete. All the land that Badger occupied was now signed over to the United States and fully paid for by the taxpayers.[165]

The process had begun three years earlier, on January 10, 1942, when U.S. Attorney John J. Boyle, acting on orders from United States Attorney General Robert Jackson, filed for a condemnation decree in the federal court for the Western District of Wisconsin in Madison. Judge Patrick T. Stone, of Tomahawk and Wausau, who happened to be President Roosevelt's first appointee to the federal bench in 1933, would preside over the case from opening round to final bell. Stone was not a man to countenance lawyerly foot-dragging. He liked to see cases proceed with a minimum of delay. As one of his colleagues remembered, "There was no dilly-dallying in his court. He moved along." [166]

Prosecutor Boyle was a native of Lafayette County, an Irish-Catholic Democrat who ran unsuccessfully for Wisconsin Attorney General in 1928 and 1930 and for Congress in 1932. He was appointed federal attorney by President Roosevelt in 1935. He would argue the government's case, assisted by federal Department of Justice Special Attorneys Ann Ruth Kanevsky, a 1940 University of Wisconsin law school graduate, and Byron B. Conway, an experienced prosecutor from Wisconsin Rapids. The volume and nature of their work would be determined by the landowners. If the owners of all the land taken for Badger would accept the government's offer of payment based on the initial appraisals announced in December 1941, the federal attorneys would not have to do much more than see that the landowners signed the necessary papers and that orders for checks were dispatched to the United States treasury. If the landowners didn't accept and chose to take advantage of their right to due process through the condemnation procedure, the federal attorneys would be busy.

That many of the landowners found the government appraisals unsatisfactory was clear from the very start. Equally clear was the Attorney General's intent to neutralize the discontent by swiftly resorting to condemnation, which allowed the land to be occupied and Badger to be built without the delays that would accompany litigation initiated by the landowners. Condemnation was not necessarily easier or faster for the government, nor less costly than other means of acquisition. It granted certain rights and protections to the landowners and the burden of proof as well. The holdouts would have their day in court but they — not the government — must prove that their property was worth more than the Federal Land Bank appraisers had originally stated.

As a result of the federal action, the "farmers committee" that had organized to resist the taking of the land in November 1941 evolved into a condemnation committee with a membership of sixty-one landowners in January 1942. Garth Premo was the president, with John Litscher, William Eschenbach and Emma Kindschi also in leadership roles. Kindschi was the secretary of the committee and, as is often the case with secretaries, the brunt of the committee's work fell on her shoulders. A Dane County native and country school teacher, Emma Schollenberger married Glen Kindschi in 1915. They farmed the former John P. Kindschi place at the southwestern boundary of the plant. Through her husband's extensive family Emma Kindschi could call about half the population of Sumpter an in-law. On her own, she came to know the township by working as an enumerator for the census of 1940. She was active in the Farmers Union and the Progressive Party and had worked in Senator Robert La Follette's re-election campaign in 1940. In late 1941 and early 1942, she traveled to Washington, D.C., where she combined family visits to her daughter Bernice, who lived and worked there, with politicking visits to the office of Senator La Follette, where she lobbied for the powder plant landowners. Emma Kindschi was committed to the cause, even though, of five Kindschi brothers and their wives who farmed on the Sauk Prairie in 1941, only she and Glen did not lose home and land to Badger Ordnance.

Among the committee's first actions was to hire the Madison law firm of Schubring, Ryan, Petersen and Sutherland, with William Ryan as leading attorney. Although he had lived and practiced in Madison since his graduation from the University of Wisconsin law school, Ryan was no stranger to farming or the Sauk Prairie. He was born just across the river in the Dane County town of West Point in 1875 and grew up on the family farm. When the new Prairie du Sac bridge was built across the Wisconsin in 1920-21, Ryan handled the legal work and spoke at the dedication ceremonies. He would not have to stretch his imagination very far to see in his new clients the faces of his parents' neighbors.[167]

He started working with his first sixty-one clients in early February and would eventually represent seventy-five clients, some of whom were also represented by the firm of Hill, Miller and Hill of Baraboo, while three other clients hired the Baraboo firm of Hill and Dithmar. Landowners who had agreed to sell their land, the Evangelical church, the cemetery associations, two of the school districts, various state and county agencies, the power company and a few landowners went through the process minus legal counsel. Ryan took the lead in the case, with the Baraboo attorneys following his strategy. He immediately hired two Madison real estate appraisers, Harry Haley and Sam Mahlkuch, who agreed to value the real estate at the rate of $50 per farm. Since almost all of the property in dispute was in farms, the appraisers had to count and measure the buildings and find, "information with respect to comparative sales, production of the farm, character of the soil, and their value considered as a part of the farm." The appraisers also agreed to testify at the condemnation hearings for a $25 per diem. At Ryan's direction, the landowners committee, i.e., Emma Kindschi, also hired photographers to supply 8 x 10 prints of the properties and began gathering information, including farm ledgers and income tax returns, which were compiled into booklets describing and declaring a value for each parcel. The material was a little more detailed, but virtually identical to the data gathered by the Land Bank appraisers. The difference was in the interpretation, of course, which is why the landowners were in court. The appraisers had to move fast to complete the on-site work. The government was taking possession of the property on March 1, with sales and removal of buildings and equipment to be completed by April 1, and construction to follow. "We are having our own appraisal made," Ryan told reporters in early February, "so as to preserve the evidence."[168]

In May 1942, Judge Stone appointed Isaac Evans, Charles Hackett and Fred Holmes as condemnation commissioners. Evans was a prominent stock buyer and farmer from Spring Green. Hackett was president of the First National Bank of Baraboo. Holmes was a Madison attorney and a Progressive

Party leader who started working for "Old" Senator Bob La Follette as the first business manager of *La Follette's Magazine*. Known for his publications on Wisconsin history, Holmes was also a devout Catholic whose book *The Trappist Silence* was a Book of the Month Club Selection in 1941. Fred Holmes was an old-school, champion-of-the-little-man Progressive who would lend a more than sympathetic ear to the appeals of the Sauk farmers for a square deal from the government. Hearings began on June 23 at the county courthouse in Baraboo. The landowners' attorneys and appraisers presented evidence to prove that the Land Bank people had undervalued their property while the government pointed out why the federal appraisals were on the money. Holding hearings that ran over ten weeks in the summer of 1942, the commissioners occupied the high-windowed county courtroom and listened to testimony. They sat through the cross-examinations, collected the documents offered as evidence, viewed aerial photos, charts and sketches, and when they had heard and seen all that was presented, the farmer, the banker and the lawyer adjourned to assess it all, based on Judge Stone's instructions. "Just compensation for the taking of property for public use must be the full equivalent of the property taken. The owner is entitled to be placed in as good a position pecuniarily as he would have been if his property had not been taken. He is entitled to receive the fair market value of his property — the sum that would, in all probability, result from fair negotiations between an owner who is willing to sell, but not compelled to sell, and a purchaser who desires to buy, but is not compelled to buy...It is not proper to estimate the value of land...or its worth to the public for the purpose for which it is taken, because it is not what the government gains, but what the owner loses, which measures the damages or compensation."[169]

Although it initiated and pursued condemnation proceedings, the federal government also attempted to sidestep them. No sooner had the Attorney General condemned the property in January 1942, then the Corps of Engineers dispatched agents to bypass the Land Bank appraisals and make higher offers to the holdouts. To the Engineers, Paul Shimniok's farm was worth $17,000, instead of the $14,300 set by the Land Bank people. Elmer Keitel's would now fetch $7,000 instead of $6,000; Paul Kirner's $10,500 instead of $8,800; Swain Mather's $21,100 instead of $18,700. Nearly all of the new prices were at least ten to over twenty-five percent higher than the original offer. The Ralph Kindschi place was revalued at $18,400 instead of $14,000. The Garth and Nora Premo farm went up from $21,500 to $26,000, Fred Waffenschmidt's from $6,900 to $12,100, Irv and Margaret Luetscher's from $30,000 to $37,900. The smallest property made the biggest jump. George Jenewein's gas station tract shot up 100 percent, from $250 to $500. Many of these offers were identical or very close to the final amount paid to the landowners more than two years later. Had they been made in December 1941, they might have been accepted, thereby sparing the landowners and the government the cost and trouble of the condemnation proceedings. But in January and February 1942, even if the offers were not too little, they were too late. The government's brusque treatment of the farmers, coupled with the low Land Bank appraisals, had locked both sides into confrontation through condemnation. Relations had so soured that when two newly assigned Army officers pulled into a farm driveway on the plant site to turn their car around, they were run off by a farmer cradling a shotgun. Even though the government renewed its higher offers in March and May 1942, there were no takers among the holdouts.[170]

This is not to say that no real estate was sold. The government continued to acquire property by direct purchase, and by March 1943 had bought and paid for approximately 2,000 acres. While owners of smaller plots were most likely to sell as soon as they could, not all of them were quick to sign on the dotted line. Ralph and Bertha Thomas, who owned one-tenth of a acre on Weigand's Bay, held out all the way to a jury trial in the spring of 1944. The owners of about a dozen farms—among them John and Ida Ehret, Ray Mather, Ben Gasser, Carl and Emma Gibbs, Frank and Ida Ruesch, Charles Eschenbach, Charles Gattwinkel, Clara Schlag and Jennie Spoors—agreed to direct purchase of their property. Families were divided, with Gassers, Eschenbachs, Schlags, Mathers and other brothers, sisters and cousins appearing on both the list of early sellers and late holdouts. The owners of about 8,500 acres accepted the government's challenge to a legal battle.

As part of the battle, the government filed a series of intimidating "declarations of taking" which stipulated that it definitely was possessing the land and paying the amount of the original appraisal. Following the declarations, the United States transferred funds to an account supervised by Judge Stone to compensate landowners. Attorney Ryan then petitioned the court so that his clients could receive at least partial compensation and Judge Stone agreed. Between July and October 1942, landowners involved in the condemnation process received payments of approximately forty percent of the first appraisal, minus any back taxes or mortgages due. Amounts ranged from $2,400 to Elmer Keitel, $5,380 to Julius and Agnes Steidtman, $7,600 to William and Amelia Burkhardt and on to $10,000 for Christ and Anna Huber. This money, about one-third of what most of the properties eventually earned, and paid five to eight months after the land was vacated, was the first compensation that the holdouts received. The delay was inconvenient but not necessarily a crippling hardship, since the government's promise to pay could be used to secure a loan. Alvin and Evelyn Hoppe assigned $17,500 of their promised payment to clear their note with the Farmers and Citizens Bank of Sauk City and so spared themselves the burden of paying a mortgage on land they had lost. Garth and Nora Premo used the original appraisal of $21,500 to secure a mortgage on their new farm, and then some. They pledged the $26,000 the government offered in January 1942, and which they did not accept, to cover the remainder of a mortgage owed on their old farm and to purchase their new farm near Columbus. The Premos and the Bank of Columbus were confident that the government would pay at least the $26,000 and probably more. Other landowners — Emma Dahnke, Lyman and Essie Kindschi, Roy and Lila Welch — secured loans of several thousand dollars at the Bank of Prairie du Sac with the principal and interest covered by the government's promise to pay.[171]

At the end of August, while the commissioners deliberated, the landowners held their first reunion picnic at the Poynette Game Farm in Columbia County. While the picnic was a social gathering, they took the time to create an organization and elect a familiar slate of officers — Garth Premo, John Litscher, Emma Kindschi — to make the reunion an annual event. They could not help but discuss the progress of the condemnation case over the cole slaw and they must have reaffirmed their commitment to stand together. William Ryan filed another set of petitions for additional payments in September. Judge Stone approved and nearly all of the holdouts received checks for another forty percent of the original assessment in October or November. They would all be paid the full amount of the original appraisal by March 1943, one year after they moved out.

The shoe that everyone had been waiting to fall hit the floor on the last day of September when the Condemnation Commission presented its findings. After due deliberation, Evans, Hackett and Holmes raised the value of the properties taken for Badger by an average of twenty-three percent. The largest percentage increases went to William and Amelia Burkhardt, whose farm value rose from $19,000 to $26,916; to Louis and Eunice Frese, from $14,500 to $20,700; Lyman Hatz, from $7,200 to $10,413; Walter Kindschi, from $18,600 to $23,907; Carl Kunz, from $7,200 to $10,239; Theodore Weigand, from 10,000 to $17,229 for the farm and resort business and from $700 to $1,552 for a resort lot. George Jenewein's gas station was still the biggest winner, with the commissioners raising its value from $250 to $516, or about 210 percent. If these prices were to hold, the cost of buying the land for Badger would rise to $1.466 million, well above the one million dollars in the Army's initial budget. Practicing lawyerly discretion, and resisting the temptation to jump up and click his heels in the air, William Ryan solemnly pronounced the new values as "reasonably sound."[172]

To John J. Boyle, the new valuations were neither reasonable nor sound. He filed an appeal as soon as the paper could be processed. Ryan countered with a petition to "quash, vacate and set aside the notice of appeal." When Judge Stone disagreed, Ryan immediately requested separate jury trials for each of the eighty tracts he represented. Hill, Miller and Hill made an identical request for the eleven tracts they represented. November passed into December and the legal system rested for the holidays, but not the legal minds. On January 5, 1943, one week before the jury trials were set to convene, Ryan received a settlement proposal from John Boyle. The U.S Attorney, "in view of advices received from the Corps of Army Engineers," asked the landowners to accept the offers the Army had made one year earlier. He also offered to pay interest on the money owed the landowners and held by the court since March 1, 1942, which ranged from a few to several hundred dollars. Ryan was to reply before January 12 and any landowner who held out for a jury trial would not be included in the proposed settlement. The amounts offered by Boyle were smaller than the condemnation commission's valuations, but Ryan saw an opening by which he and his clients could avoid the delays and costs of jury trails, and rushed through it. He asked Judge Stone to postpone the jury trials from January 12 "to a date to be set later" and Stone obliged. Ryan than asked Boyle to give him until the end of the month to reply to the proposal. Since the government had made separate offers to each landowner, and "all of said landowners live a considerable distance from the city of Madison," Ryan needed time to consult with each one of them. Boyle agreed to wait until the end of the month. Still moving fast, Ryan kept the phone line to Sauk County humming as he summoned the holdouts to Madison for one-on-one meetings. Although they "were put to considerable trouble and expense," the landowners cooperated promptly. In these sessions, held in Ryan's office in the old Power and Light Building on the Capitol Square, attorney and client weighed their options. Today's bird in hand was bigger than yesterday's, but not as big as the bird in the bush of the jury trial might be. Which was the right one to bring home?[173]

Ryan reported the decisions of the landowners to Boyle as they were made and, by the end of the month, it was determined that "there were about twenty-five cases left to be tried." Eventually fifty-four owners of sixty-five tracts agreed to accept the government's proposal, although a few went right down to the wire. Ralph Kindschi accepted the government's offer in April 1944 while in the courthouse waiting for his trial to start. The clients of William Ryan who held out for jury trials were (along with the difference between the Army offer of January 1943 and the condemnation commission valuations): Harold Thoelke ($2,892), Walter Kindschi ($1,907), William Burkhardt ($6,016), Edward Mockler ($843), Jacob Gasser ($1,279), Charles Herr ($1,920), Ralph Thomas ($402), Charles and Henry Weirich ($3,047), Lyman Kindschi ($5,001), E.M. and C.C. Waffenschmidt ($6,829).

Except for the jury trials, the matter appeared to be settled when Norman M. Littell, Assistant Attorney General in charge of the Lands Division of the Department of Justice, informed Ryan "that the Secretary of War absolutely refused to approve the settlement proposals." Boyle then hustled to file affidavits stating that neither he nor the Corps of Engineers "represented that he possessed authority to contract on behalf of the United States." Apparently the federal agents closest to the case, John O'Brien of the Corps of Engineers in Chicago and Boyle in Madison, acted without clearing the proposed settlement with their superiors in Washington. Feathers were ruffled on a higher bureaucratic perch and the groundling birds were forced to admit they had overstepped the bounds of their coop, so to speak. The affidavit filed by Boyle, in which he admitted that neither he nor O'Brien had any authority to do what they had done, presumably smoothed tail feathers in Washington because the Secretary of War did, in fact, accept the settlement.

On February 22, 1943, William Ryan filed an affidavit stating that a majority of his seventy-five clients had accepted the government's offer. On February 23, 1943, he took the first case still to be settled, that of Harold Thoelke, to jury trial in the federal court in Madison. Ten trials, covering fifteen tracts, convened and closed between March 1943 and May 1944, with the juries delighting some of the landowners and disappointing others. Here are the jury awards, listed first, followed by the rejected settlement offer of January, 1943: Harold Thoelke ($23,570-23,300), Walter Kindschi ($20,500 - $22,000), William Burkhardt ($21,500 - $20,900), Edward Mockler ($10,000 - $10,500), Jacob Gasser ($12,500 - $12,800), Charles Herr ($12,600 - $12,500), Ralph Thomas ($2,900 - $1,900), Charles and Henry Weirich ($30,500 - $27,040), Lyman Kindschi ($22,500 - $19,800), E.M. and C.C. Waffenschmidt ($26,500 - $22,350). Any trial is a gamble and a jury trial more than most. "We were unfortunate in getting the type of jury we got in 1943," Ryan acknowledged. "The situation would have been much better had we gotten the 1944 jury in 1943." In 1944, Ralph Thomas, Lyman Kindschi, the Weirichs and the Waffenschmidts were winners. In 1943, Ed Mockler, Jacob Gasser and Walter Kindschi were losers. The remainder scored a small amount more than they would have had they not gone to trial, but were not spared the disappointment of heightened expectations denied. The 1943 jury awarded Charles Herr almost $2,000 less then the Condemnation Commission had valued their property, while William and Amelia Burkhardt received nearly $8,500 less.[174]

While the jury trials were underway Assistant Attorney General Littell began filing settlements, notices of direct purchase and petitions to conclude proceedings and make final payments. Since the original appraisal amounts had been paid in full, the final payments consisted of the difference between the December 1941 and the January 1943 offers plus interest. They ranged from $284.41 to George Jenewein, to $5,578 for Louis Frese, and $9,353.56 for Irv and Margaret Luetscher. Final payments were not made until the summer and fall of 1944 and, except for Wisconsin Power and Light, all but completed by the end of the year.

By its own count made at the end of the war, the federal government acquired twenty-five of the 138 tracts — 1,574 acres — by direct purchase for a total of $187,920.00. Another 111 tracts covering 9,004 acres went through the condemnation proceeding and were purchased either by settlement out of court or jury trial for a total of $1,037,932. Two other tracts totalling two acres and worth $1,775 were taken with no payment made. In all the government paid $1,227,607 for the 10,565 acres it took for Badger Ordnance.[175]

Over a period of two years and nine months, the clients of William Ryan received a total of $956,582 in settlement and jury awards, plus several thousand dollars more in interest payments. It was $181,512 more than they would have been paid had they accepted the first government appraisal. Ryan's bill totalled $19,053.37, with $13,258.62 for attorney's fees and the remainder going for appraisals, photographs and fees paid for testimony. Emma Kindschi was paid $300 for "expenses, etc." Records of the regular correspondence she maintained with Ryan are lost but her contributions to the successful outcome of the proceedings were outlined in a letter she received from the attorney after the final jury trial closed in May 1944. "May we express to you our sincere appreciation of the very helpful services you have rendered both to the property owners and to the attorneys from the time proceedings were started...Your committee performed all the functions that could be expected of such a committee...nevertheless, the work in such cases usually falls upon the active secretary and it did so in your case...it is not likely that you anticipated all of the work, troubles and burdens which you encountered during the course of the proceedings. I am sure the property owners...appreciate your very helpful services..." To obtain just compensation for their property, the landowners required an able lawyer in Madison and a good organizer at home.[176]

Whether the owners of the property taken for Badger Ordnance truly received just compensation has been debated for nearly six decades. Often, justice is not blind but, like beauty, found in the eye of the beholder. A minority of the landowners saw justice in the government's early offer and accepted it. A majority of the holdouts found it in the larger settlement offer of January 1943. The last diehards sought justice in a jury of their peers. They found it, but it was not beautiful for all of them. What cannot be denied is the fact that the landowners who held out succeeded in increasing the amount the government paid for their property because they resisted, organized and refused to do what they were told. At the time a war was being fought to preserve basic American values of freedom from government tyranny and of rule by law. For Japanese-Americans, those values were violated. But, as they gathered around Emma Kindschi's kitchen table, sat in the benches of the summer-hot Baraboo courtroom or waited for the outcome of jury deliberations in Madison, the Sauk Prairie landowners who held out showed all who could see why the war was worth fighting. Even during a great national crisis, the constitution was not suspended, at least for citizens unafraid and able to invoke its protections.

Decontamination by Destruction

Sizzling, hissing, wreathed in pale smoke, the flame burns steadily down the rope of fuse to the edge of the grains of smokeless powder spread in a thin layer over the packed cinder surface of the Badger Ordnance "burning grounds." With a flash of intense white heat, the powder ignites and a wave of flame and smoke races from one end of the powder pile to the other. Should it meet a discarded stick of rocket propellant, the low flame erupts into an explosion that puffs a cloud of white smoke up into the clear summer air.

In a minute or so it is over. The fire smolders out and the work crew returns from its safe observation point one-quarter mile away to hose down the ashes and cool the cinders for tomorrow's burn. So the flash and smoke will be less alarmingly conspicuous, and less likely to be mistaken for an accident, the burns take place only once every twenty-four hours and never at night. After about thirty months of work, the six-man burning-grounds crew has disposed of over two million pounds of spoiled powder plus truckload after truckload of nitroglycerin "slums." It was "probably more powder than was used during the entire Revolutionary War," announced the *Badger Ordnance News*, and it was just a start. Considered to be the "safest and least wasteful way of getting rid of the explosive material" during World War II, burning became the preferred means of decontaminating materials, buildings, equipment and soil after the war ended.[177]

In September 1945, a few weeks after Japan surrendered, the War Department ordered Badger to halt operations and move into standby status. The plant had already stopped making smokeless powder. Acid, E.C. powder and rocket propellant still in the pipeline were to be finished and shipped, then all production halted. Hercules Powder had already cut its workforce by two-thirds, but kept over 1,800 people on the payroll to complete the conversion to standby. Decontamination, by burning or other means, was minimal. Instead, workers flushed powder residue out of the piping and into the drains and open ditches, greased the machinery to prevent rusting, parked the donkey engines, tram cars and angel buggies out of the rain, stored the dipping forks, ballistic gauges and microscopes in tool cribs and lockers, toted up a building-by-building inventory and sealed shut the doorways. The schedule called for the entire plant to be buttoned up tight as a drum, but still intact and able to run, by February 1946.

The world was an uncertain place at the end of World War II. Japan was defeated and occupied, but civil war and anti-colonial conflicts continued in India, Korea, China, Indonesia and Vietnam. In Europe, tensions between the United States and the Soviet Union had been rising ever since the fall of Germany and the resulting cancellation of aid from the U.S. to its erstwhile ally by Lend-Lease Director Leo T. Crowley. Added to the uncertainties of the times were the uncertainties of the future. The United States had yet to adopt a military strategy for the postwar world and did not know how much of its immense armaments establishment it would retain.

The situation had not cleared in the world nor in the Pentagon, but in November 1945, the War Department rescinded its standby order and put all of Badger Ordnance in "excess status," more commonly known as "surplus." The Army would turn over the entire facility — buildings, equipment, land — to the Reconstruction Finance Corporation, "for disposal." No one had much of an idea how the property would be disposed of, or to whom. What the Ordnance Department did know was that, by it own regulations, turning over the plant to any civilian organization — public or private — would entail an extensive process of decontamination.

"So the crews moved back into buildings completely shut down," wrote Hercules manager Louis Reuter, "and started the destruction program of removing and burning contaminated wood floors, porches, building wall linings, etc..." Wood impregnated with propellant powder, rocket paste, ether, DNT, acid, or any of the other hazardous substances used at Badger, could not be sold or even given away for reuse. It had to be decontaminated, which meant it had to be burned. Instead of only once a day, burning-grounds fires now flared all day and into the autumn and winter nights.[178]

The open air burning of apparently reusable timbers, wall studs, decking and sheathing fueled another round of talk about wasteful practices at Badger. An observer peering across the Highway 12 fence at the main burning grounds, which were located about one-quarter mile east of Gate 10, might easily presume that the Badger workers lighting bonfires from truckloads of scarce building material had to be up to no good. "There was enough lumber in just one of these truckloads to have built any veteran a four-room home," wrote Badger Village resident Ruth Hazeltine Hart to the *Capital Times*. Hart wanted to know why "our tax money is allowed to be burned up in this manner — when it is a hard pull for many of us just to provide fuel and housing, the very thing that is being wasted..." The lumber could have built many a new home had it not been contaminated and had it not been 1945, when decontamination meant destruction by fire.[179]

Building components were burned, so were boiling tubs, poaching and blending tanks, water dry tubs and other wooden production equipment. Decontamination also required the dismantling of thousands of feet of steel, stainless steel and lead pipelines, then "flashing" the metal by heating it to a temperature of 200 degrees F. The contaminants were singed off and, if the flashing fire burned at the proper temperature, the pipe maintained its shape and could be reused. If warped by excessive heat during flashing, the pipe was ruined and had to be scrapped. Centrifugal wringers, blending barrels, mix tanks and other metal components were also cleaned by flashing and, if damaged by the heat, sold for scrap. "Conductive" rubber floor coverings were torn up and burned on open fires, so was lead flooring, with the melted lumps retrieved from the ashes and scrapped. Underground sewer lines, mostly concrete, were not disturbed, but soil from "ditches and outfall sewers" was excavated and processed, "through fire." Contaminants not destroyed by the flames or released into the air remained in the burned soil, which was then used for fill, or stayed on the surface of the burning grounds until rain washed them into the soil below.[180]

"There was much urging to hasten the disposal of property and to get the government out of the business of holding and controlling the explosives manufacturing facilities which had been so urgently needed only months before," wrote a less than objective Louis Reuter, as he watched much of the plant he had helped build, operate and maintain be dismantled and destroyed. "Possibly there was a race on to see whether the armed forces could be eliminated first, or whether the manufacturing facilities could be disposed of first, and the lands returned to agriculture before the boys were all sent home."[181]

There may not have been a race, but, as months passed and the peace held, the United States rushed to demobilize both the armed forces and the arms industry. By the end of 1945, a Scrap Board composed of Hercules and Ordnance personnel was surveying the buildings at Badger for decontamination by destruction. Early in 1946, the Ordnance Department named a Building Destruction Board composed of military and arms industry personnel and dispatched it to inspect seven ordnance plants and seven ordnance works. Badger was the last stop on the tour, which ranged from Pennsylvania to Nebraska, Mississippi to Minnesota. In its report on Badger, issued in April 1946, the Building Destruction Board lived up to its name. To decontaminate the boiling tub houses, which were among the largest wooden buildings at the plant, the Destruction Board said: "Dismantle building, keeping well wetted with water, then burn all woodwork. Ground contaminated around building should be stripped and put through burning operation." The Board extended its dismantle-and-burn recommendation to the beater houses, wringer houses, nitrocellulose pump houses and the poaching and blending houses, the last of which each held forty-eight pine agitating tubs. For many other buildings, including rest, glazing, final blend, DNT screening and can pack houses, as well as the "Dehy" dehydration press houses, the Destruction Board needed only one word: "Burn." For the entire E.C. powder area the recommendation was "destroy in place." For the rocket finishing area it was "destroy in place," and for the wooden structures of the nitroglycerin area as well.[182]

Had these recommendations been acted on, virtually all of Badger's wooden buildings—over 1,000 structures—in which smokeless powder and rocket propellant had been produced, would have been destroyed. Instead of a flat prairie horizon pierced by hundreds of buildings, the grounds of Badger

Ordnance would have become a clearing littered with vacant concrete pads, dead-end pipe lines and barren roadways connecting the structures still standing. Why it didn't happen is one of the least-examined yet most important questions of Badger's history. Of the fourteen Ordnance facilities visited by the Building Destruction Board in early 1946, ten were closed and returned to civilian use, while four survived. Pantex Ordnance in Texas and the Kentucky Ordnance Works became part of the atomic weapons program. Cornhusker Ordnance in Nebraska was retained as a shell-loading facility and Badger Ordnance Works continued to make propellant. Why, when so many World War II munitions facilities were destroyed, did Badger survive?

No quick answer is possible because Badger's survival in the postwar years was, like its birth prior to the war, the child of many parents. Military, political and corporate considerations on the national and international level outweighed local concerns. Indifference on the part of war-weary public agencies swamped with surplus property, and a private sector more interested in tooling up for peacetime prosperity than maintaining the war machine, also helped keep the wrecking crews away. Badger's location in a rural area, an asset in war, was a distinct liability in peacetime, as was its sheer magnitude. If any industrial enterprise large enough to make efficient use of the plant and willing to locate in rural Wisconsin existed in the immediate postwar years, it made no claim on Badger. Policy differences between civilian and military departments, inconsistent directives and management from Washington, as well as plain bureaucratic inertia, also played a role. Eventually, time ran out. Civilian interest in demolition and conversion waned while military interest in preservation waxed. International tensions increased, crisis turned to conflict. Military action ensued and Badger was reactivated. For those who viewed the plant as the despoiler of the prairie, a great opportunity was lost in late 1945 and 1946.

For about one year, from November 1945 to October 1946, all of Badger Ordnance was in surplus status and earmarked for disposal. Decontamination through destruction was underway and property was given away or sold. About 3,000 acres of land and other real estate both in and outside the fences, including the Kingston Cemetery and farmsteads not removed in 1942, went on sale. The Wisconsin Conservation Department purchased 768 acres adjacent to Devils Lake State Park for $6,000, and four farmers who had lost their land in 1942 bought it back in 1945. Substantial quantities of tools and equipment — desks, drafting materials, lab benches, lockers, microscopes — were also distributed free of charge to high schools, vocational institutions and the University of Wisconsin. By August 1946, the value of property shipped from or destroyed in decontamination at Badger totalled $7 million. Although not destroyed in 1946, Badger Ordnance was stripped of vital equipment and materials that had to be replaced when the plant was reactivated in 1951.[183]

Salvagers purchased equipment and machinery, "cannibalized" the parts they wanted and left behind what they did not. An ammonia oxidation unit, chemical storage tanks, electric-powered compressors and other equipment that could be readily adapted for civilian use went on the block and out the gate. The city of New Ulm, Minnesota — best known as the site of an uprising by Dakota Indians that prompted an "Indian Scare" as far away as Milwaukee in 1862 — would have been the biggest cannibal of all had its bid on the No. 1 Unit of the Badger power plant been accepted by the War Assets Administration. It was not. Likewise, none of the recommendations of the Building Destruction Board were carried out. Hercules fulfilled its contracts and pulled out in the summer of 1946 but the War Assets Administration, in whose custody Badger had been placed, could not find another general contractor willing to manage the demolition and decontamination work. Even if a contractor had been hired, the destruction would have been delayed because the machinery and equipment inside many of the buildings had yet to be sold. The mixers, presses and tubs had to be stored somewhere, and the simplest way to store them was to leave them in place, which meant the buildings housing them could not be demolished. A Catch-22 situation evolved: contaminated buildings could not be demolished because they protected machinery earmarked for sale as surplus, yet potential purchasers could not be allowed to view the machinery because it was in contaminated buildings which they could not enter.

Another delay occurred when the state of Wisconsin expressed its desire to acquire the entire property. According to the hierarchy of disposal established by Washington, states ranked second behind federal agencies. If no one in Washington wanted Badger, Wisconsin stood an excellent chance of gaining title to it. In January 1946, Governor Walter Goodland, with support from Senators Robert La Follette and Alexander Wiley, informed Washington that Wisconsin would exercise its claim to federal property, i.e., Badger Ordnance, on behalf of the University of Wisconsin, which had announced plans to open a branch campus on the site. The university later reduced its plans and said that it would use Badger only for housing, but Wisconsin did not relinquish its claim. The Department of Public Welfare had already informed Goodland that it wanted the property. Accordingly, in May 1946, the Governor requested a "freezing" on all the property at Badger not already used by the university until the state could draw up a "detailed application for the transfer of said property from the U.S. Government to the State of Wisconsin."[184]

While Goodland's request worked its way through the bureaus in Madison and Washington, the Ordnance Department further complicated the process by initiating a strategic evaluation of its armament facilities. A conspiracy theorist with an anti-military bias could probably find a link between Wisconsin's claim and a revival in the Army's interest in Badger. That is, faced with the real possibility of losing the place, the Army declared it had to have it. Be that as it may, at about the same time Governor Goodland asked for a freeze on Badger, the Ordnance Department dispatched a "Field Director" to inspect and report on the premises.

Two months later, in July 1946, Colonel R.R. Klanderman reported that the N-Line rocket propellant area was "in excellent condition of preservation," with all its equipment "of the very latest and most modern design available." Little more than one year old, the lightly used rocket equipment was practically fresh out of the factory packing crate. The Ordnance Department inspector also warned that since the equipment was "highly specialized" and of little use to civilian industry, it was "ending up on WAA scrap declaration." These swords could be beaten into plowshares only at great loss to the taxpayers. The pre-mix tanks, carpet rollers and extrusion presses that cost hundreds of thousands of dollars a few months earlier were now worth no more than the iron, steel and other material that could be scrapped out of them. With his terse, telegraphic style, Klanderman was far from eloquent, but his message was clear. Keeping Badger's rocket line in "excess" status would mean destroying a brand new manufacturing plant that had been carefully mothballed in excellent condition.[185]

On the other hand, the United States was in the midst of the largest disarmament program in history. Billions of dollars worth of military and industrial equipment in equally good condition as the N-Line at Badger was headed for the scrap heap. However, tools and materials related to rocketry were as different from the battleships being beached at naval yards and the bombers being parked in the desert as tomorrow is from yesterday. Like jet airplanes and atom bombs, rockets represented the future of warfare. As Colonel Klanderman stated, "Changes in warfare and potential increased emphasis on future use of rockets, jet-propelled missiles and guided missiles emphasizes the practicability of retaining all the rocket facilities which were built late in the war..." Therefore, "consideration for retaining Badger rocket area is highly desirable."[186]

Unintentionally, the laconic Colonel would prove to be the most important visitor to the Sauk Prairie since Hercules engineer H.N. Wendle sketched out the boundaries of the powder plant in November 1941. By pointing out the excellent condition in which he found Badger's rocket area, then referring to rocketry's future in modern warfare, and concluding with a recommendation that the rocket area be pulled out of surplus, Klanderman chocked the wheel of destruction threatening the plant and respun it in the direction of preservation. Never again would Badger Ordnance come so close to a final shutdown and removal from the Sauk Prairie as it was in the spring of 1946. When Colonel Klanderman's report arrived on the desks of his superiors, Army Ordnance received the evidence and the justification to snatch the plant from destruction.

In October 1946, Ordnance drew up the boundaries of "an integrated rocket powder plant" at Badger and reclassified it from surplus to standby. The new plant within the plant consisted of the nitroglycerin, east roll, west roll, pre-dry, test, paste and magazine areas of the World War II N-Line, along with the D-Line nitrocotton, plus most of the acid area, Powerhouses #1 and #2, the river pumping station, filtration plant, reservoirs and sewage treatment plant. It was a cut-down, economy model plant but still expected to produce the same 5.5 million pounds of propellant per month as it had in World War II. Giveaways, sales and decontamination by destruction halted in the standby area. Machinery and equipment that had not been removed or cannibalized were prepared for storage once again and the buildings sealed shut.

The main administration buildings, hospital, automotive and locomotive shops, guard houses, change houses, garages, warehouses and — surprisingly — the Oleum plant, were not included in the "integrated rocket powder plant." The Oleum plant did not return to standby until March 1947, and then only after a special request from the plant commander. At the same time, the one-story administration building, the fire house, central heating plant and various shops and garages returned to standby, but not the two-story "main" administration building or the main laboratory. The new and unused F nitrocotton line, the B, C and E nitrocellulose lines and the smokeless powder finishing areas also remained in surplus and subject to cannibalizing, scrapping, destruction or transfer to new owners, should any appear. The state of Wisconsin would not be one of them. Since it was made by a federal agency, Ordnance's claim to the new rocket plant took precedent over Wisconsin's. About one-third of Badger's 7,500 acres were still in surplus and available, but Wisconsin wanted all of Badger or none of it.

Between 1945 and 1951, responsibility for the surplus areas at Badger passed through at least four civilian "disposal" agencies: the Reconstruction Finance Corporation, War Assets Administration, Federal Works Administration and General Services Administration. The Corps of Engineers supervised the minimal amount of maintenance work authorized while the Federal Land Bank conducted sales of farm land. The mission of the disposal agencies was to dismantle and sell as much of Badger as quickly as they could, but they "were slow beyond all understanding in proceeding with the task..."[187]

The mere shifting of responsibility from one agency to the other consumed months of time as new people pushed new paperwork in and out of new agency pigeonholes. Their work was not aided by the fact that, just as no general contractor wanted to take on the destruction of the surplus areas, no organization—except the Wisconsin Department of Public Welfare—had expressed real interest in taking over the entire property. The preference of everyone involved was to work with a single entity, public or private, to take over the entire property and either wreck or maintain it. Salvagers were willing to remove bits and pieces of the plant, but no one wanted the whole works either before or after the rocket area returned to standby. At a closed bid sale held in March 1947, virtually all the surplus machinery and buildings at Badger went on sale. While much of the machinery sold for scrap, no takers could be found for the buildings. "It gradually became apparent that disposal action would achieve nothing except a great accumulation of scrap material..." So Badger sat, with wreckers and scavengers knocking off what they could sell and leaving the rest of the surplus property to rot in the rain.[188]

The Corps of Engineers and Army Ordnance controlled the area that had been returned to standby and the soldiers on site consistently urged the War Department to preserve more of the plant by removing it from surplus. They viewed with genuine alarm offers, like that of the city of New Ulm, to cannibalize the plant. By the end of 1947, the Army people had succeeded in persuading the War Assets Administration to return the two-story administration building, staff village, the railroads and E nitrocotton line to standby. The rest of the smokeless powder production area could not come out of surplus because, in the eyes of one inspector, "Decontamination was performed to a greater degree than would have otherwise been done." To restore the smokeless area to standby status would mean replacing equipment and materials that had only recently been removed and sold for scrap, as well as repairing and maintaining buildings. It would cost money that neither the civilians nor the soldiers had in their budgets. Instead, Washington issued a confusing and seemingly contradictory order telling Ordnance to lease portions of the existing standby area for commercial or industrial use. The order was followed and space advertised for lease even though the prospect of finding renters for the rocket area "was not seriously entertained locally."[189]

Not entirely. One local person who seriously entertained the idea of leasing space at Badger was Jule Marachowsky. The leading figure in a family-owned retail furniture and appliance chain with a "Big Store" in Baraboo, Portage, Prairie du Sac, Mauston and other central Wisconsin communities, Marachowsky organized the Portage Warehousing Company in 1949. The company rented surplus space at Badger Ordnance for $11,500, plus $5,100 in repairs, from the Reconstruction Finance Corporation. The official in charge of the RFC's office in Minneapolis, and with whom Marachowsky dealt, was Mervyn Rowlands, a native of Cambria in Columbia County, who had served in the Wisconsin Assembly and Senate in the 1920s and '30s. Rowlands had developed a cozy relationship with an employee of the federal Commodity Credit Corporation (CCC) named Guy Cravens. The CCC of the 1940s, not to be confused with the Civilian Conservation Corps of the 1930s, managed and warehoused surplus farm commodities, such as powdered milk and cheese. Cravens reportedly kept Rowlands posted on the CCC's storage needs in return for gifts of cash, clothing, luggage and a television set—new, rare, costly and coveted in 1949, and also in stock at Marachowsky's Big Store.

The CCC had a quantity of dried milk and linseed oil in need of a warehouse and, with Rowlands' assistance, agreed to sublet space at Badger Ordnance from the Portage Warehousing Company. Here was one federal agency renting federally owned real estate from another, but paying rent to a private firm. And a whopping rent it was: for the fifteen-month period from November 1949 to January 1951 the CCC paid Portage Warehousing $215,000, a return of roughly 1,500 percent on Marachowsky's original investment. Not a bad deal, even with the forty percent of the $215,000 paid to Rowlands. The scam might have continued had Badger not been reactivated. It remained undetected until May 1951, when Rowlands was caught and fired from the RFC. A year later he was indicted, along with Marachowsky and Cravens, and charged with conspiracy and bribery before federal District Judge Patrick Stone. Another year passed in delays and postponements before the trial began and during which Stone removed himself from the case. The proceedings were moved to the federal court in Quincy, Illinois. The prosecutor was Frank Nikolay, a junior attorney in the federal prosecutor's Madison office and future state legislator. Nikolay targeted Cravens as the defendant most likely to be convicted and followed the time-honored strategy of turning the accused against one another. He offered to drop all but one of the eleven charges filed against Marachowsky and Rowlands if they agreed to testify against Cravens. They would have to plead guilty to a charge of attempting to influence Cravens, but would not be severely punished for it. Cravens fought tooth and nail, insisting on a jury trial during which he declared that not only had he paid good money for all the goods he had received from Marachowsky, he also had the receipts to prove it. The possibility that Marachowsky's receipts were not all they purported to be must not have been considered since, after six hours of deliberation, the jury found Cravens innocent of all charges. For pleading guilty, Marachowsky and Rowlands were fined $2,500 each and sentenced to two years in jail, suspended. In sentencing the pair, Judge J. Leroy Adair commented that, "It is rather difficult for me to see how these two men could be guilty and the other one not."[190]

Despite such distractions, throughout 1947, 1948 and into 1949 Badger sat in limbo, with part of the plant in surplus, part on standby. Finally, as strategic planning evolved, along with a "realistic review of international relations," the War Department determined that, in order to bring the rocket area out of standby and into production in time to supply American soldiers in combat, all of Badger Ordnance had to be maintained on a standby basis. A realistic recognition of the general lack of interest in the surplus property left at Badger must also have occurred, as well as the perception that it was deteriorating through neglect. Accordingly, those portions of Badger still in surplus, hitherto unwanted and unspoken for, rode into standby status on the back of the rocket line. In the summer of 1949, for the first time since the war ended, maintenance work was authorized on buildings in the surplus area. New asbestos shingles were laid on all the roofs in the nitrocellulose and green powder areas and the exterior walls beneath them were painted. Only the buildings of the single-base finishing areas were left as built in 1942.

In June 1950, troops from the People's Republic of (North) Korea invaded the Republic of (South) Korea. By the end of the year, after Hercules Powder said it was not interested in resuming operations, representatives of Olin Industries were negotiating the terms of an operating contract for the rocket area. In the first month of 1951, Badger's commanding officer, Lt. Colonel Oliver D. Williams, received a phone call ordering him to estimate the work necessary to reactivate the C, D and E smokeless powder lines and Colonel John Zierdt of Army Ordnance made the official announcement to the press that the plant would be rehabilitated and reopened. Badger Ordnance's first period of surplus, standby, decontamination and destruction had ended.

What to do with Badger Ordnance?

While the government officials working inside the fence line struggled to determine the future of Badger Ordnance in the years after World War II, people on the outside also looked for ways to use the site in peacetime. A few irreconcilables argued that the property should be returned to the farmers from whom it had been taken, or at least be returned to agriculture. This worthy sentiment overlooked the fact that a substantial industrial plant, with "water and sewer systems as large as Toledo, Ohio," now occupied the ground where the farms once stood. Converting the site to cropland was a very expensive proposition. It might be morally justified, but returning the land to farms could not be supported politically nor economically. Indeed, it could have been argued, but was not, that restoring the land to its pre-settlement grassland status and returning it to the custody of the Ho Chunk, was equally justifiable from the moral standpoint and equally infeasible from the political and the economic.

Discussion of postwar uses for the industrial plant at Badger began in 1943, shortly after the plant had begun making smokeless powder and before the rocket addition was announced. The Madison and Wisconsin Foundation, a consortium of business and university interests in the capital city, had assembled a committee of university and state agency people to study the situation. They helped produce a state planning commission report that examined Badger's impact on Sauk County but made few proposals for the future. In the spring of 1944, a University of Wisconsin agriculture school committee conducted its own study on future use, also without knowledge of the impact of the rocket plant. In July 1944, the ag school committee proposed the creation of "a substantial-sized modern industrial city" centered on providing agricultural services to a regional market. A central fertilizer plant, perhaps using Badger's chemical works, would serve the entire southern half of Wisconsin, so would a plant to process soybean and linseed oil. A centralized commercial feed-mixing plant and a processing plant handling milk from farms within a twenty-five mile radius would also be on site. Vegetables, dairy products and meat would be dried, frozen, cured, packed and otherwise processed. Non-agricultural industries would include paint and plastic factories, plus a sulfuric acid works. The power station would be leased to a private utility which would install steam turbines and generate electricity for the general market. The federal government would retain ownership of the property and finance start-up costs for the businesses, while charging only "nominal rentals." Within five to ten years, as predicted by Ag School Dean Fred Froker, a sufficient number of commercial establishments could be attracted to cover all or a major part of the government's operating costs." Presented to the press one month after D-Day, the ag school proposal, which bears a resemblance to the public-private "business incubators" of the 1990s, was a dead letter by V-J Day.[191]

At that time, Badger was employing 3,199 Sauk County residents, or about nine percent of the 33,700 people who lived in the county. Of Baraboo's 6,415 people, 910, or about fourteen percent of the city's entire population, worked at Badger. Comparable figures can be produced for other communities in the vicinity. Preserving as many of those jobs as possible was the logical concern of the Sauk County Federation of Labor, which had itself come into being only because of the plant. In January 1946, Abner Lee, president of the Machinists Union Local at Baraboo, and Federation Secretary August Moulton proposed that Badger be converted to a vertically integrated iron/steel/auto production plant under the management of the Kaiser-Frazer Company. Kaiser-Fraser had acquired a reputation for industrial wizardry as a builder of Liberty cargo ships during the war and the labor leaders thought the company "could turn out automobiles by the mass production method" at Badger. Their vision encompassed more than just an auto plant. They also wanted to open iron mines in the Baraboo Hills to feed a smelter at Badger. The iron would then be mixed with quartzite, also dug out of the Baraboo Hills, to make steel in a new mill, also at Badger. The quartzite could also be used to make firebrick to line the blast furnaces of the steel mill. It was all conveniently located uphill of the plant site. As Moulton pointed out, "There is what is known as the Cahoon Iron Mine near Badger Ordnance Works. All that would be necessary would be to shove the Quartzite Rock and Iron Ore down the Sauk Hill to said Steel Mill." No one can say the union men did not think large and no one would call them tree-huggers. In order to keep jobs in Sauk County, they were ready to scrape down the south face of the Baraboo Hills and construct a miniature iron smelter, steel mill and automobile factory on the Sauk Prairie.[192]

Also thinking large was Doctor William Lorenz, director of the Wisconsin Psychiatric Institute of the University of Wisconsin. He proposed a horizontally integrated, state-owned and -managed educational/social service/healthcare facility for the Badger site. As he wrote in a letter to Governor Goodland, "I am thinking in terms of a centralized community where a variety of physical and mental infirmities ranging from the child level to the adult could be properly treated, rehabilitated, or benefitted to the maximum that present day scientific methods permit." Badger would become the home of an institution for the care of children afflicted with polio, epilepsy, developmental disabilities, behavior disorders and mental illness. It would house developmentally disabled adults, the chronically insane and the elderly with serious physical infirmities. For victims of these disorders whose families could not provide the care they needed, institutionalization in either state or county facilities had been the standard treatment since the 1870s. Even the best of these facilities suffered from overcrowding and understaffing. Scandalous conditions at state facilities for the developmentally disabled at Chippewa Falls and the mentally ill at Mendota had been well-publicized during the war and amply demonstrated that needs were not being met. By bringing all the disabled into one big tent, so to speak, Lorenz hoped they would benefit from large-scale, up-to-date efficiency provided by centralized supply and nursing services, plus a 500-bed hospital. Like other institutions of its type and time, the Badger facility would also be a working farm, with the residents supplying the labor. Fresh air, sunshine and physical activity in the form of farm labor were prescribed therapies for just about any ailment, while the quality and quantity of the farm land at Badger made it a prime site for a state farm.[193]

Goodland passed Lorenz's proposal on to those he dubbed the "experts" at the Department of Public Welfare. In June 1946, the state Bureau of Property Procurement made formal application to the War Assets Administration to acquire Badger Ordnance. Many additions had been made to Lorenz's original proposal. The state now wanted Badger for university housing, a school for the blind, a health education youth camp, plus agricultural and conservation research. The healthcare facility was expanded to include care for children with rheumatic heart disease and people of all ages suffering from tuberculosis or alcoholism. There would also be a boarding school for "delinquent" boys and girls, a "Boys Town" program for orphans and runaways, and a minimum-security prison farm. Wisconsin would also hold the federal government to its own regulations concerning decontamination, which would mean acting on the recommendations of the Building Destruction Board. The state went even further by declaring that the federal government was obliged to remove not only the wooden superstructure of the buildings, but also the concrete footings, pillars and slabs beneath them so the land could again be farmed. Five weeks after Wisconsin posted its claim to Badger, Colonel Klanderman sent his report to the Commander in Chief of Ordnance. Three months later, the rocket area was removed from surplus and Wisconsin's plan for a super-human services facility passed into the realm of the what-might-have-been.[194]

The realm of reality was occupied by the University of Wisconsin. Anticipating that demobilization of the armed forces would mean increased enrollments, university administrators first visited Badger in the fall of 1945. Housing for the influx of students was their first concern, but they also hoped to use Badger's laboratories, the hospital, and several recreational and administration buildings for introductory-level classes and lab work. The university went so far as to name Professor Carl Bogholt as associate dean of the new campus. Obstacles arose both at Badger and in Madison. Citing security interests, the Army balked at allowing students, even though they were all veterans of the armed services, into the restricted hospital and lab areas. On its side, the university discovered that it could not offer all the introductory classes students living at Badger would need without moving a substantial part of the Madison campus to the Sauk Prairie. Even if all the introductory courses could be held at Badger, students would still have to commute to Madison to complete their degree work. The university concluded that if Badger students had to come to Madison for some of their classes, they might as well come for all of them. Instead of asking for more buildings at Badger, the university people requested surplus buses. When the second semester of the 1945-46 academic year began in January 1946, the first of Badger Village's student residents were already setting up housekeeping in their new homes.

"A Little Shangri-La"

In the fall of 1945, the powder plant village of Badger was a still shadow of its former bustling self. The barracks and apartments of North Badger were vacant, the new and unused community center stood empty and the school which had cared for 400 children only months earlier now held class for a few dozen students from nearby farms. A few tenants hung on at South

Badger, some still on the job at the plant, others just waiting for time and tide to carry them to a "normal" life somewhere else. It was a respite, a pause in the flow that would soon steer the village on its next course.

The still waters began to stir in November 1945, when the War Department shifted the village into surplus status. Housing was in critically short supply throughout the country and, as soon as the surplus declaration was publicized, the Federal Public Housing Administration received a flood of petitions from colleges, universities, cities and villages throughout the Midwest asking for the vacant buildings at Badger. The FPHA met a few of these requests and the first months of winter saw work crews disconnecting plumbing and electric lines, loosening the bolts holding sill plates to slabs and lifting "demountable" housing units onto moving trucks and semitrailers.

First to go were 100 pre-fab houses from the final addition to South Badger, which had been occupied for only a few weeks, if at all, in the summer of 1945. They were headed for Eau Claire to house workers at the Gillette Tire factory, which was running round-the-clock to meet the savage postwar demand for automobile tires. Close on the mudflaps of the trucks bound for Eau Claire was a fleet carrying 100 South Badger units to Madison. They would join the ninety-some trailers already removed from the Prairie du Sac trailer camp and parked outside Camp Randall Stadium. The Randall Park trailer camp, as it was named by war veteran students living there, opened with ninety-one units from Prairie du Sac when the 1945-46 class year started in September. Demand for student housing remained high so the university applied for an additional 100 units from Badger Village, which were moved to Randall Park early in 1946. Later that year, the university also took possession of the new wing of the Badger School. Completed in March 1945, the nine-classroom, four-nursery-room addition was trucked to Madison to help relieve the shortage of classroom space.

The University of Chicago trucked 125 housing units to its campus, Iowa State University took fifty more to Ames and the city of Appleton garnered 100 more to ease its housing crunch. Houghton in Michigan, Zumbrata in Minnesota, Lancaster, Neillsville, Marshfield, Milwaukee and Viroqua in Wisconsin also requested housing, but the record does not confirm that they received it. At least 375 housing units — virtually all of the single family "houses" erected at Badger in the spring of 1945 — were moved out of the village in late 1945 and early 1946. Considering the pinched national situation, all of the housing at Badger probably would have been relocated had the University of Wisconsin not expressed its desire to use it in place.

Expecting enrollments to soar with demobilized soldiers utilizing their G.I. Bill educational aids, Wisconsin administrators were prompt to visit Badger only a few weeks after the plant went into surplus. After abandoning their plans to establish a satellite campus at the plant, they concentrated on acquiring housing for married students and transportation to carry them to

Madison. Badger was particularly desirable because it had homes for married students with children, the category of student hardest to house in Madison. Badger also had a school, community center, post office and shopping facilities. All it needed was people and the university could easily supply them. Initially, Wisconsin asked to lease the vacant barracks at North Badger, along with the community building, mess hall and heating plant, plus Staff Village and the former Kindschi and Spoors farmhouses. In short time, the university also applied for and obtained a lease for South Badger.

Housing was easier to find than transportation, especially in early 1946, when automakers had yet to convert to civilian production and government-owned vehicles were still locked in administrative limbo. Buses, even old and tired vehicles like those in the former Hercules fleet, were hard to come by. The university was eventually able to strong-arm vehicles out of government hands by stating that housing was so scarce in Madison that it would have to stop admitting veterans. It was a politically sensitive, potentially explosive declaration and one that federal officials could not ignore. Wisconsin then proposed two alternatives. It could admit and house veterans in the Badger units either by moving the buildings to Madison — at federal expense — or by acquiring buses to move the students to and from Badger. Accordingly, with aid from federal agencies, the university acquired thirty-five well-used vehicles from Wisconsin and other states. They all had tired brakes, worn-out springs and odometers with six-figure readings but nonetheless went into service hauling students to campus from Badger and the university's other veterans housing project at Truax Field. For the first time in its history the University of Wisconsin became a commuter campus.

The university's lease did not start until March 1, 1946, but a small number of students had already discovered and moved into South Badger. The first housing to be occupied under university aegis was at Staff Village, where the former managers' homes were sublet to university faculty. These humble, but coveted, two-and three-bedroom residences were too precious for students and were reserved for profs and administrators. In December 1946, after completing a minimum of cleanup and repair, the university opened the South Badger apartments to student couples. Last came the barracks of North Badger, which required extensive remodeling from single to double occupancy units. Using a special State Emergency Board appropriation, the university spent $50,000 to repair footings, floors and roofs, and to rewire the entire complex. Rents were set at a just-affordable level for students living on a G.I. stipend and ranged from $18 a month for a converted two-room barracks unit with a shared bath at North Badger, to $26.50 for a three-room "apartment" at South Badger. "You couldn't touch that in Madison," said Lawrence Halle, the housing administrator who managed Badger for the university.[195]

About half the North Badger apartments were ready for tenants in March

1947, with the rest opening up a few months later. Not until the fall-1947 term was all of Badger ready for its full complement of 700 student families. At its peak in the 1947-48 academic year, the estimated population of the village was 2,700. About half were children and so many of those children were infants that if the villagers wanted a flag, they could have used a baby diaper flapping on a clothesline. One resident statistician calculated that in its heyday, Badger Village had the highest birthrate in the country or, as one wife stated, "It was rabbitville." By 1950, the population of rabbitville had declined to about 550 families with approximately 1,000 children, most of whom were preschoolers. In the final school year of the student period (1951-52) , the Badger School had 101 children enrolled in four- and five-year-old kindergarten and another 100 in all the other eight grades combined.[196]

In the course of its six-year run as university housing, Badger Village was home to several thousand students and spouses, most of whom were articulate and gregarious people who related a storehouse of memories to willing listeners. It is one of the idiosyncrasies of the Badger story that the experience of life in the powder plant housing tract has been more thoroughly recorded not by the powdermakers who lived there in wartime, but by the student families who occupied it during an interval of peace. There were differences of class, motivation and aspirations between the two groups. For the working people in residence during World War II, Badger was a means to keep the wolf away from the door and to take part in the war effort. It represented pay and patriotism. For the student families, Badger was a colorful and brief stop on their way up in the world. For the working people of the Korean period a job at the plant and an apartment in the village was an end in itself. For the students it was just a beginning. Judging by the number of Badger Village alumni who distinguished themselves in their professions in succeeding years, it certainly was a good one.[197]

Because of the accepted gender roles of the times, roles reinforced by G.I. scholarships and stipends reserved for male veterans only, life at Badger was distinctly different for men and women. While all of the men were students, hardly any of the women attended university classes. The men commuted to Madison five days a week and, depending on schedules and course load, many of them spent more weekday hours away from Badger than at the village. When they were at home, the men were likely to hole up with their books in the bedroom or at the kitchen table. With the men absent or withdrawn into studies, Badger became even more a village of women and children than it was during World War II, when women and children were in the majority but male shift workers were present at all hours of the day and night, and the women themselves left the village to work at the plant. The student wives stayed home, some not even leaving Badger on weekends, their lives centered on children born or about to be. It was almost Oriental in its gender segregation, but similar to life in the 1950s commuter suburbs where almost all the couples eventually lived, and where husbands left for work everyday while wives stayed home with the kids.

For the men the typical day, "began facing that thirty-five-mile trip into campus..." August Lemberger, who lived at North Badger from the summer of 1947 to summer, 1950 was one of them. "...for a 7:45 class we would have to catch a bus from North Badger at 6:10, at South Badger 6:20...The fellows that had problems were the six-footers...remember these were school buses for school kids...you sat crunched in a seat with your knees against the seat in front...

"In winter the buses were very cold...there was a mad rush to get a seat near heat...your legs from knees down were frozen...first choice in summer was a seat between the wheels, the most comfortable...It was very difficult to try to read or study on the buses...very difficult...pretty much you went into a cocoon [but] a lot of guys played bridge for an hour." Highway 12 presented its own challenge to the underpowered, overloaded buses at the steep, winding slope of Springfield Hill in Dane County. "In winter, you never knew if you were going to make it. If it was snowing, the bus detoured to Highway 78 and 14 to Middleton," thereby making the tedious ride significantly longer. Four buses departed Badger in the morning and six returned, with the earliest return trip at 11:00 AM and the latest at 10:00 PM, Monday to Friday. Buses also ran on a modified schedule to and from Madison on Saturdays and to Baraboo for shopping on Wednesday and Saturday. The fare was ten cents one-way for students and staff. The system was managed by student Bruce L. Solie, who, with his drivers, would get out of bed on cold prairie nights and run the engines — sometimes twice — so they would start in the morning. One bus was equipped with a padded bumper so it could push start other vehicles whose batteries had died.[198]

For the men at Badger, life revolved around getting to classes, attending classes and preparing for classes. "The husbands were under pressure to complete academic work," said Halle. The veterans tended to be older and more experienced than other undergraduates. They had seen more of the world—some of it in very unpleasant circumstances — and were impatient to get started in their professional lives. The sentiment voiced by veteran William J. Schereck was generally shared, "For me, it meant I could get the education I couldn't get otherwise." It was the promise the United States made to those who served in the war and that Badger Village helped the nation keep.[199]

Life was different for the women. "We came from the country," said Phyllis Young, who arrived at Badger with her husband Edwin and two children in the fall of 1947. "It seemed like quite a village to me...a sort of strange domesticated army base...like a house of cards...impermanent, very fragile...not slated long for this world."[200]

For Marjorie Johnson, "It was kind of fun, almost like camping...I had a sense of adventure...always challenges...learning how to operate that coal stove, washing. We all used cloth diapers and they all had to be washed...always diapers inside or out, every day of my life."[201]

"For you who haven't seen our little community, it looks like a vast area of extra-big chicken coops set kerplunk in the middle of a big prairie," wrote Nell Elperin, who moved into Badger early in 1946 before the university signed its lease with the federal government. "How you would probably wonder, do people live in those funny little boxes?"[202]

They lived as people have always lived, by coping, adjusting, improving what they could, enduring what they could not. "Diversity, " said Peggy Baime. "When I got to Badger Village, I had a whole new meaning of that word, the impact of people who normally wouldn't have gone to the university at all...There was the impact of honesty, tension, we shared only that we were women, most of us had children and most of us were poor."[203]

Baime trained for and organized a "great books" reading group where the "diversity" of the members was evident. "Anyone wanting to talk about great books could come out...women from both sides of the tracks so to speak. It was dramatic to see them confront each other...we got out of there really shaking...a sense that you had been fighting the war on the home front."[204]

The men had encountered this type of "diversity" during the war, where—as often depicted in the movies of the day—conscripted soldiers from different ethnic and class, but not racial, backgrounds were thrown together and forced to accommodate each other. For the women at Badger, the encounter with diversity came after the war, but it was still "diversity" of the 1940s vintage, with minorities in minute presence. To the best of Larry Halle's memory, no African-Americans students lived at Badger. "There were some wives from foreign countries, the Philippines for one." On the other hand, Erhart Mueller tells several anecdotes involving African-American student families living in the village and foreign students as well. "There were Greeks, Arabians, Indians, Mexicans...An Egyptian student trying to learn to drive his foreign car...A Chinese Ph.D. had married a Venezuelan girl..."[205]

Living in close quarters, separated by walls that were no more soundproof than a cloth curtain, also brought residents together. Virtually every Badger resident tells of passing the salt or the ketchup across the table when someone in the neighboring apartment asked for it. Others remember chain reaction crying, when a baby at one end of the apartment line started to bawl and set off a child in the next, and the next down the line. Almost everyone had a radio, but if one resident liked what her neighbor was listening to, she could listen as well, without turning on her own receiver. If she didn't like it, there was little escape. "We did not feel isolated, felt almost surrounded. It was too crowded, hearing through the walls was real. We kept a bit of room between us and other neighbors [but] we were living in their laps and they were living in ours." To get out of each others laps, couples learned to take walks together when they wanted privacy and to conveniently forget what they overhead through the walls when dealing afterwards with their neighbors.[206]

Sometimes, having plenty of neighbors was an asset. The A&P grocery store and the Walgreens drugstore that had operated in one wing of the school building during World War II reopened for the student era. When some of the struggling student wives found prices too high, they organized a food-buying cooperative. "A few of us set up a buyers club with the Midland Co-op. We had to buy by the case and took turns storing it." The club ran smoothly until the Heinz baby food company announced that it was about to stop packaging baby food in tin cans. Heinz offered to sell a rail carload of 3,500 cases of canned baby food to the co-op at a giveaway price the Badger Village women could not refuse. Too much of a good thing is not always good. As Phyllis Young remembered, "I ordered two cases, got string beans and prunes. After a diet of this it was another trip to the doctor in Baraboo for our little John." The university kept a nurse at the Badger school, and medical students of varying degrees of expertise were on hand, but the students traveled to the neighboring communities or to Madison for medical care. William Schereck, who lived at Badger in its final months as student housing, recalled that the medical student in resident at the time would "talk to people, let people tell him their problems, then he'd take them to the university [clinic] and get some help."[207]

Children were understandably the focus of their mother's lives, more so in a place with so few distractions and so many children as Badger. In addition to the school, which continued to run at the multi-grade level with Clarence Krumm as part-time principal, Badger also had a university-operated "nursery" for children as young as two years. "There was a great effort on the part of mothers to get them ready for preschool. Your child had to be potty-trained. I was able to get our son ready at age two, but I kept my fingers crossed everyday so we wouldn't slip under the standard." Although the nursery gave stay-at-home mothers a break from child care, it was a necessity for the minority of women who held jobs while their husbands were in school. Whether in nursery school or not, no child ever suffered from a shortage of playmates at Badger. The place was filled with children, most of them too young for school and ready to be introduced to the pleasures of squishing the fabled Sauk Prairie mud between their toes.[208]

For recreation adults looked to the community center, with its hardwood floor and small stage suitable for movies, religious services, basketball, dancing, plays or variety shows. Peggy Baime remembers a performance of the ballet Swan Lake with an all-male cast. "They began to really train and did Swan Lake incredibly seriously." On Saturday nights the students turned the community center into a cabaret where they ran their own bar and invited dance bands to perform. A speakers' forum brought university professors to lecture, and touring talkers as well. Many of the students leaned to the left politically, with liberals, socialists and communists all exercising their right of free speech. For Marjorie Johnson, "the Marxists weren't very numerous, but they were devious. The socialists were nicer, sort of soft-headed," but they did "rescue this organization from the Badger Village communists." Not surprisingly, Norman Thomas, longtime leader of the socialists in the United States, gave a well-received address in the community center, but Joseph McCarthy proved to be a lonesome fish. When he spoke at the community center, the audience was so unresponsive that he was compelled to ask, "Isn't there a Republican in the place?"[209]

If conservative Republicans were in short supply, ballplayers were not. Softball, touch football or basketball games went on most every weekend. "In Building 3 [barracks] at Thanksgiving we had the 'toilet bowl' game. The women would cook a huge Thanksgiving dinner and we'd have a building Thanksgiving...twenty-four couples," remembered August Lemberger. "We were glad we lived in North Badger. We could always find someone to socialize with, often had parties in the building. There was no keeping up with the Joneses. If you had a clean T-shirt and a six pack, you had a party...The most popular beer because it was the cheapest was Effinger...not because of the taste, because of the price."

While most of the students kept their noses to the academic grindstone, a few exercised an entrepreneurial spirit. As related by Erhart Mueller, law student Ken Hurwitz persuaded the Town of Sumpter board to appoint him justice of the peace in 1950. As such, he could call traffic court into session whenever he was short of cash and dispatch a student constable, also empowered by the town board, to collar a speeding driver on Highway 12. As justice of the peace, Hurwitz, who later became a well-known Madison attorney and television personality, could try the criminal, set the fine, and pocket it. Among the "constables" he dispatched to collar unwitting speeders on Highway 12 was future governor and chief Wisconsin law officer, Lee Dreyfus.

As typified by Hurwitz, Dreyfus and their classmates, a distinctive community evolved at Badger. The origins of its members varied, but their values and aspirations were very similar. "When we first moved into Badger, we were somewhat apprehensive," said August Lemberger, "but I can't recall a problem...It was an unusual situation, but everybody was in the same boat. We all shared common values, were approximately the same age, at the same start in life...We created an environment in which people got along...my recollection of Badger is very fond...a fond memory." So it was for nearly all the young, talented and upwardly mobile student families who passed some time there.[210]

The period of fond memories came to a close as students completed their class work and moved on. The housing shortage eased as university and non-university housing was built in Madison. When the powder plant reactivated in 1951, the federal government reclaimed its village without placing any great hardship on the remaining student families. They were allowed to stay at Badger until the end of the academic year in June 1952, when the last buses pulled out onto Highway 12 and the university's tenancy ended. "Our operation was strictly for veterans," said Halle. "When veterans dropped way down, there was no reason for us to be in business."

"I came away from it reluctantly after two-and-one-half years," said Peggy Baime. It was almost like a little Shangri-la. When I went away it was no more...Badger Village."

The Company Town

It may have been Shangri-la for the students who went away, but for the powdermakers who stayed the village was as real as the rocks on the bluffs nearby. Badger Ordnance employees began moving into the units before the student period officially ended and a few wives of students took jobs at the plant afterwards which allowed them to remain in their apartments. The transition was more gradual than in 1946-47, when the village went from vacant to "rabbitville" in about a year, and determined by the pace of hiring at the plant and necessary remodeling at the village.

Badger's reactivation meant changes on the housing front, and not just in the village itself. To meet the increased demand, the Federal Housing Administration announced that developers in the powder plant area could obtain "credit relaxation" and other aids to build seventy-five housing units. There would be fifty homes — two- and three-bedroom models — for rent and twenty-five for sale to "in-migrant personnel." In August 1952, applications were approved for thirty-six new houses in Baraboo, twenty-three in Portage, twelve in Lake Delton and four in Sauk City. Prices were set at the maximum level allowed by the federal government and uniform in all locations. A two-bedroom house cost $8,750, and a three-bedroom cost $10,000 whether in Delton or Sauk City. Rental units went for $65 per month for a two-bedroom place and $75 per month for a three-bedroom. The new housing was scattered around each community and built by several developers, including the Ritz Corners Development Corporation in Baraboo, E.M. Hines in Lake Delton and Harry F. Vogts, a Madison contractor who built the Sauk City houses.

On the grounds of the plant itself, the Army also remodeled the four World War II barracks and the mess hall constructed at Gate 10 in late 1944.

Designed to house 800 men, the Gate 10 barracks were never more than one-third filled and were not occupied for more than a few months. They went into surplus with hardly a scratch on the floor or a nick in the paint job. Although the lumber was new and the hardware shiny, the barracks had been overlooked by salvagers and survived the surplus period largely undisturbed. Near the end of the Korean period, the Army began converting the barracks into forty, two- and three-bedroom apartments for plant workers, while the mess hall became a recreation center. Gate 10 developed into its own little community, fenced off from the highway on one side and the plant on the other. Erhart Mueller lists thirty-two families in residence in March 1956. Children attended the Badger school, adults socialized in the rec hall and made the short commute to work. In 1958, after about two years of occupancy, the Gate 10 housing shut down. In the 1960s, two of the buildings were reopened for use by the Badger Conservation Club. The rest of the buildings were demolished in 1973, nearly thirty years after their initial construction.

The major construction work occurred at Badger Village, where the "temporary" structures of the 1940s were once again remodeled to meet new standards. The North Badger "barracks," which the university had converted from single and double rooms to one-bedroom apartments linked by interior hallways running to shared bathrooms, were now altered into two- and three-bedroom apartments with their own bathrooms and exterior entrances. Each of the ten buildings was divided into sixteen apartments, with fourteen two-bedroom units renting for $29 a month and two, three-bedroom flats for $39 a month. At the prevailing powder plant wage scale, these rents could be paid with two or three days' work. They were, of course, not based on a standard business model requiring a return within a reasonable amount of time of the $400,000 the federal government put into the remodeling. It would take almost eight years at full occupancy to pay back the $400,000, longer if any interest charges were included. However, Badger Village was not a business, not even after Liberty Powder took over management in November 1954. Badger was intended to supply basic housing for plant workers for the limited duration of a national emergency—certainly less than eight years. It was not an investment, but an expense charged to the taxpayers by the guardians of national security. Badger workers started moving into the new North Badger in late August 1953 and gradually occupied more apartments as they became available.

Work did not begin on remodeling South Badger until July 1953, with the contract for $300,000 awarded to the Abbott Construction Company of Chicago. South Badger consisted of twenty-nine buildings divided into one-bedroom apartments. Many of them sat on cedar posts, spanned by sagging joists and rotting plywood flooring. As one student wife remembered, "If anybody heavy came through the door, they fell right through." Windows and door jambs leaked so badly that the drafts blowing through some apartments were strong enough to snuff out a kitchen match lit to fire up the coal stove. And that was before five years of occupation by the young student families. In the remodeling of 1953-54, which was more a rebuilding, all the wooden piers were replaced with concrete, floors were reframed and resheeted, doors and windows rebuilt, roofs repaired and the walls sealed with asbestos siding. The cranky coal stoves were replaced with oil-burning units. When finished, South Badger had 393 two- and three-bedroom apartments which rented for $46.50 and $51.50 per month, respectively. Rents were higher than in North Badger, but the units were slightly larger. They were still a bargain that any blue collar or pink collar worker at Badger could afford, but relatively few of them choose to accept. By the time all work was finished in November 1954 and the village turned over to Liberty Powder, 2,600 people worked in operations and another 1,100 were in construction, but no more than 200 people lived at Badger Village.

Throughout the 1940s and 1950s, Badger was always a company town, whether the company was Hercules Powder, the Federal Public Housing Administration, the University of Wisconsin or Liberty Powder. As such, it was always an artificial creation, immune to the forces driving other villages. It did not, for example, have to levy taxes for municipal services. The plant, even in the university period, supplied water, sewer and fire protection. The town of Sumpter oversaw the village school, which had been built by the federal government and enjoyed subsidies no other public school in the county received. The federal government also constructed the community center, with its village office, public hall, post office, supermarket and drugstore. Health care was available at the company clinic when the plant was running. University medical students and nurses were on hand during the student years. The plant provided fire protection, even in the student period, when the residents organized a Keystone Kops volunteer fire corps so they could share "reciprocal" service with the real firefighters stationed across Highway 12. Police protection came in the form of the Sumpter town constable and the county sheriff, while minimal legal services came from the town's justice of the peace.

Just as it was a company town, Badger Village was also a highway town. Its existence depended on transportation or the lack thereof. During World War II, Badger filled up with people who could not reach the plant except on foot. Only a minority of Americans owned automobiles in those years. Fewer women, who made up the majority of villagers, drove than in later years, and driving was discouraged during the war anyway. In the student period, Badger would have been uninhabited if the university had not provided buses to and from the Madison campus. By the mid-1950s, Badger employees—like other Americans—were willing, able and used to traveling by car to and from work, even when the commute ate up two or more hours of time every working day. People preferred to drive to the plant not just from Baraboo or Sauk City, but from Friendship, Viroqua, Richland Center and Mineral Point, than to live in Badger Village.

With most of the newly refurbished units vacant, Liberty Powder was compelled to offer inducements to fill the village. Since reactivation, residence at Badger had been restricted to workers who paid seventy cents or more per day for travel or who needed two hours or more to get to and from the plant. Liberty dropped the restrictions and welcomed any employee to rent an apartment in the village. The company also began a "beautification" program that started with its replacing the "ugly" lathe and wire snow fence strung between the village and Highway 12 with an all metal "cyclone" fence. Liberty also gave residents paint to spruce up their apartments, distributed flowers and shrubs to brighten their yards and obtained trees from the Wisconsin Conservation Department for planting along village streets. Playground equipment was moved to the community center and lights were installed on the field at the rear for night baseball. In order to accommodate larger families Liberty started expanding two-bedroom apartments to three-bedroom units. In March of 1955, the company converted a wing of Building 273, the recreation building near the plant gate, into a chapel open to all denominations. Plant carpenter Joseph Geier had learned that a century-old oak altar and pews were available due to a remodeling project at St. Raphael's Cathedral in Madison and installed them in the rec building. Additional church fixtures came from the village Lutheran congregation. For the first time ever, Badger Villagers would not have to set up and dismantle a chapel every Sabbath or sit on folding chairs during religious services. Liberty's final inducement to recruit residents to Badger was to lower rents to $35.05 for two bedrooms and $38.55 for three bedrooms. The result of all these incentives? By the end of 1955, a year after Liberty Powder became landlord, Badger Village had about 150 residents.

The outcome might have been more favorable had rehab and construction projects, as well as operations, continued. The rocket area stopped running in December 1954, reopened in March 1955, then closed again in September 1955. Ball Powder and smokeless ran into 1956, then sputtered out. People would not live in the company town if the company did not provide jobs. Just as the plant went idle, the village went vacant. It was not Shangri-la and it was no more.

Korea vs World War II

From the start of combat in June 1950 to the signing of the cease-fire in July 1953, the Korean war lasted about three years and three weeks. The first half of the war was a blitzkrieg in which North Korean, South Korean, Chinese, American and United Nations troops stormed up and down the peninsula from the banks of the Yalu River in the north to the shore of the Sea of Japan in the south. The second half was a "sitzkrieg" during which North Korean and United Nations negotiators waged verbal warfare while seated at the conference table straddling the truce line in the village of Panmunjon. Combat continued during the talks, with small-scale infantry actions, steady artillery barrages and airborne bombs and rockets claiming the lives of thousands of soldiers and civilians. When it was finally signed, the cease-fire brought about an uneasy state of tension that was not war but not quite peace either. For the rest of the decade the threat that its opposing armies would clash again was real enough to insure that Korea would be one of the many Cold War danger zones that influenced the fate of Badger Ordnance. In turn, just as Badger brought World War II and Korea home to the foot of the Baraboo Hills, the powder plant also brought home the crises in Formosa, Berlin, Iran, Guatemala, Poland, Hungary, Lebanon and other 1950s Cold War flash points.

The reactivation of the powder plant for the Korean war had much in common with its construction and operation in World War II, but also differed in many ways. The Korean conflict called for a fraction of the commitment the United States invested in World War II and this reduction was reflected in the scope and pace of work at Badger. In addition, the United States of 1951 was not the United States of 1941. The nation was more prosperous, populous and mobile, and already settling into the middle class, suburban lifestyle, centered on the auto and television, that would define the "American way of life" for the rest of the century. These general conditions, as well as the specific demands of construction and production, influenced Badger's course in its second decade.

As in World War II, the Corps of Engineers supervised the civilian contractors performing the work of rehabilitation and new construction, while the Ordnance Department supervised the operating contractor. The rehab contractor, Fegles Construction Company, and Ball Powder construction contractor H. K. Ferguson, were good-sized firms but not in the same league with Mason and Hanger. The operating contractor, Liberty Defense Powder Company, was incorporated in Wisconsin solely to run Badger. On it own, it was but a pint-sized version of its World War II predecessor, but Liberty was also a wholly owned subsidiary of the Olin Corporation, which played on a level field with Hercules Powder.

Upon reactivation, employment figures soared upwards from the few hundred maintenance workers on hand during the standby period. However, even at the peak of rehabilitation in the summers of 1951 and 1952 and of Ball Powder construction in 1954-55, the total workforce numbered no more than a third to one-half of the top levels of World War II. At the height of production in the spring of 1953, the operations workforce rose to no more than two-thirds that of the World War II high mark. Fewer workers meant less demand for housing than in the 1940s and more housing was available in the 1950s anyway. It was not necessary to rebuild the Prairie du Sac trailer park. Construction workers did not have to board in chicken coops, nor shift workers share hot beds. Reflecting the demand for family housing over individual barracks, Badger Village was remodeled to accommodate fewer workers in larger apartments, but—as in World War II—much of the work was not completed until after it was no longer needed. The Badger workforce was still dispersed over eight counties and less concentrated in Sauk County than in World War II. A greater percentage of 1950s workers drove cars and, with fuel no longer rationed, were as willing and better able to commute to the plant than their predecessors. Badger buses ran again, but, with auto commuting easier, fewer buses ran over fewer routes. Unlike Hercules, Liberty Powder did not run its own fleet of buses, but contracted with a number of owner-operators in surrounding communities.

With many fewer men in military service and more men available for employment, Badger's 1950s workforce was more male than that of World War II, starting with the exclusively male construction crews. In operations, women made up only slightly more than thirty percent of the workforce, compared to almost fifty percent in 1944-1945. Almost all of those jobs were in traditional clerical and maintenance positions with few women on the powder lines. Badger's three in ten ratio of men to women was about average in industrial employment both before and after World War II. Also unlike World War II, all but a handful of operating employees were white, with no guest workers from overseas. Few, if any, of the black Americans who were then taking part in the "Great Migration" of African-Americans out of the southern states came to Badger. Powder plant workers also threatened to go on strike in the 1950s, an action unheard of and verging on the treasonable during World War II. Even the guard force threatened to walk off the job, then agreed to work without a contract while negotiations were underway. Construction delays due to shortages of materials were common in the 1940s, and also occurred in the 1950s, but the longest postponement was caused by the national steel crisis of 1952 which was precipitated by a dispute between labor and management.

Just as in World War II, over a year passed between the start of rehab/construction work and the full activation of the plant. In fact, it took longer to fully rehab Badger for Korea than to build it for World War II. Likewise, only part of the plant came on line initially, with additional operations phased in over time. The piecemeal activation meant that most of the propellant Badger made for Korea was fired out of heavy artillery used in the sitzkrieg phase of the war and not in the crucial blitzkrieg period, when at times it looked as if the North Koreans and Chinese would sweep the Americans into the sea. In fact, Badger ran longer after the war than during the conflict.

As in 1942, the type of explosive the plant was slated to manufacture changed before operations began. Costs ballooned from the first estimate of nine million dollars for a partial rehab in 1951 to forty-nine million to complete the makeover in 1953. Finally, near the end of the Korean conflict, just as in 1944, the Army announced it would expand Badger to manufacture a propellant type new to the plant—with a similar outcome. The rocket area was completed by the end of World War II, but finished only a fraction of the millions of pounds of propellant it was capable of making. By comparison, construction did not begin on the new Ball Powder area until after the Korean cease-fire was signed. By the time the main plant began fabricating Ball Powder in September 1955, the shooting war in Korea had been over for twenty-two months. Continuous Ball Powder production came to a halt a year later and, once again, Badger shut down with a just-out-of-the-box powder-making facility on idle. And just as the tarnish-free rocket plant had saved Badger from the wrecking ball in 1946, the barely used Ball Powder area would help keep the bulldozers outside the gate a decade later. But that was at the end of the 1950s, not the beginning.

Reactivation and Operation

At the beginning, in January 1951, Louis Reuter was a happy man. A Milwaukee native and engineering graduate of the University of Wisconsin, Reuter had come to Badger as an employee of Hercules Powder after working with the Corps of Engineers at Radford Ordnance. Nominally a production manager, Reuter made his most valuable contribution as the compiler of multi-volume reports on World War II operations. Detailed, comprehensive, they constitute the basic historical record of those years. By the end of the war, Reuter knew more about Badger Ordnance than anyone. He then topped off his copious store of knowledge by compiling an exhaustive, building-by-building inventory of the plant that included equipment lists, building dimensions, construction descriptions, site maps, plumbing and electrical diagrams, plus hundreds of photos capturing what the plant looked like at the end of its first period of operation. It was another basic document, used by the Army to designate property to be disposed of as surplus or retained in standby and by historians ever since. After Hercules pulled out of Badger in the summer of 1946, Reuter stayed on by transferring back to the Corps of Engineers. When the rocket area and other parts of the plant went into standby it became his job to maintain them on a shoestring peacetime budget. Once Badger was reactivated, everything changed. Reuter no longer had to fend off cannibals and scrappers, contend with shady schemers looking to raid the taxpayer's wallet, or petition administrators in Chicago, Minneapolis or Washington to fund maintenance on decade-old buildings designed to last five years. Instead, he would see the idling enterprise he had nurtured in its repose shift into gear once more.

Reuter was one of many veterans of the World War II military/ordnance industry partnership to return to work at Badger. The plant commanders in the 1950-'55 period, Lieutenant Colonel Oliver D. Williams, Major Ulysses G. Stewart, Lieutenant Colonel William A. Evans and Lieutenant Colonel Jewell G. Haynie, had spent the last war at home front ordnance plants. Norl Hamilton, Liberty Powder's top manager at Badger, had also served as an Ordnance Department officer in World War II. Office manager C.H. Heckman had supervised small arms production for Olin's Winchester division. Nitrocotton manager H.B. Maroney transferred in from Sunflower Ordnance, and Michael Copps came from Radford to run the acid plant. Among the World War II Badger vets were purchasing director E. W. Warrington, works engineer Ferguson Bell, technical director Robert Buell, inspection chief Hubert L. Jacobsen, protective services director Ralph Harding and guard chief Herman Johnson.

Farther down the line were hundreds of other World War II workers who came back and stayed for many more years. Among them were Virgil Wimmer, O.J. Befera, Elmer Johnson, John Troyer, Bill Clement, Forrest Sommers, Phyllis Albert, Ruth Corrao, Harold Thoelke, Robert Priebe and Walter Zechel. Also returning was Wisconsin Assembly Representative Louis Romell, who worked as a millwright at Badger when he was not representing Adams and Marquette counties in the state legislature. The workplace they all came back to was much less hectic than the one they had experienced in the 1940s. The rehab was piecemeal, focusing first on part of the rocket area, then the smokeless area, followed by the rest of the rocket area, then back to smokeless. New construction consisted of about sixty buildings for the Ball Powder plant and did not begin until after the rest of the plant was rehabbed. No more than 1,000 construction workers were needed for any of these projects, a far cry from the 12,000-man blitz of 1942.

The number of people in operations passed the 3,000 mark early in 1952 and fluctuated between 3,400 and 4,000 until the spring of 1953, when smokeless powder production peaked and pushed employment to the 5,000 mark. After the Korean cease-fire was signed in July of 1953, all powder production began to taper off, along with employment. By the summer of 1957, operations had all but halted and even with the addition of the Ball Powder plant, only about 550 people were working at Badger.

The nature of the work and the workplace had not changed since World War II. Supervisors, foremen and chief operators were hired well before operations began and sent for training at other facilities: Sunflower for rocket; Radford for smokeless; Olin's East Alton, Illinois, plant for Ball Powder. Security was still tight, with the intelligence department screening potential employees for past criminal behavior or presently dangerous political views. Safety remained paramount, with educational and motivational programs in place along with mandatory inspections and searches. The fire department spent many more hours teaching fire prevention and monitoring potentially dangerous work practices than actually fighting fires. Depending on employment numbers, the guards routinely made 10,000 to 14,000 searches of incoming workers "at the clock search booths and gates" per month. They drove up to 30,000 miles a month, chiefly on perimeter patrol, but also inside the plant and at Badger and Staff Villages. The acid plant still concocted chemicals for use at Badger or sale to other facilities. Cotton or wood pulp was still mixed with acid on the third floor of the nitrating houses, then boiled, dried, wrung, pressed and sliced into grains of the desired length for smokeless powder or sent to the premix house where nitroglycerin was added to make rocket propellant. The testing and ballistic labs still examined finished powder for uniformity and the boom of cannons making test firings again resounded throughout the Baraboo Hills.[211]

Raw materials entered the plant and powder was still shipped by rail, with the North Western and the Milwaukee Road sharing the loads about evenly. Regular work for the railroads while they were facing serious competition from the trucking industry was bound to bring a smile to the Irish eyes of Leo T. Crowley. He had retired from government service in 1945 to become the president of the Milwaukee Road, a job he held until 1969. He would have been less happy about the authorization Badger received from the Ordnance Department in April 1954 for "the shipping of all BOW powder by truck." The Army was no longer shipping by rail because trucks "eliminated up to three direct operations, including handling by powder truck and storage in the magazines. Powder shipments are loaded onto truck trailers directly in the rocket and smokeless finishing areas," with an estimated savings of $1,400 a day. Gateway Transportation of La Crosse, which had been hauling powder out of Badger since January 1954, was soon parking a "fleet of standby trailers" in the Gate 10 lot so they would be ready for loading on demand. The railroads did not take this challenge lightly and responded with the "piggyback" system of carrying loaded semitrailers on flatcars which could be off-loaded and hitched to trucks for local delivery. Piggybacks were well suited to the ordnance industry, since all the plants were connected by rail and powder from Badger could be easily unloaded at a shell loading plant in Minneapolis, St. Louis or Joliet. The Chicago and North Western hauled Badger's first piggyback shipment of 300,000 pounds of smokeless in ten trailers on flatcars in April 1955.[212]

The work was as dangerous as ever, with spot fires in the wringer houses, exposure to toxic levels of DNT, giddiness and addiction to ether fumes still occupational hazards. One notice specific to Badger and posted in the coating house of the Ball Powder line warned that frozen nitroglycerin was "more sensitive" than liquid and that workers should promptly clean up small spills with "NG killer — a mixture of water, alcohol and sodium sulphate." The killer was not to be used on large spills, since "large quantities of NG" generated heat and could cause an explosion. No fatal accidents occurred in smokeless or rocket operations in the Korean period, but Leslie Barkley died as a result of an explosion and fire in the Ball Powder hardening house in November 1955. Forty-five years old, with a son on duty in Korea, Barkley was a chief operator in the hardening house and also the barber at Badger Village. Construction, always a dangerous occupation, took its toll as well. Rehab worker Roy Elmer fell through the opening left in the second floor of a boiling tub house where a tub had been removed during decontamination and died as the result of a skull fracture in July 1952. Electrician Henry Grunke died beneath a load of light poles that unexpectedly rolled off a tram car in August 1954. Also in July 1952, a fire started in a rework roll house in the rocket area and spread to an open tub of carpet rolls which then exploded. The operator of the rolling machine and others working nearby immediately fled from the building and beat the explosion out the door. Their quick action and the design of the building restricted the damage and prevented injuries. She wasn't involved in this incident, but the quick thinking on the part of the rework roll operator illustrates a notion expressed by Ruth Corrao, also a Korean-era rocket worker: "One memory of the rework roll house job was that you learned to move quickly when you heard the distinctive click of the sprinklers, which was caused by a flash from the reworked powders, or you got an unscheduled shower."[213]

The training and emphasis on safety paid off in the rework roll house fire, but some workers took a more relaxed attitude about the hazards of life at Badger than Ruth Corrao. In September 1953, sixteen employees were fired for smoking cigarettes inside the plant. Dismissal was automatic in such cases but, since Badger was a federal installation supervised by the Army, the commander could pursue additional punishment. Accordingly, four of the sixteen — Willard Suppy, Delzie Leigh, Clark F. Wilkinson, Edwin Quinn — were indicted on charges of "smoking within the exterior confines" of the powder plant. Federal attorney Frank Nikolay prosecuted the defendants with Judge Patrick Stone presiding. In his deposition for the trial, plant commander Colonel William Evans expressed his "grave concern over increasing carelessness" at Badger. He was so alarmed that he had hired two private detectives to go undercover at the plant. It was on the word of the detectives that sixteen workers were fired and four of them faced criminal charges. Evans said the defendants were found smoking in the carloading and magazine areas while cars were being loaded with powder. He also reported that the men even smoked "while sitting on a powder keg." The "keg" was actually a special powder can with a non-sparking liner and lid sealed airtight, yet a hot ash or glowing match tip dropped on some spilled powder could set off a whole car load. The maximum penalty for smoking in an ordnance plant, as set by federal law, was a $5,000 fine and a year in prison.[214]

While not going for the maximum, Judge Stone did slap Suppy, who was a senior foreman, with a $50 fine and a year's probation. Leigh, a foreman with less experience, got a year on probation and a $25 fine. Wilkinson and Quinn were fined $25 and sentenced to six months probation. The situation revealed by the prosecutions was serious enough for the Chemical Workers Local, one of the largest unions at Badger, to pass a resolution in support of Evans and the plant safety program. In the 1950s, at least two of every three adult American males smoked cigarettes and many workers addicted to tobacco must have had a hard time getting through a full shift without a hit of nicotine. Men working outdoors in the loading and magazine areas, away from production lines, guards and supervisors, like the quartet hauled into federal court, could easily succumb to temptation. Considering the pervasiveness of cigarette smoking at the time, it was fortunate that Badger did not have a serious accident caused by a smoker who could not endure eight hours without a fag.

Although the gap was not as great as in the 1940s, Badger still paid higher wages for equivalent work than other employers in the area. When operations resumed in September 1951 Liberty paid wages ranging from $1.20 to $2.00 per hour. This nominal rate was rendered meaningless in December when Badger starting running on a "continuous basis," with four shifts each clocking forty-two hours for five days. Powder makers worked five days, changed shifts, worked five more days, then took four days off. They came back to spend five days on a different shift, then five on another, then had four more days off, and so on. Wages were calculated on the basis of "constant overtime," which resulted in pleasantly enlarged pay checks.[215]

As a federal contractor, Liberty Powder had to accept a unionized workforce, which was the norm for any sizable industrial enterprise in 1950s Wisconsin anyway. Signed up by county Labor Federation Secretary August Moulton, a beefy and blunt archetype of the labor leader of his day, Badger workers were organized into eleven individual trade unions affiliated with the American Federation of Labor (AFL). Negotiations for a wage and benefit compact began in March 1952 and dragged on for seven weeks with no agreement. In May, Moulton and Vito Mazzacano, an AFL official from

Chicago, called a meeting of union members. More than 1,600 workers filled the Badger Village community center to overflowing and voted overwhelmingly to give their negotiating committee the power to call a strike. Armed with the club of solidarity, the union committee sat down for a marathon session with Liberty's industrial relations director, A.E. Giertsen. The unions wanted a minimum hourly wage of $1.40, a maximum of $2.36, plus twenty-five cents more for the head tradesman in each department. They also wanted two weeks of paid vacation after one year on the job, three weeks after three years and seven paid holidays a year. Liberty wanted to keep the $1.20-$2.00 wage range, along with less vacation and fewer holidays. After twenty-five hours of talks, Mazzacano and Giertsen agreed to a $1.35-$2.25 wage range with a fifty cent increase for shift work, paid lunch periods, one week of vacation after one year and two weeks after five years. In addition, Liberty agreed to pay one-half the cost of the woolen protective clothing worn by workers handling chemicals or powder. Meeting individually in the following week, the Badger locals voted to accept the agreement and work continued without interruption. Mazzacano was quick to point out that the settlement meant that an additional "one or two million dollars would be paid out by Liberty," and spent locally.[216]

The Badger unions and Liberty Powder had re-enacted a familiar 1950s scenario. Labor productivity was rising, making it easier for management to raise wages, but not without a struggle. Corporations, while not necessarily enthusiastic about collective bargaining, were more accepting of it, and less prone to union-busting than they would be a generation later. Liberty Powder was, of course, more willing to grant concessions than it might have been since it could pass the additional costs of labor on to the taxpayers. There was also a war on and, although it did not seem to be much of a talking point during the negotiations, American soldiers were in combat and the peace talks were in stalemate. If patriotism was not incentive enough to settle the dispute, mutual self-interest was. Badger's output was in demand at the moment, but for how long into the future? An ordnance plant did not operate according to the usual laws of supply and demand. The market it served depended on continually developing arms technology, which could make the latest weaponry obsolete overnight. The political situation was equally unpredictable, with both the national and international situations as fluid and as testy as nitroglycerin on an angel walk. It was in the interest of both labor and management to "make hay while the sun shines" and sign a contract without a work stoppage. When the labor agreement came up for renewal in 1954, the climate had changed. Production was winding down, a layoff was underway and, when Liberty refused to accept the unions' request for an across-the-board raise of ten cents an hour, the workers voted not to strike. They settled for a compromise that raised pay for shift work and time spent changing clothes but left their wages flat.

Liberty Powder and Badger's unions could agree on a compensation package, but final approval had to come from the federal Wage Stabilization Board. This World War II relic agency took the rest of the year to study the May 1952 contract, while wages were frozen at the pre-settlement scale. In February 1953, the Board approved the pact and personnel in the payroll department at Badger began laboriously calculating the back pay owed for a year's work. After all the time cards were checked, the ledgers filled and the arms on the mechanical adding machines pulled forward a few thousand times, a month had passed. Then on one fine day in March, Badger workers received a total of one million dollars in freshly printed paychecks. "It was like Christmas..." said Curt Page of the *Baraboo News-Republic*, and not just for the wage earners. "Business in Baraboo received quite a stimulant... employees came in with their checks, paying up bills and making purchases so that more than one merchant found himself scurrying around for currency as the employees cashed their checks. The retroactive pay, like the regular payroll, makes a nice business for Baraboo." The economic impact of Badger's multi-million dollar payroll on Sauk County had never been so vividly illustrated.[217]

The threat of a strike in 1952 and the increased cost of labor at Badger did not deter the Army from expanding operations there. A few weeks after the unions accepted the contract with Liberty, Army Ordnance announced that it would spend another $26 million to complete the rehab of the entire rocket area and restore the B and C smokeless powder lines as well. Employment would increase from the 3,800 currently at work to "more than 6,000" in the spring of 1953. While the money was spent and the powder lines rehabbed, the 6,000 jobs did not appear. By the time the rehab was completed, the war had wound down and with it demand for ammunition.

The Army's original schedule for Badger called for the activation of one-half of the World War II N-Line rocket area, the nitroglycerin plant and the E nitrocotton line. Of necessity the power, water and sewer systems were restored, along with administrative, maintenance, testing, security and transportation facilities. Everything was to be up and running and rocket powder shipped in six-to-seven months. Reactivation started not at the head of the rocket line but with the roll, press and finishing areas at the end. Since the N-Line could start rolling, pressing and finishing rocket paste from other powder plants, just as it had in World War II, rehabbing the front end of the line, where the nitroglycerin and nitrocotton components were formulated, could wait for a few months. Putting the tail of the horse before the head, so to speak, would enable Badger to ship propellant to assembly plants sooner in the hope that finished rockets would get to the soldiers, sailors and pilots who needed them as quickly as possible.

The first order from Army Ordnance commissioned Badger to make the propellant designated as T-6, with its smooth, 4.5-inch diameter cylindrical "sticks," instead of the 5-inch-wide Mark 16, 18, 21 and 22 cruciform grains the plant manufactured for the Navy in World War II. A month before production was scheduled to start, Ordnance cancelled the T-6 and ordered Badger to make the World War II propellants. "This change was very welcome to all the former personnel who were again at Badger and filling key positions in technical and production capacities," Louis Reuter reported. Doing what they knew how to do all over again was easier for Badger workers than mastering a new process and product. Indeed, producing T-6 contradicted one of the Army's justifications for reopening Badger: that experienced rocket powdermakers were on hand and ready to work.[218]

The people were present, but not the equipment — or at least not all of it. The supersonic testing devices used during World War II were experimental and had to be returned to their manufacturers when Badger was declared surplus. They were not needed to make T-6, but now had to be purchased again for the "Mark" powders. Other equipment specific to cruciform powder production and removed from Badger five years earlier also had to be replaced. The Excello brand saws, used to cut the extruded propellant to "stick" size, had all been cannibalized out of the plant. Single-purpose, highly specialized devices, the saws had not been manufactured since 1945. Badger's rehabbers were able to find replacement saws idling in the standby inventory at the Indiana Ordnance Works. Indiana also supplied extrusion presses and rolling equipment to replace gear that had been stripped from Badger in the 1940s. The Volunteer Ordnance Works in Tennessee and the Missouri Ordnance Works, also former Hercules World War II plants, supplied equipment that had been removed from the acid plant. All of this searching, finding, removing of T-6 equipment and installing of cruciform machinery, plus training of new workers, pushed back the date when production could resume.[219]

Badger had always been a giant pipe farm where the buildings were secondary to the plumbing. Accordingly, pipe fitting made up the bulk of the rehab work. Miles of pipes and fittings that had been destroyed by decontamination or stripped away by salvagers had to be replaced. Shortages of plumbing materials, as well as stainless steel boiling tubs for nitrocotton, delayed the reactivation of the N-Line. Work fell further behind schedule in June 1951, when the pipe fitters abruptly walked off the job and did not return until their work week — with overtime pay — had been increased from forty-eight to fifty-three hours. In 1952-53, the rehabilitation of the smokeless lines was delayed for a good six months because of the national steel crisis. Starting as a labor-management dispute over wages for steel workers, the conflict drew in President Harry Truman, who seized the steel mills, and the Supreme Court, whose judges declared the seizure unconstitutional. In the meantime, the people restoring the B, C and D nitrocellulose lines at Badger waited for the flanges, couplings, tubs and valves they needed to finish the smokeless pipe farm.

Using paste trucked in from the Sunflower Ordnance Works in Kansas, the N-Line processed its first propellant of the Korean period in September 1951, only three months behind schedule. By the end of the year, E-Line nitrocotton and the nitroglycerin plant were up and running and Badger could stop importing paste from Kansas. Even so, the plant was running at far less than half-speed. In four months Badger produced two million pounds of propellant, the equivalent of about ten days' operation at full capacity. The rocket line continued to run at a fraction of capacity for all of 1952, even though rehab work on all rocket powder facilities was completed in August of that year. The N-Line would not produce the same volume of propellant — fourteen million pounds — that it manufactured in World War II until September, a full nineteen months after rehabilitation work began. In fact, the plant never did operate at its design capacity of 5.5 million pounds a month in the Korean period. By the spring of 1953, the N-Line had filled all its current orders and rocket workers who could not shift to the then-expanding smokeless operation faced a layoff. Once again, as in World War II, by the time the rocket line was ready to hit on every piston, the race was all but over.[220]

Badger's limited mobilization for Korea mirrored the country's, where the war surfaced as more a nagging presence Americans were duty-bound to heed than an acute crisis in need of resolution. Despite its low profile, Korea was a real war requiring real sacrifice. For example, more Wisconsin people served in the armed services during the Korean war than in World War I. More than 2000 of them were wounded and over 800 died. Korea was also a concentrated artillery war, especially in its final eighteen months when the armies dug in, held their lines and fired round after round of cannon shells at each other. By the spring of 1952, the artillery war had consumed so much ammunition that Army Ordnance ordered the reactivation of the C and D smokeless lines along with the remainder of the rocket line. In December 1952, the Army placed an order for an additional sixty-three million pounds of cannon powder that would require the activation of the B smokeless line. In the summer of 1953, just as the Panmunjon negotiators were about to initial the cease-fire agreement, smokeless production peaked at Badger. "If the present truce in Korea turns into a genuine peace then there will be a cutback in production," said Colonel William Evans. "However, many months of production are necessary to build up the ammunition reserves sufficient to take care of the defense needs of the nation." He was correct. Badger continued to make both smokeless and rocket powder until September 1955.[221]

In four years of operation, much of it after the Korean cease-fire was signed, Badger produced 35,862,600 million pounds of rocket powder and 235,832,900 pounds of smokeless propellant. By comparison, in World War II, Badger made 13,394,700 pounds of rocket and 257,968,900 pounds of smokeless. The plant that was saved from the wrecking ball because of its rocket line actually produced only about two-and-a-half times as much rocket powder for Korea as it did for World War II. It was the equivalent of about seven months production at full capacity. Although Badger started its Korean rocket run with Mark 16 cruciform, about two-thirds of all its rocket production in the 1950s came in the form of the larger Mark-18 rocket mortar grains. The N-Line also finished a relatively small amount of T-16 rocket mortar sticks. By comparison, the smokeless lines that had been left to rot and ruin for several years came back to make almost as much powder for Korea as they had for World War II. In total, Badger produced about 271 million pounds of rocket and smokeless for each war.[222]

Reflecting the demands of combat, all of the conventional smokeless Badger made for Korea was cannon powder. Most of it was for heavy artillery, ranging from the 105 mm to the 240 mm howitzers, with the 155mm howitzer at the top of the list, followed by the 8-inch howitzer. Badger also made propellant for the lighter 75 mm and 90 mm cannons, but the closest it came to making World War II-style smokeless for small arms was the powder it turned out for the 105 mm and 75 mm recoilless rifles. Smaller than a cannon, but larger than a bazooka, the recoilless rifle was known as the infantryman's artillery piece since it could be handled by a crew of two men and fired a true cannon shell that could demolish a tank.

In both World War II and Korea, Badger also manufactured "fuming sulfuric acid," better known by its brand name, Oleum. Not itself an explosive, it was used to make compounds that were explosive, including smokeless, rocket and TNT. The Badger Oleum works was one of the largest in the United States and was originally designed to be part of the TNT plant which was never built. During World War II the Oleum works produced all the sulfuric acid Badger needed and tens of thousands of tons that were shipped in tank cars to TNT plants at Radford, Barksdale, Kankakee and elsewhere. Badger started shipping Oleum to the Kankakee and Radford Ordnance Works in September 1952 and later received orders for 90,000 tons, which was two-thirds of the volume produced in the 1940s. The plant continued to operate throughout the Korean period and, although TNT was never manufactured on the premises, the volume of acid shipped made Badger one the nation's most important suppliers of the chemicals required to formulate the high explosive.[223]

New Powder, New Plant

A young engineer not long out of Cornell University, Franklin W. Olin founded his first explosives company in 1892. He was in East Alton, Illinois, atop the vein of coal-bearing rock that stretches across central Indiana, Illinois and Missouri, and his Equitable Powder Company got its start making "black" blasting powder for mining companies there. In 1898, Olin organized the Western Cartridge Company to make black powder ammunition for duck and deer hunting, target shooting, police firearms and other non-military uses. In World War I Western built its own brass mill to manufacture cartridges, loaded them with powder made in its plant and sold finished ammo to the United States. Diversified and adaptable in peace and war, Olin/Western continued to prosper in the 1920s and was solid enough to expand during the Great Depression. In 1931, Western acquired the failing Winchester Repeating Arms Company, the legendary gunsmith whose quick-firing rifles had "won" the West in the years after the Civil War.

While Western Cartridge was completing the acquisition of the nation's best-known rifle maker, its research and development people perfected a new small arms propellant that took its trademarked name from the spherical shape of its grain — Ball Powder. Here was powder that actually looked like gunpowder was supposed to look — fine-grained, round and shiny black when coated with graphite, as it usually was. Ball Powder proved to be exceptionally uniform, stable in storage and reliable in varied combat conditions. It was ideal for the semi-automatic and automatic weapons carried by modern infantry soldiers who slogged through steamy jungles, snowy mountaintops or bone dry deserts. Since it was double-base powder, made of nitrocellulose impregnated with nitroglycerin, and could be tucked more compactly into a cartridge because of its spherical shape, Ball Powder packed much more punch, load for load, than Du Pont's smokeless and the antique lover's black gunpowder. Western introduced Ball Powder in 1933, well before World War II but, despite its superior characteristics, the new propellant was not accepted by the United States military until 1944. Through a subsidiary organized for the purpose, Western Cartridge participated in the government-owned, contractor-operated system by loading Ball Powder manufactured at its East Alton works into cartridges at the St. Louis Ordnance Plant. Between St. Louis and East Alton, Olin/Western produced over fifteen billion rounds of ammunition during World War II.

Most of those rounds were .30 caliber loads for the standard military rifle of World War II, the M1 Garand semi-automatic, of which Western's Winchester division was a principal manufacturer. In 1942-45 Winchester and the Army's Springfield Armory together manufactured over four million M1 rifles. In all its operations, from powder to cartridge to rifle making, Olin/Western/Winchester had 62,000 employees in World War II and was as much a part of the military-industrial partnership that won the war as any company in the United States. Together, Olin's three divisions constituted a one-stop small arms shop. It manufactured its own patented powder—the only propellant fired in the standard American military rifle—packed it into cartridges cast in its own foundries, then loaded the ammunition into rifles assembled in its own plants. In 1951, the one-stop arms company expanded by purchasing timberlands in Louisiana and a paper mill in North Carolina. It now owned the trees and the equipment to supply wood pulp for nitrocellulose should it choose to do so, which it did not.

When it signed the contract to operate Badger, Olin had no direct experience at manufacturing rocket or conventional smokeless powder. It was, however, the only maker of Ball Powder in the United States and well up to the task. Olin was also contracting to manage high-explosive manufacturing operations at the Wabash River Ordnance Works in Newport, Indiana, and to turn out finished ammunition at the reactivated St. Louis Ordnance Plant. The company began studying Badger for a Ball Powder plant in the summer of 1952. With its ample infrastructure, acid works, magazines and especially its nitrocellulose and nitroglycerin facilities, Badger was a logical place for a plant to make double-base propellant. After a year of study and negotiation, the Army authorized a multi-phased construction project that included a new extraction plant where smokeless powder could be reprocessed, a "pilot" Ball Powder plant and a full-fledged Ball Powder line. Sixty new buildings connected by miles of process pipeline, tram lines, roadways and railroad spurs were in the plans. In addition, the F-Line nitrocellulose would be rehabilitated to make NC for the new plant. Completed in the summer of 1945, the F-Line had not run in World War II. It went into surplus unused, remained there for five years and, since it had not been contaminated, was severely cannibalized, which only added to the cost of its rehabilitation.

The "pilot' plant came on line in February 1955, about eight months earlier than the main plant. It was used for experimental runs of new powder formulations and for regular production and was soon renamed the "auxiliary" plant. This designation was consistent with the customary Ordnance Department policy of redundancy, which held that two of any facility needed for national defense was better than one. The auxiliary plant continued to operate along with the main line until Badger shut down in 1957. Built in the north central section of the grounds near the Oleum works, the auxiliary plant took over space occupied by the old E.C. powder area. The main Ball Powder line was laid out on the vacant ground east of the F-Line, north of the Gate 10 barracks and west of the rocket area. The initial public announcements, which proved to be wildly optimistic, said that Badger would start making Ball Powder in the spring of 1954. In fact, the first call for construction bids was not made until February 1954, and the main line did not start running until September 1955.

The new structures followed the forms required by the powder process as well as progress made in construction since the 1940s. Ball Powder line buildings were framed with steel, floored with concrete and sided with asbestos cement, compared to the wood and concrete structures of World War II. They tended to be larger than the smokeless line units and multi-storied, with the wet screening house, for example, covering nearly 20,000 square feet. The demands of safety had not changed, so individual "houses" were spaced according to the 1950s version of the table of explosive distances. Many were barricaded and had no-spark flooring. The roll and dewater houses, for example, were lit with externally mounted lights and heated with air warmed in and circulated from an adjacent room. In addition, greater use was made of automatic fire suppression systems, using both water and carbon dioxide to smother flames.

The production process for Ball Powder differed from that of smokeless and rocket in that it was "wet" instead of "dry." The individual grains of Ball Powder were shaped in a chemical bath, unlike the grains of smokeless and rocket, which were mixed wet, then dried to a putty that could be forced through presses to take on a variety of sizes and shapes. Ball Powder making relied on vats and tanks, tubes and tubs, agitators and rollers to mix and form propellant into a dozen different-sized grains. With highly explosive materials being heated, churned, beaten, stirred, sluiced, flushed and piped, the entire operation had to be closely watched and tightly controlled to prevent not just an accident, but a heart-stopping disaster. The process began in the hardening weigh house where nitrocellulose was mixed with water, agitated into a slurry, weighed into 5,000-pound charges and pumped to a hardening "still" in the hardening house. Calcium carbonate (chalk), ethyl acetate, diphenylamine and sodium sulphate (salt) were added. The whole mixture was gradually heated and agitated until grains of powder formed. The resulting slurry was then piped to the solvent handling house where excess chemicals were washed out for reuse. The washed slurry was then pumped back to the hardening house and on to the wet screen house where the grains were separated according to size. The largest building in the Ball Powder area, the ninety-foot-tall wet screen house had two full stories, plus a top floor connected by five levels of platforms and ladders. It housed six concrete slurry tanks, each capable of holding 60,000 pounds, seven tiers of shaker screens through which the slurry passed, pycnometer tanks for weighing and measuring the screened powder, and, finally, thirty-six 60,000-pound concrete storage tanks on the ground floor. Pumped as a slurry to the top of the screens, the powder grains were sorted into three sizes. Too large for processing, the "overs" were sorted out for rework into another charge in the hardening weigh house. The "fines" were too small, and also contained impurities, which required disposal at the burning grounds. The "cut" was the size of grain desired and it continued down the production line.[224]

The wet screened "cut" powder was pumped to the coating house where it received more than a coating. Next to each coating house stood four nitroglycerin transfer houses. Each one was about the size of a one-car garage, but with a few added features, like grounded sheet lead flooring, two inches of asbestos-cement siding, a carbon dioxide fire suppression system, thick steel doors and double-rivetted, earth-filled barricades on four sides. Into one of these garages was wheeled a transfer buggy bearing the nitroglycerin necessary to charge 23,775 pounds of raw Ball Powder with additional explosive might. After all workers had left the transfer house, the nitroglycerin was pumped out of the transfer buggy and into the still tank containing the Ball Powder. It was carefully heated and agitated until the NG fully permeated the powder. It was at this point that the Ball Powder became a double-base propellant, just as nitrocotton first became double-base rocket powder when it was charged with NG in the premix house of the N-Line. The Ball Powder grains were then coated with a "deterrent" of dibutlyphthalate to control the rate of burning.

At this point, the coated powder, still in a water-based slurry, was piped to the roll and dewater house where its moisture level was reduced to twenty-five percent. Depending on the desired type and size of the powder, most of the grains then went through a rolling machine that perfected their shape and reduced them to the exact thickness required, measured to a tolerance of five ten-thousandths of an inch. Out of the roller the powder was divided into 1,000-pound lots and loaded into "circle buggies," which were wheeled down covered walkways to sit in the wet powder rest house where heated, circulating air reduced their moisture content to a level of five to ten percent. Still in the buggies, the powder was then loaded into tram cars and taken to the tray dry house where the grains were dumped into hoppers that deposited them in a thin layer on a series of wooden trays stacked seventy-five to a rack, six of which were then wheeled into a dryer, with air circulating at 72 degrees, summer or winter. After sitting in the dryer for up to five hours, the powder was loaded back into a buggy and transferred to the salt coat and glaze house. Dumped into a "sweetie barrel" mixer, the powder first acquired a thin layer of potassium nitrate, which aided its property of ignition, followed by a coating of slippery graphite, which reduced friction and static electricity. As with smokeless and rocket powder, the Ball Powder was next sent to a preblend house where different batches were mixed into a uniform blend. The process was repeated in the final blend house where blended batches were mixed even further. Then it was on to the testing labs, the can pack house and the magazine before final shipment via rail or highway to a cartridge loading plant.

In their brief period of operation, Badger's Ball Powder plants made nearly four million pounds of .30 caliber powder for the M1 and almost two million pounds for the 20 mm machine gun, plus tiny experimental runs of other grains. The plant began operation in September 1955, ran continuously throughout 1956 and fitfully in 1957, until slowing to a halt at the end of the year. By then, the Soviet Union had lofted its Sputnik satellite into outer space, B-52 bombers armed with nuclear bombs were stationed in Upper Michigan and Nike antiaircraft missiles nestled in launchers on the lakefront near downtown Milwaukee ready to bag Russian bombers coming over the North Pole. The international situation had hardened into the atomic showdown of the Cold War and editorialists pondered the state of America's readiness for war. With Badger Ordnance as a major contributor, the nation's arsenal of conventional weaponry was full again. None of the Ball Powder made during the "Korean" period—nor the plant itself—was finished prior to the 1953 cease-fire. It went into storage ready for the next war. The entire plant went into standby with its four production areas—acid, smokeless, rocket, Ball Powder—all ready to run should the call to arms sound once more. "Our flag flies on," wrote plant manager H.B. Maroney in a final message to Liberty Powder employees, "and attests perpetual alertness in protection of the liberty we love..."[225]

On Hold Between Wars

When Badger shut down at the end of the Korean period of operation in 1958, the attitude of its neighbors and the plans of the Army differed considerably from those of the post-World War II years. In 1945, Badger Ordnance was still an intruder in Sauk County, not unlike the weekend guest who is dutifully welcomed by his hosts on Friday and happily ushered out the door on Sunday. Nor was the Army unfailingly hospitable to the plant, with its designations of "excess," deferrals of maintenance and tolerance of "cannibals" inside the gates. Thirteen years later, the atmosphere was much more welcoming. The tens of millions of dollars in payroll and purchases Badger had brought to Sauk County between 1951 and 1958 prompted its hosts to extend a permanent invitation to stay. At the plant, no one posted declarations of excess or initiated decontamination by destruction. Instead, when they shut the gates to the production areas in 1958, the Army and Liberty Powder treated Badger as if they knew it was only a matter of time before the powder lines would run again.

Badger's fate after 1958 was determined by its status in the Army's national mobilization plan. Revised annually to meet current defense needs, the "MOB Plan" applied not only to Badger as a whole, but to its parts as well. The 1958 plan, for example, placed the entire facility, except for the tank farm and the magazines, on high-maintenance standby status. In 1959, rocket propellant from Badger was dropped from the mobilization plan, but the N-Line did not drop into the excess category, just into a lower priority of maintenance. Heating lines to the rocket area were capped as an economy measure, but equipment was not dismantled nor were the buildings allowed to fall into disrepair. Instead, the 1959 plan called for the conversion of the rocket area to the manufacture of "cast propellant" solid fuel for the guided missiles now entering the national arsenal.[226]

As mobilization plans changed over the next few years, Ball Powder and cast propellant shifted in and out of the schedule of "requirements." Badger never did manufacture cast propellant, but it might have had it been reactivated in 1960, one of the years when "cast" was in the MOB Plan. Likewise, had Badger come back on line in 1964, the Ball Powder plant would not have been reactivated because it was not in the plan for that year. Rocket powder was removed from the plan in 1959, and did not return until reactivation in 1965, because defense strategists had decided that the "facilities would not be needed in the event of future mobilization." During intervening years, maintenance slid and a considerable amount of equipment was dismantled, given away or sold. The East Rocket Area, for example, "was completely cannibalized for equipment." Other production areas — NC, NG, smokeless, acid — were kept in a higher state of "readiness," so they could remobilize as quickly and inexpensively as possible. Except for the rocket area, Badger was kept ready to roll again in three to six months.[227]

As always, the military was not the only organization with an eye on the powder plant. Two weeks after it was announced that Badger would shut down by the end of 1957, Baraboo news editor Curt Page wrote a letter to United States Representative Henry Reuss. Echoing the prevailing sentiment in the county, Page wrote, "It would be fine if this huge facility could be converted to peacetime use and in that way give much needed employment." Reuss, along with Gaylord Nelson and William Proxmire, belonged to the postwar generation of Democrats that had revitalized the party in Wisconsin in the 1950s. An up-and-comer in the Democrat-led Congress, Reuss represented a Milwaukee district, but he had greater ambitions and he displayed more public interest in the Badger closing than Sauk County's Representative Gardner Withrow, who is remembered for faithfully visiting the plant every two years in the month prior to election day.[228]

Reuss was the author of a measure in Congress to provide economic development services to "distressed" rural areas throughout the United States. With the plant shut down, every community within a forty-mile radius qualified as distressed and Page thought Reuss could help. The Congressman responded by contacting Baraboo lawyer and county Democratic Party leader William Hayes, who had traveled to Washington to lobby for the siting of Badger on the Sauk Prairie in 1941. Reuss urged Hayes and fellow

Democratic activist Elaine Smith to organize a county-wide economic development committee. He also offered the services of his own development specialist, Roderick Riley, who would come to Baraboo to get the local ball rolling.

Reuss also went to work on Thomas S. Nichols, chairman of the Olin-Mathison Chemical Corporation, which had been formed out of the merger of the separate Olin and Mathison companies in 1954. As a result, O-M was the sole owner of Badger's operating contractor, the Liberty Powder Defense Company. A major manufacturer of arms and chemicals for military and civilian markets, Olin-Mathison owned forty-three plants with 36,000 employees in twenty-four states and foreign countries. To "Dear Tom" Nichols, Reuss wrote that as "one of the great corporations of the country," Olin-Mathison should "negotiate promptly for a lease that would enable commercial production to begin" [at Badger] "and the idleness of extensive plants and a splendid labor force to be ended." Three weeks later, Norl Hamilton, former manager at Badger and now a vice president at O-M, penned a formal application to the Army to lease an ammonia oxidation unit, nitric acid concentrator and enough of the nitrocellulose area to produce at least 400,000 pounds of NC a month for three years. O-M would also require steam, water, sewer and electrical services. Although it would use only a portion of the acid and NC works, O-M would still create several hundred jobs and make monthly payments of $24,000 to the government. Nichols then told Reuss that his company had made an earnest effort to utilize "at least a small part of the Baraboo facilities" and promised "to come up with other ideas or suggestions for maximum utilization of this vast facility."[229]

Olin-Mathison had scored points with an influential Congressman, but could not close the deal with the Army. Although it was advertising for leases "on all or any portion of the facilities" the Army's stated preference was "to lease all of the facilities to one lessee." It was an echo of the 1940's rental policy and equally unsuccessful. Other offers went unaccepted as well, including a proposal similar to O-M's from Soils Research, Inc., Sun Prairie, which expressed interest in the acid plant and the C-Line nitrocellulose, and from the Farmers Union Central Exchange (not yet Cenex), which wanted four acres and a building for a fertilizer plant and rail loading facilities. Except to allow the University of Wisconsin to set up a seismograph in one tunnel of the cannon proving grounds, the Army would not lease pieces of the plant until the 1980s.[230]

Reuss's county economic development organization did not get off the ground either. In the 1950s, business and civic leaders who thought about economic development saw it in purely local terms. Baraboo had its group, as did Reedsburg, Wisconsin Dells and the Sauk villages. Although the county supervisors had an "industrial development committee," Sauk County did not create an agency to work for the entire county until after Badger shut down for the third time in the mid-1970s.

In 1957, as Curt Page was writing to Henry Reuss, Baraboo was in the midst of its own development crisis. During World War II, Badger had shown the Gem City what a major industry could do for a small town and also what happened when that industry left town. A group of businessmen — Ted Mandt, George Martiny, Roy Lindgren, Henry Moeller, John Lange, Curt Page — organized the Baraboo Industrial Expansion Corporation and raised $65,000 to purchase fifty acres for an industrial park on the southwest side of the city. They succeeded in bringing thirteen new industries to Baraboo and helped several older businesses expand. By 1951, as a result of Expansion Corporation efforts, a total of $500,000 in payroll was entering the Baraboo economy annually. It was a fraction of the money Badger Ordnance could bring to Baraboo, but it was steady income, unlike Badger's, which could boom briefly and bust just as fast.

By 1957, Flambeau Plastics, one of the Expansion Corporation's larger post-war catches, had outgrown its quarters. Flambeau's officers, William and Norman Sauey, announced that the company had to have another 35,000 square feet of plant that would cost about $200,000. They also pointed out that several other communities had offered incentives to relocate and asked what Baraboo was willing to do to keep Flambeau. With Badger shutting down and cutting approximately $2.5 million out of the county economy, the Flambeau situation became even more urgent. The Expansion Corporation agreed to borrow $100,000 to build a building for Flambeau and to raise another $100,000 through the sale of corporation stock at fifty dollars a share. So small investors could buy stock, the corporation issued stamps that could be purchased for one dollar each, with fifty stamps the equivalent of one share of stock. This grass roots effort in a community with an average annual income of about $4,000 and whose largest employer — Badger Ordnance — was laying off hundreds of workers, succeeded. Flambeau got its building, stayed in Baraboo and has been a steady employer and community investor ever since. The consequences of the effort to keep Flambeau in town go far beyond Baraboo. In the 1970s, Norman Sauey's son Ron came home to Baraboo with fellow ornithologist and close friend George Archibald. They set up headquarters at the Sauey farm north of Baraboo and organized what they called the International Crane Foundation. It has since become the leading organization in the world for the protection and propagation of the planet's oldest extant species of bird.

Sauey and Archibald would have founded the Crane Foundation anyway, but probably not near Baraboo if the Sauey family had moved Flambeau Plastic out of town in 1957. Flambeau stayed with help from the Industrial Expansion Corporation, which was organized to create a steady growth alternative to the bust and boom of Badger. The links between the ammo plant and the conservation foundation are not straight-line, but clear and indicative of the course of history in Sauk County since 1941. Not all roads in the county lead to Badger Ordnance, but nearly all the pathways of local history in the last half of this century wind their way back to the plant on the prairie.

One path that led to a dead end was the industrial development trail Henry Reuss attempted to blaze, but not for want of trying on the part of the Congressman. Once it became clear that he could not bring a non-military enterprise to Badger, he tried to bring Badger out of the military. In 1960, at the behest of David Carley, director of the Wisconsin Department of Resource Development, Reuss joined Senator William Proxmire in a joint request to the Army asking that Badger be removed from "standby basis and given to the Governmental Services Administration to dispose of as federal surplus property." Carley had informed Reuss that he had "a specific industry now interested in purchasing a plant of this type...to manufacture solid fuel propellants."[231]

The Army was itself proposing to convert the rocket area to solid fuel production upon mobilization and its plan for 1960 called for Badger to extrude four million pounds of cast propellant per month. However, when Pentagon spokesman General H.A. Gerhardt replied to Reuss and Proxmire, he did not mention solid fuel. He did say that Badger "has the only ball powder capability in the military establishment." It could also produce "single and double base propellants" and "the condition of the physical plant is such that reactivation costs would be low." Badger was not available for civilian use and would remain on quiet hold in the Cold War arsenal of the United States.[232]

In the Cold War year of 1960, the United States was embroiled in the U-2 spy plane confrontation with the Soviet Union. At Badger the heating lines to the rocket area were shut off, along with the boilers of the Number One power plant, and twenty tons of unused coal were shipped to the Argonne National Laboratory in Illinois. At the initiative of County Agricultural Agent Richard Halbach, an informal committee of Sauk County supervisors, farmers and business leaders met in Baraboo to discuss the possibilities of bringing the state fair to the grounds of Badger Ordnance. The idea generated little interest at the plant or anywhere else in Wisconsin.

In 1961, the Berlin Wall was erected and, in the ensuing crisis, the Sauk County unit of the National Guard was summoned into active service. Responding to an inquiry from Senator Proxmire, Secretary of Defense Robert McNamara said that Badger would be reactivated only in the event of war. He did not say that many Kennedy administration officials considered the possibility of a nuclear exchange over Berlin to be imminent, in which case Badger could not be activated because it would likely be destroyed by a Soviet bomb. The crisis wound down, and with it the maintenance budget for Badger. Work crews reported that hundreds of the World War II-era wooden poles carrying electrical, telephone, steam and other lines were decaying and should be replaced. When no money was budgeted for new poles, work crews searched for solid specimens in the low-maintenance rocket area, uprooted and replanted them in the higher priority smokeless and Ball Powder areas. Robbing Peter to pay Paul was a short-term solution for the problem of rotting utility poles, but it would not work for the wooden revetments of the barricades. Work crews had also discovered that the posts, beams and sheathing on about one-third of the hastily constructed twenty-year-old wooden walls would rot to powder in five years. In ten years, the wood in all but a few of the newer barricades erected in the Korean years would decay and collapse.

In 1962, the United States and the Soviet Union again faced nuclear Armageddon for the eleven days of the Cuban missile crisis. The Army was in the midst of a reorganization plan that revamped the Ordnance Corps. Up the chain of command from Badger, the Ordnance Ammunition Command became the Ammunition Procurement and Supply Agency (APSA) while the Chief of Ordnance became the Army Material Command (AMC). At Badger, further budget cuts forced Liberty Powder to reduce employment to 116 maintenance people and to close the cafeteria which had served meals every day since 1951.

In 1963, the regime of South Vietnamese president Ngo Dinh Diem was overthrown in a bloody coup in Saigon and the war that had been grinding on since the 1940s entered a new and perilous phase. President John Kennedy, who was himself assassinated several weeks after Diem's fall, had already approved a steady buildup of American troops in Vietnam that totalled 13,000 soldiers at the time of his death. At Badger, all hands were trying to orient themselves to the fact that, as a result of the Army reorganization, the facility was no longer designated an "ordnance works" but an "army ammunition plant." Despite new letterheads, badges and signs, most folks continued to call it "the powder plant" as they had since 1941.

In 1964, Lyndon Johnson was elected President by a landslide in large part because voters believed he was more likely to keep the United States out of war in Vietnam than his opponent Barry Goldwater. The United States Congress, including Senator Gaylord Nelson, Senator William Proxmire, Representative Henry Reuss and Sauk County's Representative, Vernon Thomson, had already voted in favor of the Gulf of Tonkin Resolution which empowered Johnson to escalate the war in southeast Asia. At Badger, the Army continued to keep the Ball Powder, smokeless and acid areas on a high level of maintenance, but also asked for bids to remove more than 20,000 running feet—about four miles—of "surplus" steam and air lines in the rocket area. According to the MOB Plan, all the rocket propellant American forces would need in the next war could be supplied from other sources.

In February 1965, the United States Air Force launched Operation Rolling Thunder and B-57 aircraft began the bombing of North Vietnam. At Badger, staff from the United States Fish and Wildlife Service, the Wisconsin

Conservation Department and the plant began to implement the new wildlife management plan they had jointly drawn up the year before. In late fall, John Troyer, the Army's top manager at the plant, who bore the title of Commanding Officer's Representative, was ordered to prepare cost estimates for the reactivation of each area of the plant, including the rocket line. Secrets about Badger were hard to keep, so Sauk County buzzed with quiet talk about the plant's reopening. A few job seekers stopped at the gate to ask about work, but not until the end of the year were the rumors about reactivation confirmed.

At War Again

In the early 1960s, while the Cold War threatened to boil over, Vietnam simmered and Badger sat on a back burner switched off, the Army was in the midst of a research and development program to find a replacement for the legendary M-1 rifle as the standard combat weapon of the American foot soldier. It was the real-world equivalent of a cool, new generation actor like Steve McQueen replacing rugged old John Wayne in a gung-ho war movie, and about as public and controversial. After several years of ballistic tests and firing-range trials, a rifle initially designed by the Colt firearms company emerged as the leading candidate to succeed the hallowed M-1. Designated the M-16, the new rifle proved itself in combat with the Army of the Republic of (South) Vietnam in 1962, and was touted as the modern infantryman's dream weapon. Lightweight and automatic, the "little black rifle," as it was known in Vietnam, weighed about as much as a 1960s glass gallon milk jug (filled) and carried a clip loaded with high-powered 30 caliber ammunition that a soldier could trigger off in rapid-fire automatic bursts that were accurate and deadly for any target within 200 yards. Unlike the M-1, which soldiers were trained to aim from the shoulder and to fire one round at a time, the M-16 was held at the waist and fired multi-shot volleys which sprayed bullets like a garden hose spitting water. From the point of view of an ammunition manufacturer, soldiers equipped with M-16s were better customers than soldiers firing M-1s. They consumed more product.

Trials of the new rifle continued in 1963 and '64, but when the Vietnam buildup intensified in 1965, the fate of the M-16 had to be settled. In June 1965, General William C. Westmoreland, U.S. Army commander in Vietnam, reported that he would need 150,000 troops to fight the war. By the end of the year he asked for another 25,000 men and predicted he would need

100,000 more in 1966. To meet the demand for manpower, the Selective Service ordered local draft boards to call up 340,000 men in 1966, and 300,000 more each year until 1971. Along with his call for more soldiers at the end of 1965, Westmoreland placed the Army's first order for 100,000 M-16s.

As any powdermaker knows, a gun is only as good as the ammunition it fires. In the years it took to develop and test the M-16, the Army also experimented with a variety of propellant loads for the "233" cartridge the rifle fired. In 1964, the nation's three leading powdermakers — Du Pont, Hercules and the Western division of Olin-Mathison — each offered trial loads for the Army to test in its M-16s. After tens of thousands of test firings, the Army accepted both the Du Pont propellant and the WC846 Ball Powder manufactured by Western and authorized cartridge manufacturers to load either one into M-16 ammunition sold to the U.S. The ammunition industry was dominated by the big three of cartridge manufacturing, Olin-Mathison, Federal and Remington. Western was a division of Olin-Mathison, so, to no one's surprise, O-M chose WC846 Ball Powder for its M-16 ammo. Federal, the smallest of the three and a division of no one but itself, also chose Western Ball Powder. Remington was a division of Du Pont so it chose Du Pont powder for its M-16 ammo, at least until the end of 1964, when it too opted for Western WC846. In 1965, just as the United States was preparing to put several hundred thousand soldiers on the ground in Vietnam, and arm them with M-16s, the only ammunition being manufactured in quantity for the new rifles came from one company in one place, Olin-Mathison of East Alton, Illinois.

Olin-Mathison, with its brass foundries, Ball Powder works and shell loading plants, anticipated orders for tens of millions of cartridges for the M-16. As the East Alton powder works expanded to fill those orders, management found itself locked in a drawn-out dispute with its union workers. The situation was reminiscent of the negotiations at Badger in the spring of 1952. The difference was that O-M and the unions could not reach an agreement in 1965 and the workers went on strike. Despite the efforts of a presumably labor-friendly federal mediator from a Democratic administration, O-M's workers walked out on December 1, 1965, and were still out at Christmas.

Late in the afternoon on December 29, Curt Page called John Troyer and asked what he knew about the reactivation of Badger Army Ammunition. When Troyer replied that he had received no word from Washington on the matter, Page told him that an aide to Representative Thomson had called the paper and said that Badger would definitely reopen in January. Senator William Proxmire's office also alerted the media and news reader Walter Cronkite scooped his television rivals by reporting it on the "CBS Evening News" that night. NBC's "Huntley-Brinkley Report" did not air the news "in color" until December 30. Cronkite, Huntley and Brinkley were the original celebrity television newsmen and the mere mention of Badger and Sauk County on their shows was itself treated as a news item by local journalists. The morning after hearing from Page, and probably watching Cronkite, Troyer contacted his superiors at the Ammunition Procurement and Supply Agency, who must not have been watching TV. They confirmed the news and expressed their surprise that someone in Washington leaked it prior to the official announcement, which they had planned to make after returning from a long New Year's break on January 5.

As reported in the press on December 30, Badger was being reactivated because the strike at East Alton was threatening the supply of ammunition to the troops in Vietnam. Secretary of Labor Willard Wirtz "admitted in theory" that ammunition could run out without a settlement but in fact the situation was far from critical. President Johnson had accepted the military's recommendation to escalate American involvement in Vietnam early in 1965 and the number of bombing runs and troop shipments had increased steadily afterwards. As Troyer also told reporters, he and his staff had been preparing cost estimates for reactivation "for at least four to six weeks." The mere demands of the war, as well as the customary and commonsensical defense policy of redundancy, would have prompted the Army to reactivate the second of only two Ball Powder plants in the country no matter what the labor situation in East Alton. Reopening Badger did help settle the strike at East Alton and force the strikers to accept less than they initially demanded. Both plants remained open and made hundreds of tons of propellant and profits for their operator.[233]

Olin-Mathison received official notification that the Ball Powder plant, along with supporting acid, nitrocellulose, nitroglycerin and other necessary facilities, was ordered out of standby on January 3, 1966. O-M would be the contracting operator and not its subsidiary, the Liberty Powder Defense Company, which was no longer active. One month later, the Army informed O-M that the hitherto forsaken rocket area would also be reactivated. Weapons development had come full circle. What once was old and obsolete was new again. Since 1959, when the Army had deleted rocket powder from its Mobilization Plan, it had expanded the use of helicopter-mounted machine guns, cannons and rockets in support of ground troops. Already, by the end of 1965, it was clear that the helicopter gunship would play as large a role in Vietnam as the tank had in World War II and the machine gun in World War I.

There seems to have been a sinister warlock haunting the Pentagon, an evil genie of inefficiency who decreed that whatever weapon the military planners deemed superfluous at the end of the last war would become a combat necessity at the start of the next. Just as the Army left Badger's smokeless powder area unmaintained after World War II only to find that it needed all the cannon powder those lines could produce for Korea, it found that the rocket area it had let slide into low priority in the early 1960s had to come back on line for Vietnam in the latter half of the decade. Thanks to the evil genie, production of ammunition in time of war would be delayed and the cost of rehabilitation would far exceed that of the maintenance deferred.

In order to fund the mobilization of which Badger's reactivation was a part, President Johnson asked Congress for $12.8 billion in 1966 dollars, worth at least twice as much as the greenbacks of 2001. Rehabilitating Badger would cost about $20.7 million, with the rocket area claiming $9.4 million and smokeless $8.7. The Ball Powder area was a comparative bargain at $2.77 million, including the reconstruction of the hardening house destroyed in the explosion of September 1966. As the bill indicates, peacetime maintenance paid for itself in the Ball Powder area. However, since it was the newest, smallest, least-used and most modern production facility at Badger, the Ball Powder plant should have cost less to rehab. By comparison both the smokeless and rocket areas were older and physically much larger than Ball Powder, and the rocket plant had been on low priority for maintenance since 1959. The cost of bringing back the rocket line was also driven up by the purchase of new equipment to convert the facility from the Mark 16 and Mark 18 cruciform sticks that made up nearly all of its output for Korea to the smooth Mark 43 sticks on order for Vietnam.

Badger produced its first propellant for the Vietnam war in mid-May 1966. By the end of June, the plant had manufactured 100,000 pounds of 7.62 mm Ball Powder for use in lightweight machine guns and in rotating, multi-barrel, rapid-fire Gatling guns mounted on helicopters. In July Badger produced its first 50,000 pounds of 5.56 mm Ball Powder for the M-16 rifle. By the end of the year the plant made about one million pounds of 5.56 mm and 1.7 million pounds of 7.62 mm powder, as well as nearly 290,000 pounds of 30 cal. powder for heavier machine gun loads. In less than eight months Badger made 3.1 million pounds of Ball Powder, more than one-half of its total Ball Powder output in the Korean period. Production surpassed the Korean total of 5.7 million pounds early in 1967, a year in which the plant manufactured over 6.7 million pounds of Ball Powder. Production had kicked up so quickly it threatened to outstrip the supply of packing containers.

The initial runs of Ball Powder were made with reclaimed nitrocellulose processed at the extraction plant and glycerine nitrated with imported acid. By the end of August nitric acid manufacturing and sulfuric acid concentration were underway, followed by start up of the C-Line nitrocellulose in preparation for activation of the rocket line. The fabrication of Oleum for shipment to the Joliet Army Ammunition Plant and production of rocket paste and carpet rolls began in September 1966, but no Mark 43 rocket powder was extruded and finished until June 1967. Rehabilitating the rocket area, finding, purchasing and installing new equipment, training personnel and working out the inevitable glitches took its toll on the time clock.

However, the rocket area started finishing rocket powder for Vietnam in less time than for Korea.

All work stopped on September 8, 1966, when the plant experienced an explosion and fire on the Ball Powder line. Start-up periods, with inexperienced work crews and untried machinery, were notoriously dangerous. Fatal accidents were most likely to occur in the opening months of operations and had occurred when the E.C. Line opened in 1943, the N-Line in 1945 and Ball Powder in 1955. Part of the first production line reactivated for Vietnam, Hardening House One contained a series of hardening still tanks in which 5,000-pound charges of nitrocellulose slurry were mixed with ethyl acetate, diphenylamine, chalk and potassium nitrate, then heated and agitated. The process was underway at a few minutes before ten in the morning on a sunny, late-summer day when an explosion in Hardening Still #4 blew out about one half of the two-story building and the ethyl acetate storehouse nearby. Even though it occurred a mile away inside the plant, with other structures in between, the blast shook the guard shack at the main gate, and rocked mobile homes on their piers in the Bluffview trailer park another mile away. Windows rattled in Baraboo, crockery clattered in the Sauk villages and a grey mushroom cloud of dust and debris rose over the plant. Jim Mattei, a fresh-faced twenty-five-year-old junior manager for Olin was in a meeting with two plant veterans when he heard the blast. "I didn't know what to make of it," he said, "but I looked at those two guys and saw their faces just go white." The three men jumped in a car and rushed to the blast site where firefighters were already in action. Fire Chief Percy Loomis was in charge and directing the efforts of his own crew and firefighters from Baraboo, Sauk City and Prairie du Sac. "I was new to the job," Mattei continued, "and I used to wonder just what it was Percy Loomis did for his pay. But after watching him that day...if Percy Loomis never did anything else for the rest of the time he worked at Badger, he earned all his pay that day." The fire crew had the blaze under control in about an hour, but it was still smoldering until 1:00 PM. Spectators in cars, some caring, some curious, lined the shoulder of Highway 12 as smoke rose from the embers and the air filled with the model-airplane-glue odor of ethyl acetate.[234]

Thirteen men were at work in the Hardening House, two escaped unharmed and eleven were injured, but no one was killed. "I was walking up the stairs when all of a sudden, Bang!" said operator John Karasik. He and twenty-one-year-old David Hutchens were blown off the stairway and about thirty feet into a wall which had already burst into flames. They remembered to stay low to the ground and crawled out of the building as "flames shot into the air." John Adams was knocked out by the blast and when he woke up "everything was lying on top of me." He was on the second floor but was able to escape by crawling down the stairs and out a doorway. All the men escaped from the burning buildings, some with help from other workers who bravely came to the scene instead of following the standard safety protocol of getting as far as possible away from an accident. Thomas J. Schwichtenberg was the most seriously injured with a compound fracture of his right leg, facial burns and fragments of debris imbedded in his abdomen.

"We have absolutely no idea what caused the blast," John Troyer told reporters. One of the workers confidentially told another newsman that he thought sparks from a "defective machine" touched off the blast, which was entirely possible given the nature of the materials in the hardening still and the state of the equipment. According to the official operating report for 1966, "the Ball Powder plant was not operated at a high enough rate after construction in 1954-55 to reveal latent deficiencies found in new plant design." The machinery was not fully broken in and then idle for a decade. The blast sent hot, flaming debris cascading in all directions. An electric motor the size of a bushel basket was blown into the air and hit so hard when it came down about 200 yards away from the hardening house that it drilled itself through the asphalt and completely buried itself below the surface of the pavement. By design, the hardening house was far enough away from neighboring structures so damage was reduced to a minimum and the fire did not spread. It was not barricaded, but the machinery inside was designed to direct a blast upwards, a feature that launched the electric motor into space but also probably saved the lives of the work crew. The force of the blast dissipated upwards instead of concentrating inside the building where it could pulverize the men between ungiving beams and berms, as in the fatal accidents of 1943 and 1945.[235]

Since redundancy was also built into the design of the Ball Powder plant, the explosion at Hardening House One delayed but did not halt production. Two days later, work crews began reactivating Hardening House Three. They brought it on line in about three weeks and made up for lost output in September by manufacturing greater volumes of powder in October, November and December. It was needed, too. By September 1966, the United States and allied forces in Vietnam had fired approximately eighty-nine million rounds of 5.66 mm Ball Powder.[236]

Production continued to increase in 1967 as the Army ordered more and different types of propellant requiring the activation of more production areas. Demand for artillery ammunition brought smokeless line B back into action, along with nitrocellulose B to feed green powder into it. By the end of the year, in addition to four types of Ball Powder, Badger was making four types of cannon powder. It was processing 2.75-inch, Mark 43 rocket grains from the first mixing of NC and NG to the final capping and wrapping of finished sticks. Rocket paste not used for Mark 43 was shaped into carpet rolls for shipment to other plants, as was nearly all the output of the Oleum plant and a substantial portion of newly formulated nitric acid. The acid, nitrocellulose and nitroglycerin works, which were the first steps on the line to finished propellant, were all up and running hard. Output was so great that, in addition to substantial tonnages shipped out, finished powder filled over 200,000 of the 250,000 square feet of storage available in the magazine. Over it all, and symbolizing the bustling activity at their feet, stood the five stacks of the power plant, idle since 1959, now burning pulverized coal by the hundred ton, delivering pressurized power to pipe lines and machinery while pumping so much smoke into the sky the prairie smelled like a miniature Pittsburgh and on damp, foggy mornings looked like smog-choked Los Angeles.

All wars are nasty, foul and demeaning to those caught up in them, but the Vietnam conflict seemed to be better able than most to besmirch fine people, honest work and noble aspirations. Badger's impressive start-up in 1966 and '67 was marred by the public controversy over the performance of the M-16 rifle. Reports coming back from Vietnam in the first months of the buildup stated that the guns jammed alarmingly often, disarming and endangering men in combat. A House of Representatives Armed Services Subcommittee investigation directed by Representative Richard Ichord of Michigan revealed that in 1965, the Army's Combat Development Command had tested the M-16 and found that the fouling which caused the gun to jam was "caused primarily by qualities of the propellant used in standard ball 5.56 mm cartridges." Badger had yet to be reactivated, so the propellant the Army fired in those tests was made in the Olin-Mathison plant in East Alton, but it was the same powder later made at Badger. One of its objectionable "qualities" was that Ball Powder deposited a grimy residue that built up and jammed the firing mechanism of the M-16. As a double-base explosive, Ball Powder was also powerful enough to increase the rate of fire of the M-16 from 850 to 1,000 rounds per minute and this accelerated rate of fire also added to the grime and scarred the rifle's firing chamber.

After a period of embarrassment and denial, the Army accepted the revelations of the Ichord Committee and ordered that the steel firing chambers of the M-16 be replaced with sturdier chromium-lined units to resist scarring, that buffers be installed to reduce the rate of fire to 650 rounds per minute and that the necessity for keeping the guns clean be drummed into the soldiers firing them. The marriage of the M-16 and 5.56 mm Ball Powder took place even though the rifle was designed to fire single-base propellant. After the specifications were changed and Du Pont decided it could not produce the new powder the Army wanted, Ball Powder was drafted to fill the breech. That the resulting marriage of convenience cast doubt on the performance of an otherwise highly praised weapon while it opened the defense establishment and the war effort to suspicion and criticism from yet another source was typical of the times. It was another thing wrong in the 1960s. It was Vietnam.

The M-16 controversy was resolved by the end of 1967 without any repercussions at Badger other than some fine-tuning of the ball-powder formula to reduce deposits of residue after firing. To the contrary, powder production increased and diversified. In 1968, the plant manufactured eleven types of Ball Powder, from the controversial WC856 5.56 mm for the M-16, to the WC860 .50 caliber for heavy-duty machine guns. In May the Ball Powder auxiliary plant was activated to manufacture propellant for blankfire ammunition and work began to refit the C-Line smokeless to meet larger orders for cannon powder. The E-Line nitrocellulose was reactivated in August, giving Badger three lines nitrating cellulose that was processed into smokeless, Ball and rocket powder. Between mid-1968 and early 1970, during which American troops saw some of their heaviest combat, Badger produced powder like it had never produced powder before. The peak production month for smokeless was October 1968, 5.8 million pounds; for Ball Powder it was July 1969, 1.6 million pounds; and for rocket, it was January 1970, 178,000 pounds. In this period Badger exceeded the 271,000 pounds of powder it made for World War II and the 277,000 pounds it packed out for Korea. Demand for powder and production started to decline in the latter half of 1970, about one year after the United States made its first significant withdrawal of troops from Vietnam. The gradual winding down followed, leading to the inevitable shutdowns: smokeless in August 1973, Ball Powder in May 1975 and rocket in June 1975.

In sum, Badger manufactured 486,724,800 pounds of powder between January 1966 and June 1975, its longest period of operation. The largest volumes by type were 160 million pounds of smokeless for the 155mm howitzer, 71 million pounds of propellant finished into 10.2 million individual sticks of rocket and 37 million pounds of 5.56 mm ball for the M-16. Despite these impressive numbers, no more than one-half the production area had been activated and it was operating at about one-half of its capability. If, instead of running at one-half of one-half, Badger had run at full capacity, it could manufacture three million pounds of Ball Powder per month, sixteen million pounds of smokeless and five million pounds of rocket propellant. Had Vietnam escalated into the massive, non-nuclear, Asian land war that many Americans feared, Badger could have supplied an equally massive quantity of ammunition for it.

War is always hell and it is never cheap. Vietnam's toll of destruction, suffering and death was levied not only in Southeast Asia. After the explosion in Hardening House One in September 1966, Badger suffered a string of minor but destructive explosions and fires. In September 1968, for example, a flash fire in the solvent store and pump house on Smokeless Line B subjected Anthony Dugan and John R. O'Keefe to burns on their hands and faces. Their injuries might have been more serious had they not been wearing safety glasses, flameproof coveralls and the hard hats that had only recently become required headgear — and a political token — for American workers.

While many workers turned the hard hat into a symbol of blue collar support for American involvement in Vietnam, others considered it to be an uncomfortable nuisance. "When the OSHA men came through, they said you have to have your hard hats," remembered Walter Schuck, who worked in a mix house where NG and NC were blended. "Don't drop your hat in there, it will blow up," the government man also warned Schuck. "That's why we didn't like to wear them."[237]

With or without hard hats, the accident toll started to rise in 1970, which began with the attempted bombing of the plant by the self-styled Vanguard of the Revolution, Karleton and Dwight Armstrong. Except to litter the grounds with broken glass and metal trash, the Armstrongs left the plant unscathed, but accidents during operations were serious and the damages real. The peace of a calm overcast night in April 1970 was shattered by what was likely the largest explosion ever at Badger. "Eight tons of powder in a drying house at BAAP went up in an orange ball of fire about 7:20 Saturday night," reported the *Baraboo News-Republic*. A tray dry house in the main Ball Powder Line, where grains of powder were stacked on racks with heated air circulating through them like shelled beans in a food drier, went up not once, but twice. "I saw one huge ball of fire go skyward," said Russ Potratz, who was driving on Highway 12 near the southwestern corner of the plant. "The entire area was lit up by the bright orange flames which kept shooting skyward. Debris was burning several hundred feet in the air." Other witnesses reported one distinct explosion followed by another and fountains of orange flame. The flashes were visible in Baraboo and one motorist as far away as Wisconsin Dells wondered why he could see the flash of the setting sun since it had not broken through the clouds all day. A single-story building, the tray dry house was all but invisible behind its heavy barricades, which directed the force of the blast and flames upwards with such spectacular results and also prevented the fire from spreading through the plant. Fortunately, once the powder was loaded in the drier, tray drying was an automatic process so no one was in the building when it exploded and no one was injured. The most powerful of several serious explosions at the plant since its reactivation, the tray dry house blast led APSA to give Badger its least favorable safety rating and to order "considerable effort toward eliminating unsafe acts and conditions at the plant." As a result, the plant operated for nearly three years, from December 1971 to November 1974, without a lost-time accident. Not only were there no major accidents, but also no minor mishaps that harmed workers enough to halt their work. While Olin placed more emphasis on safe operation, the Army took measures to tighten plant security, "including modification and intensification of anti-penetration measures," just in case the accidents were not as accidental as they seemed to be.[238]

Despite the increased emphasis on safety, accidents continued to occur. In October 1970, Victor R. Schommer, a carpenter reshingling an activated-carbon recovery building, encountered a blast of live steam—penetratingly wet and terrifyingly hot—that severely burned his upper body and respiratory tract. After lingering painfully in the hospital for two days, he died from respiratory failure. After the accident it was discovered that Schommer was working on the wrong building and that he should have been on another building where the steam lines were shut off. Construction work always went on during operations at Badger and was responsible for almost as many injuries and deaths as powder making. A few months after Schommer's death, "sweetie" barrel operator Richard Hardy was working alone in one of the bays of a salt coat and glaze house in the smokeless area when a "low order" explosion destroyed the building. Several hours passed before firefighters discovered his body and it was not known if he died immediately after the blast or lingered before passing away. Hardy was the first worker to die in an accident on the powder production line at Badger during the Vietnam period. The death of Leo Turner exactly one month later was not the direct result of powder making, but of making power. He was working at the bottom of a hopper car when its load of frozen coal thawed and shifted, smothering him in coal dust. In the terse works of the *Badger World* plant newspaper, Turner "lost his life in performance of his duties at the powerhouse."[239]

On one rare occasion the danger zone was outside the plant. In April 1971, a freight train on the Milwaukee Road main line from Minnesota was on its ways to the plant via Portage. It was hauling a box carload—about 50,000 pounds in 640 fiber drums—of blankfire propellant from the Twin Cities Ammunition Plant on its way to Badger for reprocessing into Ball Powder. As the boxcar approached the overpass carrying Highway 16 over the tracks, it exploded. The embankments along the tracks acted like barricades and directed most of the force of the blast upwards, but at least ten cars were derailed and "gigantic pieces of boxcar, plants, powder kegs and other debris could be seen falling through the air." Witnesses as far away as Madison to the south, Randolph in the east and Pittsville on the northwest, as well as National Guard fighter pilots in the air over Stevens Point, reported seeing the flash and smoke. No one was injured and the Highway 16 overpass was undamaged, but it was a near miss. The train had passed through the heart of Wisconsin Dells about twenty minutes earlier and was only moments away from the center of Portage, not to mention the many other communities large and small that it had passed through on its way from Minneapolis-St. Paul. On the other hand, Badger had shipped out and received millions of tons of explosive powder and flammable acids in tens of thousands of railcars over the years and this car was the only one to explode.[240]

The fatal year of 1971 ended with a final fatal explosion. In the paste breaker/blender house on the rocket line, bags of dried rocket paste were emptied into a barrel for reblending into a uniform mix prior to extrusion.

Standard procedure called for the operators to pour 720 pounds of paste into the blending barrel and flip the switch to start the barrel spinning. They were then to leave the building and take shelter behind a thick concrete wall. Somewhere in the course of this procedure on December 8, "a low order explosion and fire occurred." Although officially characterized as "low order," the blast "shook the foundations of nearly every building in the plant," as one worker told a reporter. It destroyed the paste breaker/blender building, inflicted lacerations and bruises on operators Dennis Pechan and Avery Hartley, seriously burned Leroy Graves on over one-third of his body and killed John F. Mitchell. Pechan and Graves were on the first floor and made it out the door. Hartley, who was on the second floor with Mitchell, sprinted for the escape chute and survived. Mitchell didn't make it. He was the second Badger employee to die on the powder line during the Vietnam period.[241]

In 1971, 2,357 of the 58,193 Americans who died in Vietnam met their end. Tens of thousands more Vietnamese, Cambodians and other Southeast Asians died that year. At Badger the toll for 1971 was three men dead. The war was not so far away from home. Those who made the tools of war suffered along with those who used them.

Working For The Boys At War

In Sauk County the news of the reactivation of Badger prompted immediate interest. On December 31, 1965, the day after official word of reactivation leaked, job seekers started lining up outside the plant gate at 7:30 AM. By the end of their shift, Guards Bruce Lindgren and Gerald Sabby reported that they had issued over 350 entrance passes to applicants. Although not even John Troyer knew when hiring would begin and how many jobs would open, and despite a chilling winter rain drizzling on the parking lot, go-getters lined up at first light to put their names at the top of the applicants' list. They knew Badger would pay higher wages for unskilled and semi-skilled work than most other employers in the area. If the past was any proof, there would also be plenty of overtime at higher pay, bonuses for shift work and other fringe benefits. In line that morning when a reporter from Milwaukee arrived were Frank Hein, Baraboo, a construction worker unemployed for the winter, and Ronald Hanko, Cazenovia, a veteran of Badger's World War II days. Both of these men and their wives, who also filled out applications, wanted jobs at Badger because it offered "a better opportunity." From this very first day of what was not yet reactivation, the personal impact of the Vietnam war, and its potential divisiveness, were evident. Sixty-year-old Charles Wilson of Avoca, who had spent over four years at Badger during World War II and was now working as a salesman, said he wanted to come back because, "I figure this is more important." A couple from Richland Center, who would not identify themselves because they did not want their current employer to know they were ready to leave, said they would quit their jobs to come to Badger because they had two sons in the Air Force and "it might help" them. Also in line was eighteen-year-old David Elder of Rock Springs. He was prime draft bait and knew it. He wanted to hire on at Badger not only because it paid better than his current job but also because it could earn him "a deferment from the army."[242]

Job seekers, no matter what their motivation, were happy to see Badger open again, but many employers were less than pleased. "It was the worst thing that could happen to the city at this time because it might stop expansion of other industries," said Ted Mandt. He believed that the local economy should be based on the slower, steadier and more stable growth fostered by his Industrial Expansion Corporation. "I don't like the boom and bust aspect of this particular thing." Mandt's associate John Lange, who had worked hard to bring Badger to Sauk County in 1941, agreed. "It was too bad. It might put a crimp in the industrial expansion. Local industries might lose workers because they can't afford to meet the wages." On the other hand, more jobs in Sauk County meant less commuting to Madison. "Without a doubt, a lot of people who are working in Madison will come back to Baraboo," said auto repair and service station operator Ben Rotzoll, who also remembered staying open round the clock during the World War II boom.[243]

The population of the area within a reasonable commuting distance to Badger had so increased and commuting itself had become so much the norm that the status of housing near the plant was irrelevant. Since it considered Badger's reactivation to be "temporary," the federal government took no action to build or improve public housing nor did it offer any extraordinary financial incentives to private housing developers. After the experience of the Korean years, when most of Badger Village remained vacant even after the federal housing authority spent nearly $800,000 on improvements, the Army's vision was sufficiently clear-eyed to note that public housing had become unnecessary at the powder plant. In 1960, the General Services Administration sold Badger Village to a partnership of local businessmen headed up by Forrest "Woody" Zantow and Wayne Hatz. They had remodeled the old former H-Barracks of North Badger into housing for senior citizens, and had no more than a few vacancies for powder plant workers. When the plant reopened they developed a mobile home tract with 110 sites, many of which were occupied by Badger workers, but it was nothing on the scale of 1944 or even 1954, nor should it have been. The automobile had made the company village obsolete. Busing resumed, but on a reduced basis, with contractors running lines to communities north and west of the plant: Adams, New Lisbon, Hillsboro and Richland Center.

A little more than a year before the reactivation order, the federal government had also sold off the Staff Village housing to Mazomanie Attorney Thomas Winch, to the consternation of the military officers assigned to the plant. Available housing was either below standard for the soldier executives or too expensive for their housing allowances. They had to stretch their not overly elastic paychecks to make up the difference. Its isolation from other military installations was also a negative for the soldiers assigned to Badger. The subsidized shopping, health care and recreation stateside soldiers enjoyed elsewhere were not conveniently present. The nearest military housing was in Sun Prairie and the closest base-exchange shopping facilities were at Camp McCoy or Fort Sheridan, Illinois. However, no provisions for special housing were made for the newly appointed commander, Captain Sidney I. Davis, or his successors.

Except for degreed professionals with special technical, scientific or managerial skills, who were recruited throughout the United States, filling employment slots at Badger was not difficult in the Vietnam era. At the end of 1965, prior to reactivation, the Olin-Mathison workforce numbered about 150, and the Army's totalled three people. Within a year, O-M had more than 1,800 people on the payroll and by the end of 1967, nearly 3,600. Among these were at least ninety-seven men and women who had been on the job since the Korean activation and were thereby qualified to receive 15-year service pins. Experience was important and Jim Mattei, who started his thirty-year career with Olin in March 1966, had the job of tracking down Badger alumni and getting them to come back. "It wasn't too hard," he recalled. "I'd call one guy and he'd give me ten more names." Many workers already earning a good wage were reluctant to jump back to Badger, "then we had to offer carrots, incentives, appeal to their patriotism to get them to come back." Growth continued in the wake of the escalating American commitment to the war, with every increase in troops sent overseas eventually followed by a rise in hiring at Badger. The workforce rose to and remained above 4,000 throughout 1968 and peaked at 5,362 in August 1969. As American ground troops were gradually withdrawn, the line on the Badger employment chart turned downwards: below 4,000 in early 1970, to 3,000 in 1971, 2,000 in 1973 and 1,000 in early 1975. When production halted in June 1975, about 400 people were on the Olin-Mathison payroll.[244]

Throughout the 1960s, Badger workers stood with the majority of Americans who supported American policy in Vietnam. As Robert Quick, a Mule Operator who started commuting from Mineral Point in 1967, told the *Badger World* plant newspaper, "It's a good place to work. I have a wife and six children. My three boys are all in the service, one a nine-month veteran of Vietnam...I know that I can help my boys and everyone in the Armed

Forces. I know I have to produce here and all of us have to produce here because if we don't there'll be no powder for them." Quick was one of many Badger production workers with a family member in Vietnam. Young men from working class and lower income families were much more likely to be drafted than men from the middle or upper classes who could more readily obtain educational deferments or find openings in the National Guard. Draftees were much more likely to be sent to Vietnam than volunteers, to serve in combat and to become casualties. In Wisconsin, working class and lower income men were overrepresented in the draft by a proportion of two to one in relation to their numbers in the population. U.S. Representative Alvin O'Konski, an old McCarthyite whom no one would call "soft" on communism, surveyed 100 inductees from his economically depressed 10th District in northern Wisconsin, and found not one from a family earning more than $5,000 a year. "They say the poor are always with you," O'Konski commented, "If the draft goes on as it has, they may not be with us much longer."[245]

This fact of political and economic life in the 1960s gave Badger workers a doubly personal link to the war. It existed at work during the day and at home at night. "I have a son who's in the army in Germany, but is going to be transferred to Vietnam next month," said Lab worker Ruth Alberts shortly after she was hired in 1967. "I look forward to working here and I feel it's my patriotic duty to help as much as I can." About one-third of Badger workers during the Vietnam period were female and a good number of them were women who were re-entering the workforce because their children had grown old enough to drive, drink in a beer bar and be drafted. So numerous were these women at Badger that they could have organized a club called "Mothers With Sons In Vietnam." Had the club—which did not exist—included female cousins, aunts, sisters, wives and sweethearts, it could have filled the largest recreation hall at the plant for meetings—and without inviting any of the men with family in Vietnam. "He's proud of what I'm doing," Ruth Alberts said of her son. "He said, 'I'll be watching for cases of ammunition from Badger.'"

More than one son wrote home to say he had used ammunition bearing identifying marks from the Badger Army Ammunition Plant. "I have two sons serving in Vietnam," said LaCretta Eno, Reedsburg. "I feel I'm helping them by what I'm doing here. My boys tell me in letters how important our jobs here are." PFC Roger A. Cormican, who graduated from Baraboo High School in 1965, enlisted in '66 and was in Vietnam in '67, wrote Badger Commander Sidney Davis, "I have a great deal of confidence in the rounds that I have loaded in my magazines. This confidence comes from the knowledge of knowing that they are loaded with powder that comes from BOW. I have the privilege of having both of my parents working there for you." Roger Cormican's father, Elmer, was a chief operator in the Ball Powder extraction plant and his mother, Marjorie, worked in a rocket powder milling house. Their direct connection to the men on the ground in Vietnam gave Badger workers a double motivation to work hard and well. They wanted to give their soldiers top-quality ammunition to show their support and to help them win the war so the "boys" could return home alive and well. Politics did not matter to a mother whose son was being shot at. When asked what she wanted from her job early in 1970, Elaine Hellenbrand, an NC worker from North Freedom, said, "Security for all the boys in Vietnam. I like my job because it's challenging. It's a good way to let our boys know we're really backing them. What I want is a feeling of security — that we're doing all we can. It's a small thing for a woman to do, but I see it's my way of contributing." Or as Isabell Carpenter, another NC worker from Baraboo, said, "My son is on his second year of duty in Vietnam. I want to do the best I possibly can. To me this is important."[246]

About one-quarter of the Badger workforce — especially in the buildup years of 1966 through 1969 — was under twenty-five years old. June high school grads and college students in need of summer work were welcomed because, thanks to the baby boom of the 1940s, there were a lot of them around. They were trainable, didn't mind shift work, nor care about job security beyond Labor Day. The work of powder production had not changed since the 1930s. It was a relatively simple, labor intensive process the young people could readily master. "I'm a University of Wisconsin student majoring in math," said Ralph Thiede in June 1968. "I'm working my way through school and I need the money." Laurel Lange came to Badger that same summer because she "was just graduated from Baraboo High School...I'm anxious to earn money to buy school clothes and help defray college expenses." Badger also participated in work-study programs for students in relevant technical fields. John Kinzler of Prairie du Sac spent the summer of 1968 at the plant. "I'll be a junior at Stout State University in the fall. I'm majoring in Industrial Education and I plan to teach...I'm getting two credits toward graduation in a work-study program."[247]

If they took jobs at the plant, those who felt uncomfortable with or opposed the war parked their opinions at the gate. "I worked at Badger in the summer of '69, the height of Vietnam [and] the summer they hired the most college students," said Susan Mossman Sweeney. " They told us whatever your feelings on the Vietnam war, you need to forget about your personal feelings...I was kind of in the middle on the war, as a student in Madison, I was aware of what was going on...when I went to school I didn't want to tell anyone I worked here." As a student on a campus rife with anti-war sentiment, Sweeney faced the dilemma presented by Vietnam a bit earlier than other Badger workers. As the war continued into the 1970s and opinion polls recorded that a majority of Americans opposed involvement, more Badger workers expressed their own second thoughts. Al Gallagher, Chief of

Administration in the Vietnam period, recalled workers who expressed discomfort about how the products of their labor were used. "I remember one fellow who said he felt ashamed to work at Badger." After completing nine years on the line at the plant in 1975, Howard Harmon described what it was like to watch the evening news from Vietnam after coming home from work: "It made you sick." His coworker Thomas McCullough agreed: "Sure, it bothered me to make powder to send over there. The wife and I — she works at the plant too — used to talk about it at night. Our boy was over there in Vietnam and we worried about him all the time."[248]

Unlike Harmon and McCullough, many Badger workers — like most Americans — did not reflect on the world situation or their place in it, most did not have a choice. "But what are you going to do?" continued Harmon. "It was a good job."

Joan Lenerz, who worked in product improvement in 1968-69 expressed the reality of the situation. "My husband Howard, a former employee in ballistics, is now in Vietnam and the allotment checks I get just don't reach. I didn't come for patriotic reasons. It was a matter of economics."[249]

"I'm working here to help support my family," said material lab worker Arline Malone. "We have six children and you know what shoes and clothes for the kids cost today. It takes a lot of money. I've been working most of my life and I worked here back in 1942."

All Badger workers hired in because they needed a job, but for many women the job was more than a means to pay the bills. Much was made of the need and right of middle class women to work outside their homes in this period, but those sentiments were not new among working women at Badger. Rose Saunders of the metrology lab said, "Originally I wanted to help earn money to further our children's education...I am very interested in the quality control field. When I began work...it was to be on a temporary basis, but I found it so interesting that I stayed with it." Jane Denman, a ballistics lab worker from Portage, said, "My children are all in school and, frankly, housework bores me stiff. I enjoy being with and working with people...I've been working since we were married and I worked here in the last operation." Nancy Turdot, Chief Operator in the boiling tub house on the C-Line was a veteran of all three operations. "Women I feel have better job opportunities now than in any other operation. If any woman possesses the necessary ability, opportunities for advancement are better than ever."[250]

The plant newsletter, where most of these interviews were published, was not likely to print opinions not in agreement with company policy and the government line, so workers critical of the way Badger's products were used, of American involvement in Vietnam or of policies and practices regarding women, were not given a voice in the *Badger World*. However, just because a worker agreed with the company and the government line, happened to like the way she was treated at Badger, wanted to make powder to win the war and therefore was published in *Badger World*, does not invalidate her opinions. Abundant evidence supports the contention that Badger workers, like the majority of working class people in the United States, supported American involvement in Vietnam until even the Nixon administration acknowledged the futility of the enterprise in late 1970.

For female wage earners, the starting pay scale at Badger averaged at least fifty percent higher than on their previous jobs. Men, who usually started a few rungs higher up the wage ladder, averaged twenty percent better at Badger. With regular raises, a wage earner who stayed on the job for five years could expect to see a paycheck eighty percent larger than when he or she hired on. Blue- and pink-collar workers at Badger earned between $2.21 and $3.00 per hour to start. After five years, their pay ranged from $3.41 to $4.60 per hour straight time, plus overtime and shift bonuses. It was possible for a frugal person to support a family on those wages in the late 1960s and early 1970s. A family with two such wage earners could enjoy all the standard middle class comforts of the time: a comfortable home, a car or two, a color television, plus a nest egg for retirement or an education for the kids. In the economy of the Vietnam era, with inflation rampant, the regular wage increases at Badger helped workers keep up with rising prices, which made a job at the plant even more desirable. "I was working on a farm," said Ronald Labrenz, a Mule Operator from Westfield, "I've got two brothers, two cousins and a sister-in-law working here. I'll be making more money here than I've made in a long time."[251]

And so it was with Badger. It was a workplace where people traded time for money and the personal satisfaction derived from employment. Political opinions, in this most intensely political of times, were, like street clothes, shed in the change room. However, Badger could not remain a busy but serene island in the tempestuous 1960s sea. The impact of the plant on its neighbors in the county, its presence near the college city of Madison half a county away and its direct link to the Asian war half a planet away could not be ignored, shrugged off or hung on a hook in a change room locker.

From Badger Village to Bluffview Courts

Wayne Hatz was a boy when he first encountered the powder plant. His parents' farm, about two miles west of the plant, was not taken, but his Uncle Lyman's place off of Highway 78 was, and Wayne was old enough to help with the packing and the relocation. His friend Forrest Zantow, who could not escape the nickname "Woody," had a similar background. His family farm was up on the Bluff northwest of the plant and was spared, but the Zantows also had family and friends who were displaced. Both men grew up and stayed close to home in postwar Sauk County. Hatz became an insurance agent who cut his teeth in sales by making door-to-door calls on his neighbors, farm by farm. Zantow, as if to confirm his nickname, went to work at the Lachmund Lumber Company in Sauk City. By the late 1950s, both men had acquired business experience and an entrepreneur's ear for opportunity's knock.

About one year after the end of the Korean period of operations, the Army turned over Badger Village to the General Services Administration for disposal as surplus property. Despite numerous remodelings, improvements and upgrades in the 1950s, the company town had not been fully occupied since the last of the university students moved out in 1952. As the plant wound down between 1953 and 1958, the village moved from a state of semi- to full-fledged status as a ghost town. Zantow himself had helped lay $25,000 worth of new floor covering in one set of apartments, only to see the linoleum remain unscuffed by human feet since the plant closed before the apartments could be rented. The Badger School was still open after the 1958 shutdown, but, like other country schools in Wisconsin, it had been integrated into an area-wide district. After the final eighth-grade graduation ceremony in May 1960, the doors were locked, the equipment removed and the students bused to the consolidated Sauk Prairie schools.

Following standard procedure, the GSA offered the village's 111 acres, its eighty-two buildings, its roads, water and sewer lines to federal, state and local governments. None expressed interest so the property was advertised for sale with sealed bids to be opened at the federal building in Chicago on March 10, 1960. "We should buy it and advertise it as the largest motel in the world," Zantow joked as he drove past one day with his curling partner Harlan Packard. The "motel" consisted of the original ten H barracks of North Badger built for construction workers in 1942, now divided into 160 units, plus sixty-eight smaller buildings of the original South Badger built in 1943, now parceled into 272 apartments, and the community center. All the living units were furnished with gas cook stoves, oil-burning space heaters, refrigerators and hot water heaters. [252]

Packard was the local representative for Skelgas LP gas and appliances in Baraboo. One evening not long after he heard Zantow's motel joke, he was sitting in his shop with Zantow and Ron Nichols, who was also with Skelgas. When the talk turned to Badger Village and the possibility of putting in a bid, Packard called Hatz and told him to "Come down here and put your money where your mouth is."[253]

Hatz did and the quartet tried to guess what Badger Village was worth. When that proved to be a futile exercise, they looked at how much they could afford, both as a final bid and as an initial ten percent down payment.

Each man could put up $1,000, so if they had a fifth partner they could lay $5,000 down on a $50,000 bid. Zantow knew that the group couldn't go wrong with a plumber on its side so he invited Bill Fuchs of Prairie Plumbing and Heating to come in and Fuchs accepted. Although none of these men was a shrinking violet, the prospect of owning all that property, remodeling it and bringing it to life as a viable business prompted more than a few sleepless nights. They comforted themselves by mumbling, "well, we can always sell it for scrap." Undaunted by insomnia and without telling his partners, Zantow raised the ante. Having heard that another prospect was bidding $52,000, he borrowed $250 from Lachmund Lumber so he could put $5,250 down on a bid of $52,500.

The five partners agreed that the bid would be made in Zantow's name and that he and Hatz should attend the opening in Chicago. Of the thirty-some bids made, several were from other local people. Walter Ring, Baraboo, bid $13,000 and Richard Hegerfeld, Prairie du Sac, offered $33,514. James J. Delaney, of King's Corners, who wanted to buy the village and sell it for salvage, thought $36,666.66 would do it. The bid that almost won was a $51,000 offer from the Osterhus Publishing Company of Minneapolis, which planned to use it for a boys camp. When Zantow's bid was opened, a man sitting behind him exclaimed, "Jesus Christ, didn't they look at it?"

"If there had been a mouse hole, I would have crawled into it," Zantow recalled. He would have needed an extra-large mouse hole to contain the property he and his friends were now destined to own. He was further embarrassed when a reporter asked what he planned to do with the property. Speaking first and thinking later, Zantow answered, "I don't know," which made a provocative headline in next day's *Chicago Tribune*. The partners had about one month to think about it, while the GSA passed final judgment on the deal. In the meantime they were contacted by job seekers, former villagers hunting for souvenirs and at least one shady character who claimed he could invoke the influence of Congressman Vernon Thomson and Vice President Richard Nixon to pressure the GSA, in return for payments of only $25,000 a year and a new car for each of the politicians. The partners, who were good at fooling people into thinking they had come to town on a turnip truck, said they would take their chances minus inside influence.

On Good Friday in April, a GSA representative called Zantow and said the bid had been accepted. On Easter Monday, John Troyer called him and said that the sign still identifying the village as government property was now inaccurate and misleading and what was Zantow going to do about it? He alerted Hatz, who rushed out with a can of paint and brushed over the offending words. If the impact of what they had taken up had yet to soak into the partners, it soon would. They had purchased a tract of housing and infrastructure that the federal government had put several million dollars into, yet in the eyes of any other interested party was worth no more than its value as used lumber, hardware, roofing and plumbing. To the naked eye Badger looked like the low-rent motel Zantow joked about, or maybe—with a long stretch of the imagination—like one of the recreational subdivisions of the type appearing with increasing frequency along the lakeshores and riverfronts of Wisconsin; but there was no water frontage at Badger and little, if any, potential in the recreational home market. In addition, although Badger looked like a village, it depended on the powder plant for basic municipal services such as drinking water and sewage disposal. When the school term ended in June 1960, these services were cut off, leaving the partners with several hundred uninhabitable apartments. Still, they decided not to scrap the place. It was another of those seemingly simple yet historically pregnant turning points in the Badger story. Had any of the other bidders bought the place, or if the five partners had fallen prey to their legitimate doubts, the buildings would have been torn down and trucked off in pieces, leaving behind a weed-choked grid of blacktop, gravel and cement atop a network of decaying water and sewer lines. Left vacant, the grounds, which could not be farmed, would likely have become a parking lot for junk cars and scrapped machinery, leaking oil and rusting in the prairie rain.

Instead, Zantow, Hatz, Packard, Nichols and Fuchs tried to make something out of the place. In their efforts they were aided by a demographic process that would bring about the greatest change to the face of rural America since the white pioneers displaced the Indians. Demographers called it "turnaround" migration because it represented a reversal in the decades-old pattern of rural people leaving the country and moving to the city. It was led not by the much-publicized baby boomers of the 1960s, but by their grandparents. Born at the turn of the 20th century, these were the people for whom the term "senior citizen" was coined. They were in better health and would live longer than previous generations. They had contributed to the social security system long enough to rely on it for an income base, as well as on their own pensions and savings. After 1964, they could also rely on Medicare for aid with health care expenses. They were mobile, used to jumping in the car and traveling for work and recreation. They had vacationed in the country and looked forward to ending their days there, even if it was hours away from their extended families. For many of Wisconsin's turnaround migrants the changing racial composition of Milwaukee, Chicago and other midwestern cities was also a factor. "White flight" sent migrants not only to the collar suburbs around the cities, but to Sauk County as well.

The countryside itself had also changed. It was more friendly to people who did not farm but wanted to live beyond the sidewalks, unlike the 1940s when so many of the older farmers displaced by Badger Ordnance had little choice but to follow the traditional pattern and retire to the comforts of village life. By the 1960s, rural roads had improved, as had the cars traveling on them; so had telephone, electric, television broadcast and health care services. Rail passenger service was all but dead, replaced by the interstate highway system, which reached Wisconsin Dells in 1964 and made long distance auto travel more convenient and faster than ever.

All of these broad historical forces were at play, and waiting for the five men who purchased Badger Village to take advantage of them. It did not take long. After a brief and fruitless attempt to market their apartments as recreational property, the partners discovered senior citizens. "It was Wayne's suggestion that we use North Badger as a retirement center," said Zantow. They had found their market. Now all they had to do was generate income to pay for the place and locate financing so they could replace the disconnected water and sewer systems.

To pay the mortgage, the partners decided to sell South Badger, buildings and furnishings, and were peddling it piece by piece throughout the summer and fall of 1960. A toilet stool, uncracked, sold for $45. Cracked, but still usable, stools sold for five dollars less. Hatz tells the story of one out-of-town buyer who wouldn't pay the full price for an uncracked stool, but was willing to drive home, call back to ask if a cracked one was available, then drive back to pick it up — a penny-wise and pound-foolish gesture even in the days of twenty-nine-cents-a-gallon gasoline. An apartment shell large enough for a farm machine shed, dismantled and loaded, sold for $1,000. Zantow happened to be minding the store on the day when a farmer came in looking for chicken coops. He was able to lift the mortgage substantially and pay for a drilled well by selling twenty buildings at once.

The partners also raised funds by renting out the community center to the "National Institute of Concrete Construction." This federally funded organization trained approximately 300 Peace Corps volunteers in the art and science of making concrete. Here is yet another of those wonderful turns in the Badger story, that a facility built to house those who made an instrument of war was now used to train people enlisted in the cause of peace. If they could make war with wood pulp, why not peace with concrete? Space was also rented to a diesel-truck-driving school and to farmer Ed Knutson. He fenced a few acres, grazed sheep on the lawn and sheltered them overnight in vacant apartments.

After reading an article in the *Reader's Digest* and consulting with the Department of Natural Resources, the partners learned they could replace their severed link to the plant treatment facilities by digging a seven-acre holding pond and running the village sewer lines into it. Since the pond had to be at least one-quarter mile away from the housing, another forty acre parcel was purchased on the western edge of the village. After a hectic summer of selling salvaged materials, negotiating with the DNR for permits, acquiring real estate and pursuing financing for it all, the partners held their first open house. Wishing to clearly separate themselves from the powder plant, they had rechristened the place Bluffview Courts and incorporated themselves as Bluffview Acres. A total of 2,500 visitors, mostly Sauk County people, attended the first open house. They liked the model apartments which had been furnished by Gattschall Furniture in Prairie du Sac and they liked the location on the sunny prairie next to the idle powder plant, but they did not like the insecurity and lack of equity of a rental property.

Hatz then consulted with Madison attorney and future Wisconsin Supreme Court Justice Roland Day, who devised a system whereby Bluffview residents would purchase their apartments and contract to pay monthly fees for maintenance and services, including lawn care and roof repair. Both Zantow and Hatz are quick to point out the own-and-lease agreement Day drew up was one of the first, if not the first, of its kind in Wisconsin. "We invented the condominium," said Hatz. "Wisconsin did not have the word condominium in their vocabulary," said Zantow. Be that as it may, the condo was new to Sauk County, and Bluffview was where it was introduced.

The first units were "pretty spartan" and priced accordingly, at $2,495-$2,595, plus a monthly fee of $43. Zantow then selected one unit to serve as a model, installed about $1,000 worth of improvements, added the cost to the price, and sales picked up. After placing ads in the Milwaukee and Chicago newspapers, Hatz and Zantow followed up with phone sales calls. "We rented a hotel room in Chicago and got on the phone." They were hitting the right market at the right time — people of retirement age or soon to be who wanted to get out of the city — stereotypical turnaround migrants. A few of them thought Bluffview was too good to be true, their fears justifiably fuelled by reports of unscrupulous rural real estate developers who hoodwinked city folk into buying mosquito ranches on swampland. Bluffview was a bargain for a senior selling a $25,000 house and relying on social security and a company pension to finance retirement. "Many prospects were impressed but when they told their families back home the price was so low they were afraid we were dishonest. However, we were able to sell out in a couple of years."

By the end of 1967, Bluffview had a population of nearly 200 retirees who were becoming part of a genuine community. The layout of the place, with the units closely connected and arranged in U-shapes around park-like clearings, served to bring people together. It was an oddly urban landscape transplanted to the country, one whose appearance improved as the trees transplanted from the Zantow woods matured, the lawns took hold and the gardens flourished. To further enhance the property, the developers traded the gravel beneath the surface for dragline services from Wilbur Deppe, who excavated a small lake with an island in the center. Students from the concrete school built a dam and floodgates to maintain water levels, as well as casting letters spelling out the name Bluffview, which they placed on the island. Over the years, a grocery store, barber shop, beauty salon and cafe came and went. A chatty newsletter kept everyone up to date on what the neighbors were doing. At a time when cities and villages all over Wisconsin, and the United States, had just begun to establish councils, commissions and agencies to serve their newly discovered senior citizens, five entrepreneurs had created a model retirement community out of barracks hurriedly constructed for powder plant construction workers.

Among the stereotypical turnaround migrants who took up residence at Bluffview in later years was Dr. Donald Kindschi. An environmentalist and patron of the arts and history, Kindschi had left Sauk Prairie for medical school in Madison and a practice in California. He retired to Bluffview to live on land which, by coincidence, was formerly owned by his family. Not all the residents of Bluffview came from big cities. It also became the home of Erhart Mueller, who took up residence after retiring from the farm in 1970. Instead of moving to a village farther away from his old home, Mueller was able to stay on the prairie and find the community he needed at Bluffview. Had nothing else of note happened there, Bluffview would be worthy of honor as the site where Erhart Mueller crafted his chronicles of Sumpter. There was little in the powder plant that he looked on with favor. It had destroyed the old settler's community that he had loved. Yet, it had also given rise to a new home which nurtured him as he re-created what had been lost.

When the plant reactivated in 1966, the Bluffview partners expanded to provide housing for the latest round of workers by developing what had been "Middle Badger" for mobile homes. No longer the humble trailers or "demountable houses" of the World War II era, the mobile units of the 1960s were more spacious and comfortable. In time they made way for the "double-wide" modular homes that became the largest single-family dwellings in the park. The partnership continued to own and manage Bluffview until 1994. Of the original quintet, Harlan Packard passed away in 1981. Ron Nichols is retired in Florida. Bill Fuchs lives in Prairie du Sac. Wayne Hatz and Woody Zantow are active retirees and still in Sauk County.

Focusing On The Obvious

As a consequence of the transformation of Badger the company town into Bluffview the independent village, the powder plant acquired neighbors. The countryside had changed and the plant's position in it was now similar to that of the old farmer who, after dutifully working his land for many years, suddenly finds himself surrounded by suburban homeowners who complain about the noise, dust and bad smells wafting over his fence line. During the

Korean period, except for a minor tiff or two involving agricultural leases or access for pheasant hunting, and while the plant was inactive between 1958 and 1966, the old farmer at Badger Army Ammo got along with his neighbors. When operations resumed, with coal smoke pouring out of the power plant chimneys, grey "mist" floating out of the Oleum plant vents and brown puffs of nitrogen oxide pluming out of the acid works stacks, the situation changed.

"The most significant community relation interaction encountered by Badger relates to the emission of environmental pollutants," states an Army report penned in the summer of 1971. The first complaints had been lodged a few years earlier by Bluffview villagers miffed over soot soiling wet laundry hung out to dry. They also did not like the smog that hung low over the prairie on still mornings when the barometer was low nor the rotten egg smell of sulfuric acid fumes. None of this was new and had occurred whenever the plant was running. What was new was the presence of people who lived near the plant, did not work there and were vocal about their interests.

The passing of time caught up with Badger in other ways as well. Built in the 1940s, and using 1920s or older technology, the entire plant—especially the power plant, acid works and waste disposal systems—were ripe for renovation or replacement. The operations reports for the Vietnam era recount numerous minor but troublesome breakdowns, delays and rush-job repairs necessitated by age and wear. Badger's pollution control equipment was equally antique, which was the same as saying it was non-existent beyond basic safety and sanitary requirements. Reducing smokestack emissions and sewer pipe discharges to harmless levels was not on the drawing board when Badger was built in the 1940s or when it was renovated and enlarged in the 1950s. "It's simple," said Lt. Colonel Eugene Belcher, not long after he assumed command of the plant in 1970, "most of those years few people even thought about pollution...The emphasis has been on production, jobs, a growing economy."[254]

If attention was paid to the environment prior to the late 1960s, it was in the form of tree planting and "wildlife" management. Planting trees was what conservation was all about then, as it had been since the days of Theodore Roosevelt, and about 125,000 seedlings had been set out at Badger between 1952 and 1967. The plantings, along with trees and brush that had sprouted on their own, had transformed acres of Badger's unworked ground from grassland to woodlot. Northern forest trees were now growing in places where they had not grown since they had been replaced by the prairie-savanna grasslands 6,000 years earlier. Taking place gradually in the 1950s and '60s, the conversion from grassland to woodland was a less dramatic change than the construction of the plant itself but as extensive and as profound in its impact.

The rise and fall of Badger's pheasant population illustrates the process. In the early 1950s Badger supported a rich population of pheasants the hunting of which was enjoyed by the farmers who leased fields inside the plant as well as the poachers who wormed their way through the fence. In 1953, Lt. Colonel William Evans, the plant commander, became a very unpopular man and created a small stir in the newspapers when he prohibited the farmers from hunting pheasants on their leased fields. Evans contended that it was hazardous to discharge firearms inside an operating powder plant—at least for the farmers to do so. He did not mention the parties of Army officers and business executives he routinely invited for wing shooting excursions on the grounds. "I personally control this matter and am responsible for who is allowed to hunt there," an unapologetic Evans was quoted in the papers. In time, as the rules governing farming on federal property required the Badger leaseholders to switch to crops less favorable to the birds and trees increasingly reduced their grassland habitat, pheasant numbers declined. At the same time and also as a result of larger woodlots, the whitetail deer population increased. In the farming days of the 1930s, deer were as scarce on the Prairie as airplanes. By the late 1950s, the herd grew to "huntable" size and continued to increase.[255]

In the fall of 1967, a heavier than normal release of sulfuric acid "mist" from the Oleum plant damaged pines and other trees near the water reservoir and across the fence in Devils Lake State Park. The dead trees were the canary in the mine shaft whose death alerts the proprietors that something is wrong. The trees damaged inside the fence were mostly pines planted for old-style conservation purposes. White pines were most vulnerable to the mist, with red pines, oaks, maples and others less troubled. No more than five percent of the white pines were killed, about as many as a pack of hungry rabbits would gnaw down in a typical winter—but trees poisoned by a chemical release are a more ominous and potentially dangerous sign of trouble than trees girdled by rodents, and were taken as such in 1967. The Oleum mist that killed the planted pines delivered a symbolic message appropriate to the times. Well-meaning, old-style conservation work was superfluous unless accompanied by wider environmental awareness.

Even before the acid mist fell on the pines, Badger engineers were working on a pollution control plan. "I think we are ahead of most of our critics," said John Troyer, and in 1967-68 the plant was ahead. In the summer of 1968, a $100,000 Brink's Mist Eliminator was installed at the Oleum plant that would all but eliminate sulfuric acid emissions. No more would the mist defoliate the pines. Rarely had the solution to a problem not directly tied to production been planned, approved, funded and implemented so swiftly. Badger's engineers had also completed plans to combat air pollution by converting the power plant to burn fuel oil and/or natural gas and by installing catalytic reductors on the acid plant to cut down nitrogen oxide emissions. Surface water pollution would be reduced by building a plant to neutralize

100 percent of waste acids before they entered the plant sewers and, ultimately, the Wisconsin River. Discharge of solid wastes into the river would be lessened by the construction of a secondary treatment plant, aeration spillways and a twenty-five-acre sediment pond. Then the pace slowed to normal and Badger fell behind. Months, years passed while the plans—which were ambitious in the budding days of environmental awareness—worked their way up the Army chain, through the Pentagon maze, into the Congressional budget and onto the President's desk.[256]

In the meantime, Wisconsin environmentalists won a momentous victory that led to the banning of the pesticide DDT. In a well-publicized, trial-like Department of Natural Resources hearing in Madison, the Conservation Natural Resources Association of Wisconsin and the Environmental Defense Fund presented evidence that led the DNR to identify DDT as a pollutant. Within two years state and federal bans on the use of the chemical were enacted. The local significance of the DDT ban is vividly displayed by the growing number of bald eagles that have wintered on Lake Wisconsin ever since.

The success of the DDT hearing, which was a court trial in all but name, encouraged environmentalists and those who sought their favor to use the legal instead of the political process to fight pollution. As the operator at Badger, the Olin Corporation was charged as the defendant in two anti-pollution suits filed in that familiar Badger Ordnance battleground, the federal court for the Western District of Wisconsin. Congressman Henry Reuss filed a suit arguing that, by discharging pollutants into the Wisconsin at Badger, Olin was violating the Rivers and Harbors Act of 1899. Invoking a law written to promote water-borne commerce for environmental protection was a stretch and Olin's lawyers had little trouble convincing the federal judge to dismiss the case—but not until after November, when Reuss was safely reelected. A few months ahead of Reuss, and on firmer legal ground, Wisconsin Attorney General Robert Warren had also filed a suit against Olin. The Wisconsin suit charged "in general terms that the operation of Badger Army Ammunition Plant creates air and water pollution." The DNR then officially found that Badger's coal-fired boilers were in violation of the state's air quality regulations and ordered Olin to correct the condition by September 1, 1971.[257]

Olin's attorneys countered by pointing out that, although it was the operator of the powder plant, it was not the owner. It could not compel the United States government to modify procedures or physical plant at Badger; nor, as Olin continued, could the DNR, since federal facilities, especially defense plants, were not subject to state environmental regulations. Attorney General Warren and Olin were not able to bring their arguments to judgment because Congress and President Richard Nixon unintentionally made them irrelevant by approving funding for the Badger anti-pollution plan. The 1971 federal budget appropriated $1.6 million to implement the measures previously drawn up at Badger. As a result, Warren dropped Wisconsin's suit against Olin. The funding constituted a tacit admission by the federal government that the powder plant was contaminating its neighbors' air and surface water. By inaugurating pollution control measures the federal government was also accepting responsibility to halt further pollution and to clean up the negative remains of past practices. It was the start of thirty years of often controversial but methodical pollution control at Badger. It was also incomplete. Focusing on the obvious symptoms — black smoke in the air, oily discharges in Gruber's Bay — the pollution control planners of the Vietnam period ignored the very ground beneath their feet. Not until the end of the 1970s would the Army conduct its first survey of groundwater contamination inside the plant and not until the end of the 1980s would it recognize that pollutants had been migrating through the soil under the fences and into the drinking water of the plant's neighbors. The "significant community relation interaction" due to "the emissions of environmental pollutants" would continue and intensify for the rest of the century and beyond.

Target BAAP

"On this date an unsuccessful attempt was made to disrupt operations by dropping homemade devices on the plant site from a stolen aircraft. It is believed that this is the first and only aerial bombing of a military installation in the continental United States." So read the typically laconic official report of the 1970 New Year's Day bombing of Badger written a few months later by administrative manager and plant historian Al Gallagher. It is accurate to the pinpoint, with no unnecessary words, the exact opposite of the bombing itself, which was well off target and upon which volumes of verbiage have been expelled.[258]

The New Year's bombing was a contemptible crime, treasonable sabotage in time of war, a reckless endangerment of life and property and a terrorist act that could have inflicted casualties unrivaled in American history until Timothy McVeigh parked a truck crammed with explosives outside the Murrah federal building in Oklahoma City.

Or...

The bombing was a sincere statement of outrage against an evil government ignoring the will of its people, legitimate direct action taken against the war-mongering military-industrial complex and a righteous rattling of the cage in which Americans had locked their consciences in regard to wrongs perpetrated in their names in Southeast Asia.

Or...

The bombing was a merry prank perpetrated by hippies with minds in a chemically altered state who were spared from doing harm to themselves and to others by fool's luck and the kind of thing you'd expect to come out of Madison in the 1960s.

Or...

The bombing was a composite of all three points of view, a whacked-out, seriocomic adventure undertaken for high purpose but marred by its deadly criminal potential — like the Vietnam war itself.

The image of a airplane coming over the Bluffs with a bomb-bearing revolutionist in the cockpit stands in high contrast to the first anti-Vietnam demonstration at Badger, which took place in June of 1966. Martin Luther King and Robert Kennedy were still alive. Lyndon Johnson spoke proudly of the war he wanted to fight — against poverty. The terms hippie, hard hat and Haiphong had yet to become household words. In these summery times, a group of about forty-five men, women and children, in a scene reminiscent of the Civil Rights movement, marched up Highway 12 from the University of Wisconsin campus at Madison. Led by John White, a crew-cut graduate student from Chicago who wore a sports coat, button-down shirt and tie, they sang "We Shall Overcome" and "Blowin in the Wind," and carried signs reading "End War Production" and "Badger Brings Death Not Peace to Vietnam" as they trudged through the sun-dappled cow pastures of Dane County. At the end of a day's walk, they would drive back to Madison for a night's sleep, then return to their stopping point on the next morning to take up the march again. The announced goal of the "Badger for Peace" group was to make a "non-violent entry of the plant" in order to persuade the workers to walk off their jobs.[259]

"They have no business there," was the no-nonsense reaction of plant commander Captain Sidney I. Davis. "No one is allowed in the plant unless they are on official business." Undeterred, the marchers kept marching and, after three days on the road, they reached the main gate. They were greeted by Sauk County Sheriff Merle Spencer, whom everyone knew as Mike, plus six deputies equipped with riot helmets and long-handled nightsticks. Dane County Sheriff Jack Leslie and six deputies were also on hand, along with twelve additional deputies on horseback, plus a beefed-up force of plant security guards. The ratio of law enforcement officers to marchers was about one to one.[260]

Mike Spencer was the last of the old breed of Sauk County sheriff and the first of the new. Initially elected in 1957, Spencer, like his predecessors, lived with his family in the 1890s red brick county jail in Baraboo. Mike's wife, Dorothy, whom everyone knew as Dolly, cooked meals for the prisoners just as sheriff's wives had cooked meals for prisoners since the 1840s. On

Saturday nights the Spencer children gave up "Gunsmoke" on TV to peep over the top of the Dutch door separating their kitchen from the jail to watch Dad hustle live lawbreakers into the lock-up. Because of a state term limit law, Mike and Dolly took turns holding the title of sheriff, and it was she who officially held the job in 1961 when Mike became the most famous law officer in Wisconsin. Sheriff in fact, if not in title, he organized 500 deputies into a county-wide manhunt for three fugitives accused of gunning down county traffic officer James Jantz and then bagged the killers himself by disabling their speeding getaway car with an ace shot from his handgun. After that feat, Mike Spencer couldn't have been more popular in Sauk County if he walked across the surface of Devils Lake in his cowboy boots.

In 1966, he was sheriff in his own right and ready to handle the "Badger for Peace" marchers with common sense and restraint. He told them they could "peacefully demonstrate and picket" as long as they wanted to, but they could not enter the plant. "I'm not allowed in there either," he said. The marchers themselves did not attempt to break through the police blockade, just sat down on the gravel, waved their signs in front of the news cameras and chanted "Johnson, Hitler, General Ky, they all sound the same to me." A few of them exchanged angry words with plant workers and other bystanders, but no one shoved or punched anyone else. Ever the politician, Spencer circulated among the dozen or so children accompanying the marchers and gave each one a junior deputy sheriff ID card, which they all accepted, and then told the youngsters to be "good citizens."[261]

It was a summer kind of demonstration. The marchers did not rush the police. The police did not attack the marchers. Having stayed out in the hot sun long enough to insure they would make the front page of the Sunday newspapers the next day, the marchers climbed into cars and got back to Madison in time to spend Saturday night on the town. The deputies and security guards dispersed in time to be home for supper. Somewhere today, a middle-aged man or woman probably has a Sauk County junior deputy sheriff ID card tucked away with other childhood souvenirs. Mike and Dolly Spencer stayed on as sheriffs of Sauk County until 1970, when he assumed command of emergency police services for the state of Wisconsin and supervised troopers attempting to control anti-war demonstrations in Madison that made the 1966 Badger march seem like a summer afternoon in the park.

Badger returned to the slumbering sidelines of anti-war action. If University of Wisconsin students came into contact with the plant, it was to seek out employment. The politically minded among them had plenty to do in Madison, which was on the front line of the anti-war battleground. Sauk County people engaged in the struggle there either as student demonstrators on one side of the police lines or as deputies, state troopers and National Guardsmen on the other. As the war escalated, so did anti-war activity in Madison. In October 1967, student opposition to the presence of recruiters from the Dow Chemical Company, the manufacturer of the napalm loaded into incendiary bombs dropped in Vietnam, was so intense that, for the first time on an American campus, police used night sticks and tear gas to control demonstrators. After the Tet offensive in early 1968, which the Viet Cong lost on the battlefield but won in the media, anti-war demonstrations and disruptions occurred so often in Madison they could have been listed in the university catalog. In February 1969, Governor Warren Knowles and University Chancellor Edwin Young, who had spent his student years living at Badger Village, called out the National Guard to contain thousands of students participating in the "Black Strike." Over 2,000 soldiers equipped with machine guns, automatic rifles and grenade launchers deployed on the campus at the foot of Bascom Hill. In a televised address the following November, President Richard Nixon coined the term "silent majority" to refer to those Americans who did not march, demonstrate or publicly complain and therefore, so he said, must agree with his policy of gradual disengagement from the war. However, ever since Tet, opinion polls had been registering that a majority of Americans—silent and unsilent—believed that America's involvement in Vietnam was a mistake and should end quickly. The number of those in opposition would only increase when, two weeks after the "silent majority" speech, the *New York Times* published the first accounts of the action in the village of My Lai, in which American troops firing M-16s loaded with 5.56 mm Ball Powder killed over 300 Vietnamese civilians, most of them women and children.

Karleton Armstrong was already convinced of the immorality of the Vietnam war, of mainstream American culture and of the university's participation therein. Born in Madison in 1946, he was raised on the northeast side of the city and graduated from East High School. His father, Donald, was a skilled machinist at Gisholt Machinery, his mother, Ruth, an attendant at the Central Colony state facility for the developmentally disabled. They were the kind of people who, had they lived in Sauk County, would work at Badger. As it was, the Armstrongs were well acquainted with the place. Ruth's brother Ellsworth Goff had died in the nitroglycerin storehouse explosion of July 1945. Karl himself would apply for a job during the Vietnam buildup, but was not hired. He later said that he first started to think about the Vietnam war after hearing about the 1966 "Badger for Peace" march. His biography for the years after he enrolled at the University of Wisconsin in September 1964 reads like a standard late-1960s resume. He drifted in and out of school, lost his student deferment and reported for his induction physical, but was classified 1-Y and escaped obligatory military service. He held a variety of jobs, hitchhiked around the country, took part in anti-war demonstrations in Madison and Washington, D.C., had his share of sex, drugs and rock and roll. Along the way, he filled his head with a smattering of leftist polemics, anarchist rhetoric, old and new revolutionary jargon.[262]

By the end of 1969 he had decided that American society was so corrupted by the war that it could be redeemed only by what radicals of an earlier era called "direct action," but what most everyone else knew as vandalism, assassination and terrorism. Targets directly connected to the war and therefore suitable for direct action were numerous in Madison. The Pentagon maintained a presence on the University of Wisconsin campus through the ROTC officer training program and by funding dozens of research projects. The Air Force trained its officers in a surplus property quonset hut off Babcock Street on the west end of the campus which was designated as T-for temporary-16. In the pre-dawn hours of December 28, 1969, Karl Armstrong walked across campus from his room at the Phi Kappa Sigma house off Lake Street carrying a brown paper grocery sack containing two glass gallon jugs filled with gasoline. When he reached T-16 he walked to the back of the building, away from the street, pulled a hammer out of his pocket, broke a window and tossed in the gasoline, followed by a match. Gasoline slithered over the floor, then burst into flames, and so began the "winter offensive" of the "Vanguard of the Revolution." Over the next two months the Vanguard — Karl Armstrong, his brother Dwight and Karl's girlfriend Lynn Schultz — took direct action four more times. In early January the trio attempted the aerial bombardment of Badger, then Karl firebombed the university's Old Red Gym and the Primate Research Center. After laying low for over a month, he and Schultz attacked Badger again by attempting to dynamite the Wisconsin Power and Light substation at Prairie du Sac.

For an anti-war radical in Madison, Wisconsin, there could be no target more deserving of direct action than the Badger Army Ammunition Plant. It was an integral part of the war-making machine, the enemy incarnate. On a more practical level, by knocking out Badger, Armstrong could hamper the war effort. Even if he failed to damage the plant, he would make a statement to the world at large and bring the war home to the people making powder for it. Yet Badger was not an easy target. As the summer-of-'66 peace marchers had learned, it was difficult to penetrate from the ground. By air, however, it was as vulnerable as a skyscraper in Manhattan.

Dwight Armstrong had worked as a line boy at the Morey Airport in Middleton for the first six months of 1969. He had received some informal flight instruction in a Cessna 150 airplane but had never soloed and was not licensed. Nonetheless, as the last hour of December 31, 1969, turned to January 1, 1970, Dwight and Karl Armstrong stood at the doorway of a hangar at Morey Airport. Earlier that day, Karl had persuaded Dwight to join him in stealing an airplane and heading off in the night to "take out Badger Ordnance." He had also enlisted Schultz, a nineteen-year-old Wisconsin Dells native working as a secretary at Ray-O-Vac in Madison. She was an unlikely revolutionary, but persuadable, and Armstrong talked her into supplying and driving the getaway car on several bombing forays, including his raids on

Badger. After dropping off Karl and Dwight at Morey Airport, she was to drive to Sauk City, make a warning call to the plant, then pick up the bombers after they landed their plane at the Sauk Prairie Airport.

Karl had prepared three bombs out of a pair of commercial-size mayonnaise jars he had taken from Gino's Pizza, for whom he made deliveries, and a hollow metal ash tray stand he found at the Sigma Kappa frat house. The explosive was a mixture of ammonium nitrate crystals and kerosene. It was the first time he used ammonium nitrate and kerosene in a bomb. The second was at the Army Math Research Center the following August. There was no detonator and they would not explode on impact, but if the bombs landed in the right place, say, down one of the chimneys of the power plant, the damage could be serious. Knocking out the power plant would shut down the entire operation. More dangerous than a mayo jar bomb was the plane itself. Manned by an unqualified, inexperienced pilot flying in uncertain weather at night over unprotected wooden buildings filled with flammable, explosive materials, even a tiny Cessna could turn into a bomb.

The Morey hangar was locked, but when Dwight yanked on the padlock, it ripped free, along with the door to which it was attached. The Armstrongs stepped inside, found the 150 at the rear of the hangar and a Cessna 172 in front. Dwight had never flown the 172 so he and Karl pushed it out of their way and rolled the 150 onto the runway. A light snow was falling and the temperature was in the teens.

The Cessna's ignition key was "hidden" under the engine cowling, just as Dwight remembered. Since he had never flown this or any other plane in the dark, Dwight could not remember how to turn on the instrument panel lights. The brothers settled into their seats in the tiny cabin, with Karl cradling a mayonnaise jar under each arm, and the ashtray stand propped between his legs. Dwight started the engine, but he could not read the instruments indicating oil pressure or the amount of fuel in the tanks. Dwight was able to steer the Cessna down the runway because Karl had remembered to bring a flashlight which he managed to haphazardly flash onto the instrument panel while hanging onto the mayo jars. Suddenly Karl's door popped open. He had forgotten to latch it and, with his arms filled with slippery, explosive-filled glass jars, he could not do it now. With snow swirling into the cabin, one door flapping and no panel lights, Dwight, who had never soloed, never flown at night and never in a snowstorm, got the Cessna into the air.

Karl freed one of his arms and closed the door. A minute or two passed until Dwight announced that he had neglected to set the gyrocompass, which meant they were already lost. The only thing Dwight could think to do was to drop through the clouds until he could sight a recognizable landmark. Once again, the bombers' luck held. They did not fly into a hilltop or radio tower and soon spotted the lights of Cross Plains. Staying low, Armstrong turned the plane back towards Middleton until he sighted the lights of Highway 12. He followed 12 north and west to Sauk City, where he recognized Schultz as she got out of her car and entered the phone booth at the corner of Philipp's and Water Street. She was having her own problems. The first phone booth she stopped at was occupied by a caller who would not give up the phone. She drove to another phone, but it was out of order. When she came back to the first phone — just as the plane was flying over — she put in her dime but nothing happened. Although she had not noticed the plane overhead, she thought she would be late for the pickup so she rushed to the airport without making the warning call.

Apparently she did not give much thought to what the call could mean if Karl's bombs were real. He later said that he wanted Schultz to make the call because it might save lives. It might also help him appear to be less cold-blooded than someone who attacked without warning. Be that as it may, a five-minute warning was worse than no warning at all. None of the operations in the plant could be shut down in five minutes. Had they heard that a bomb was on its way, the workers would have to face the choice of staying at their stations and waiting for the bomb to fall, or leaving their stations while potentially explosive processes were underway. A blending barrel full of nitrocellulose and nitroglycerin left untended was many times more deadly than a mayo jar full of ammonia and kerosene.

As they flew out from Sauk City, the Armstrongs saw Badger light up the Prairie like a small city. They then skirted the plant and continued north to Baraboo, where they turned back to approach Badger from over the hills just west of Devils Lake. Karl told Dwight to head for what he thought were fuel tanks, but were actually the chemical tanks of the acid area. Dwight wanted to keep the plane at 1,500 feet, well above any blast, but Karl ordered him to drop to 750 feet. He then opened the door and kicked out the ashtray bomb. It fell harmlessly into an open area well away from any building or chemical tank. Dwight then banked the plane around and prepared to make a lower dive over the smokeless powder area. Plumes of vapor rose from the power plant stacks directly ahead, as Karl, without even looking where they were headed, threw the mayo jars out of the door. They fell to earth and shattered in the snow. No blast, explosion or fire. Karl later said that he never intended to do any damage to Badger and that merely demonstrating the plant's vulnerability delivered his message of protest against the war.

Below, the plant hummed serenely on. Dwight continued to fly south over the Ball Powder plant, the burning grounds, the magazines and on to Sauk Prairie airport. He planned to land on the unlighted snow-covered runway aided only by the sputtering glow of the two railroad flares Schultz had lit and stuck in the snow. A tailwind pushed the Cessna too hard and fast on the first approach so Dwight banked, turned, came around again and dropped the plane into a near-perfect landing. As soon as the plane stopped, the Armstrongs jumped out and made for the trees where Schultz waited with the car. They drove off and were back in Madison within an hour.

"Whoever flew that plane in here last night had to be one hell of a pilot," airport manager Milton Zech told reporters the next day. "No one ever lands here from the north." Dwight had unwittingly made his approach across a power line and missed the wires by what had to be no more than a few feet. He then brought the Cessna to a skidding halt on the snowed-over runway without flipping it onto its nose.

By intention or not, the powder plant bombing, like other actions in the "winter offensive," was crudely executed and, fortunately for both perpetrators and potential victims, not as destructive as it might have been. Karl's firebomb at the T-16 building did only slight damage to the Quonset hut. At the Red Gym, the only part of the building not damaged by the firebombing was its target, the ROTC office. The Primate Research Center burned out because Karl mistook it for his real target, the office of the Selective Service, located across the street. The Prairie du Sac substation bombing failed because Karl forget to bring along tape to connect the detonators to the dynamite. While he was fumbling with the explosives in the cold he was spotted by a watchman and had to flee. During their revolution Russian Communists coined the term "Vanguard of the Revolution" to describe the hard-edged, no-nonsense position they had carved out for themselves as leaders of the peoples' struggle. In Madison, Wisconsin, in the winter of 1969-70, the Vanguard of the Revolution was more like the gang that couldn't shoot straight. They never got it right, even in their final direct action against the Army Math Center at Sterling Hall. The van load of explosives they parked next to the supposedly vacant building wrought over $2 million dollars worth of damages to property, destroyed years of invaluable scientific research, injured six people and killed a young post-doctoral researcher named Robert Fassnacht, but inflicted only minor damage on its target, the Army Math Center.

Had the bombs detonated at Badger, had they caused explosives already there to go up or had the plane crashed into the right — wrong? — building, the destruction could have been catastrophic. Badger had millions of pounds of propellant and other flammable, explosive materials on site. Because of the New Year's holiday, the regular shift of approximately 1,000 workers was reduced, but several hundred people were at the plant and hundreds more men, women and children lived in Bluffview directly across Highway 12. The plant was designed and operated first to prevent explosions, then to minimize the damage should an explosion occur. Although there was some talk about the need for aerial defenses in the early panic-filled days of World War II, none were ever installed because everyone knew that an attack from the air could never happen in the United States.

Although a guard reported that a small plane with markings matching those on the aircraft discovered at the Sauk Prairie Airport had flown over the plant "at 0213 hours, 1 January 1970," no one at Badger was aware that a bombing had occurred until several days later. The Madison "underground" newspaper *Kaleidoscope* reported that it had been contacted by one of the bombers who described what had happened at Badger on New Year's. Radio reporters repeated the story and, on January 5, the FBI contacted plant commander Lt. Colonel Russell Enoch and told him that the *Kaleidoscope* was usually accurate in such matters. In light of the three firebombings that had taken place in Madison in the past few days, it was quite possible that Badger had been attacked. A device timed to explode at any moment, or an unexploded bomb, might be lurking in the plant, so Enoch, plant security chief John Probert and the FBI very quickly organized a search. An Army bomb squad, the 259th Ordnance Detachment, was immediately dispatched from the Savanna, Illinois, Army Depot. On January 6, the Ordnance men, civilian federal officers and every firefighter, safety tech and guard on the roster began the search. Starting in the cannon range, they walked a careful grid pattern through the ankle-deep snow. After two hours, they found the remnants of two glass jars with screw-on lids, along with a white crystal-like substance impregnated with a liquid that smelled like kerosene. The jars had landed in the northwestern part of the plant between the acid area and the power plant and shattered in the snow. The metal ashtray stand, which was discovered in the grass north of the acid area, looked more sinister than the mayo jars so the bomb squad handled it with care until they saw that it too was harmless.[263]

Reaction within the plant ranged from amusement to anger. "It was a joke," said engineer Frank Wolf. Afterward, whenever Fire Chief Percy Loomis was about to drive through the plant he would ask his passenger in a sarcastic voice, "do you want to go by the bomb site?" The official response was cautiously vigilant. "I would like to believe that any tendency on your part to dismiss the event as the inept effort of 'some nut' was offset by an inner bitter anger at anyone who would endanger your lives by any means, amateurish or callously malicious," wrote Colonel Enoch in a memo addressed to employees. "It should be obvious that we must take the 'bombing' seriously." He ordered "increased security measures," but there was little he could do beyond adding more guards, severely restricting access to and increasing the number of guard patrols on the perimeter road, persuading federal and state aviation authorities to designate the air space above Badger as a no-fly zone and urging the employees to be more vigilant. If another airplane, one carrying a real bomb, suddenly appeared over the Baraboo Bluffs, nothing short of military aircraft on constant patrol could protect the plant. Even then, could an Air Force fighter or National Guard helicopter risk firing at and downing an airplane over a munitions plant? Even from outside the plant, terrorists could still disrupt operations without crossing the fence line, and would have if Armstrong had not flubbed his attempt to dynamite the Power and Light substation in February.[264]

Radicals continued to take direct action in Madison and other places in the United States, but not at Badger, although Karl Armstrong still talked about it. He considered using primacord explosive fuse he had stolen from a mine in northern Minnesota and stashed in the old quarry at Devils Lake to topple power poles carrying electricity into the plant. After working briefly for the Chicago and North Western railroad in the spring of 1970 and bringing trains into and out of Badger, Karl wondered aloud what it would take to hijack a box car of propellant, park it on the Wisconsin River bridge and dynamite it. After the bombing of the Army Math Research Center in August 1970, the Vanguard of the Revolution fled to Canada. Karl Armstrong was captured and returned to Madison in 1973. He did not stand trial. Instead, the state prosecutors and his defense attorneys struck a bargain by which the bomber pled guilty to the charge of second degree murder in the death of Robert Fassnacht, to four charges of arson, the attempted bombing of Badger and the dynamiting of the Wisconsin Power and Light substation. After an extensive sentencing hearing, Judge Robert Sachtjen sentenced Armstrong to twenty-three years in state prison. District Judge James Doyle Sr. added ten years for federal offenses, the two terms to be served concurrently. Karl Armstrong was out of jail and on the street again in 1980.

A New Powder Plant

At the end of April 1975, the helicopter carrying the last contingent of United States Marines on duty in Vietnam lifted off the helipad at the American embassy in Saigon. It was a sad, dramatic moment, but anticlimactic as well. The United States had been cutting its forces in Vietnam for over five years. A peace agreement had been signed and American prisoners of war repatriated. The winding down in Vietnam had been mirrored on the powder lines at home. Badger's sister rocket powder plant, Sunflower Army Ammunition, shut down in 1971. After steadily declining for three years, smokeless production halted entirely at Badger in August 1973. The Ball Powder and rocket lines were still running when the Viet Cong marched into Saigon, but in a month Ball Powder was shut down, followed a month later by the rocket line. The Vietnam closing, like the final act of the Korean period, was accompanied by protests from Wisconsin congressional representatives, promises from the military to re-examine the question, then a final statement declaring that the arrival of peace meant the exit of Badger from the active duty roster.

The standard procedure was interrupted by Wisconsin State Representative Kenyon Giese of Sauk City, who also worked at the plant.

Even though the war was winding down in the 1970s, the Army had ordered and contractors had completed a new rocket powder production area at the Radford Army Ammunition Plant in Virginia. Word had somehow reached Giese that Badger was being shut down so rocket production could be shifted to the new works at Radford. "Virginia politicians are perhaps more powerful in defense matters than Wisconsin politicians," he charged. The prowess of senators and representatives from the southern states at garnering Defense Department spending for their districts had been well recognized for decades, while Wisconsin Senators William Proxmire and Gaylord Nelson were hardly hawkish when it came to Pentagon spending. Robert Kastenmeier, who had replaced Vernon Thomson as Sauk County's representative after Wisconsin's congressional districts were reapportioned in 1972, was also known as a man who never met a defense appropriation bill he could like. Those who did not give, did not receive. Further stirring the publicity pot was a multi-million dollar order for rocket powder, said to arrive sometime in the future, from the Shah of Iran. He was awash in oil revenue and itching to expand his military to world power status. Defense contractors like Olin and Hercules Powder, which ran Radford, were in fierce competition in both the shrinking home and the expanding overseas markets, and Giese suggested that Hercules must have had smoother lobbyists than Olin. Congressman William Wampler, who represented Radford's 3,000 powder workers and their neighbors in southwestern Virginia, stated, "There are no political overtones whatsoever to the possible transfer of production." Representative Kastenmeier also declared that politics was not involved in the Badger closing and accused Giese of "trivial thinking." After joining Kastenmeier for a military briefing in Washington, Giese's thinking expanded. When he came home he recanted and no longer spotted politics at work in Virginia or Wisconsin.[265]

Nonetheless, Kastenmeier formally requested that the General Accounting Office of Congress compare the cost of producing powder on the cost-plus basis proposed by Hercules at Radford and the fixed fee charged by Olin at Badger. After having produced finished rocket powder for eight years, Badger personnel were convinced then and still are today that they could make powder with fewer flaws and at a lower cost than the inexperienced crew at Radford. After a month of inquiry the GAO announced its findings. Setting aside the question of which plant could make more powder for less money, the Congressional investigators announced that, "The military services' total inventory of 2.75-inch rockets on hand are more than twice as large as needed for the war reserve stockpile." Neither Badger nor Radford were needed and, following this investigation, the Army did not request funding for the Radford rocket plant. Except for test grains, it went into standby without pressing a stick of powder. The order from the Shah also did not materialize, which was fortunate, since his regime was overthrown in 1979 by the

Islamic revolution of the Ayatollah Khomeini.[38]

After this final flurry of non-political action, Badger closed its rocket line, but the plant remained in "active" status for another two years. "It's as if they wanted to be able to come back and turn it right on again," said Olin engineer Dennis Thurow. He worked at the plant from 1971 to 1976 and witnessed Badger's high state of readiness in those years. Because it kept a few hundred people employed, Badger's "active" status also tempered the economic shock of the shutdown, which, in addition to slashing employment at the plant, would remove $5.5. million and 2,500 jobs from the regional economy.[266]

Employment at Badger had been falling steadily since it peaked at 5,300 in the summer of 1969, to a low of 400 in the spring of 1975. More significant than the peaks and valleys was the fact that Badger had kept at least 3,000 people employed from the summer of 1967 until the spring of 1971 and over 2,000 people from late 1966 until the summer of 1972. The painful impact of the loss of jobs on Sauk and neighboring counties had been felt for several years. In December 1974, for example, the unemployment rate for the county approached eleven percent, a figure inflated by seasonal layoffs in construction and tourism. Six months later, with tourism and construction in full swing, unemployment hovered near eight percent, which was the historic and unpleasant post-World War II norm for Sauk. As Badger wound down in the early 1970s, employment compensation claims rose to the highest levels since the last plant shutdown.

Applications for welfare in Sauk rose only slightly, which reveals much about the nature of employment at Badger and the people who worked there. The government-mandated safety net was firmly in place. Durlin Janzen, chair of the labor council representing eleven unions at Badger, pointed out that his people's adjustment to the closing would be eased with back pay for unused vacation days, six weeks of severance pay and a year of unemployment compensation. He was underestimating since all laid-off wage earners at Badger were eligible for the maximum level of unemployment compensation of $108 per week for sixty-five weeks. "All that adds up to more than you can make in other area plants," he said in April 1975. "It's going to be a year from July when the glut comes in the job market and damn it, somebody better come up with a good idea."[267]

No one did come up with an idea as good for working people as a job at Badger. Salaried workers, professionals and skilled tradespeople would fare well, as they always had, but "the man who pushes the button and holds his finger there until the gauge reaches the right point...is out in the cold. It is the second breadwinner who has no place to go." Few of those "second breadwinners," women who moved in and out of employment as the demands of family and opportunity allowed, would find work for comparable pay. "I guess the only job I'll get is babysitting," said Geraldine Rossing,

Lone Rock. She was a single parent and the first breadwinner for herself and a teenage son. "A lot of people will have to move to Madison. There are just no jobs here." Many of the men who worked at Badger were farmers who crammed a shift on the powder line into a day of fieldwork and barn chores, with a lot of help from wives and children. They or their wives would find other jobs and accept less pay. Limited employment options and low rates of pay had long been a part of life in rural Wisconsin. Badger was the golden exception and when it shut down, the entire area was poorer. In 1969 Badger's payroll totalled $32.5 million in dollars worth about three times as much as the greenbacks of 2000. How Sauk and its neighboring counties would have developed—what homes would be built, farms sustained, schools improved—had Badger continued to pay out even fifty percent of that amount for the rest of the century is only a matter of dreamy speculation. A decade or more would pass before the rural economy would develop sufficiently for Sauk, Columbia, Marquette, Adams, Juneau and Richland counties to make up for the loss of Badger's economic punch. It was the price paid for peace. As laid-off Badger worker Betty Ulrich said a few months after the last Marine helicopter took off from the embassy in Saigon. "I'd rather have it this way than having a war going on over there."[268]

Although it was certainly not in their minds when they began drawing lines on paper in 1967, the engineering staff at Badger proposed its own economic stimulus package for the post-shutdown era. They wanted to build a new Badger Army Ammunition Plant. "It was built in World War II and used technology from World War I," said Frank Wolf, who spent much of his twenty-year career at Badger at work on the "Modernization Plan." Wolf was exaggerating to make the point that industrial processes had progressed in the three decades since Badger was built. For example, much of the equipment used at the plant was no longer manufactured and spare parts had become impossible to purchase. To keep the old machinery running and, as was often necessary, to make otherwise unobtainable parts, Badger had an extensive machine shop and employed skilled smiths, machinists and toolmakers. They took justifiable pride in keeping the equipment in excellent shape but could not overcome the fact that it was aging and obsolete.

Industrial work had also become less labor intensive and more automated. Just as the backhoe had replaced the crew of shovel-wielding ditchdiggers on construction sites, continuous production systems were supplanting one-at-time or single-batch processes in chemical plants. In modern nitrating houses, instead of a crew of operators feeding cellulose into a dipping pot by hand, wringing the nitrocellulose by hand and then forking it into the slurry line to the purification house, a single operator opened and closed valves on a multi-chambered nitrator where ingredients were mixed, wrung, washed and spent acid removed continuously. Not only had operating procedures become more efficient and less costly, they were safer for workers. People pushing carts and buggies were replaced by pipelines in which, for example, a nitroglycerin charge could be measured precisely and was less likely to explode. The pipelines also reduced worker exposure to ether, DNT or other hazardous chemicals, another feature of modern production methods that received only minimal attention when Badger was built.

The demand for environmental protection, largely unheard of in the 1940s and '50s but stated loudly and clearly in the late 1960s, would only grow in the 1970s. Every day it ran, Badger's 1940s ammonia oxidation plant discharged nearly 32,000 pounds of nitrogen oxides that later fell out of the sky as acid rain. State-of-the-art equipment in 1972, for example, discharged about one-tenth the amount of nitrogen oxide pollutants.

The need to update the plant was obvious to those working most closely with it, seemingly from the day the plant was reactivated. As they brought the old equipment on line once again they laid plans to replace it. The engineers started with a five-year timetable, that soon grew to twelve years and then, as funding was postponed or denied, to twenty years. Had it been fully executed the Modernization Plan would have completely reversed Badger's course for the rest of the century. Instead of a facility, if not in decline, then at best maintained in its antiquated status, Badger would have been an expanded, diversified, cutting-edge operation whose mere construction would have brought hundreds of millions of dollars into the local economy. Rarely has the Badger story followed the simple logic applied by third-party observers, but it seems safe to say that had the new Badger been built as planned in the last twenty-five years of the 20th century, it would not be decommissioned in the 21st.

The plan started at the source of all production at Badger, the acid works. As designed in World War II, acid formulation was divided between the acid area in the northwest corner of the grounds near the nitrating houses and the Oleum plant, which was located farther east and adjacent to the aborted TNT facility it was built to serve. The acid area consisted of the ammonia oxidation/nitric acid manufacturing works and the nitric acid/sulfuric acid concentrator, while the Oleum plant produced highly concentrated "fuming" sulfuric acid, i.e., Oleum. In the Modernization Plan all acid production was moved to a new area located on the central rail line, due east of the administration building and almost exactly in the center of the production area. It would be the hub of the plant, with individual manufacturing areas stationed around it like spokes in a wheel. On a spoke running east was a new nitroglycerin plant and, beyond, a new Ball Powder plant. South from the new NG plant was a new rocket area, utilizing about one-half the old paste area and about one-third of the finishing area. Directly to the west of the new acid area, just across the railroad tracks, was a new nitrocellulose area, while to the north would sit the smokeless powder finishing lines. Also in the plan was a completely new water distribution system and other support facilities.

The magazine area would remain unchanged, but in total, the new Badger would require a little more than one-half of the 7,500 acres of the old plant. It would be a concentrated facility, clustered in the center of the grounds, and produce somewhat smaller quantities of some propellants than the old Badger, but still meet military requirements for propellant production as set forth by the Defense Department.

The cost in dollars in 1973 was estimated at $319 million, more than twice as much as the Badger of the 1940s — on paper. However, after thirty years of inflation, the plant as estimated in the mid-1970s would cost about the same as the original model. Benefits not in the budget came in the form of reduced air and surface water pollution since the new plant would meet all the standards of the newly enacted federal Clean Air and Water Acts. Operating costs would be lower because the new plant, with its highly automated, continuous processes, needed fewer people. The old nitric acid plant, for example, required twenty-eight operators, compared to eight in the new plant. It may not have been good news for potential job seekers, but those who would be hired could expect higher pay as skilled technicians and a safer, healthier working environment. The schedule Olin's engineers presented to the Army and Congressional auditors in January 1973 had the new Badger built and ready to roll in 1981.

Of course, it did not happen that way. Funds were appropriated and actually paid to construct the new acid and nitroglycerin plants and to purchase equipment for the first phase of the new nitrocellulose facility. The acid and nitro plants, state of the art in the 1970s and, except for computer upgrades, state of the art in 2001, made a few test runs and shut down. They were completed in 1976 and 1978, respectively. The NC equipment is in storage at the plant. Since 1976, money has been appropriated to study, design and draw sketches of the rest of the Modernization Plan, but no ground has been broken, concrete poured or wrench clamped to a pipe. "I left in 1976 as product improvement manager," said Dennis Thurow. "We were very involved in the modernization...working for about two years, setting up, designing a pilot plant to be used to study the technology that would be used in the modernization of the Ball Powder plant. We had just received funding to put in the pilot plant when I left. When I came back in 1991, funding was just received to do the design for the modernization of the Ball Powder plant...It was all there but the Army, rather than start to put it in, they said, 'when we need it again, we'll start it up.' From a taxpayers' standpoint it really makes sense. Spend the money when you need it."[270]

They almost thought they would need it for the Persian Gulf War in 1990-91. Even though fifteen years had passed, Badger had been kept in a higher state of maintenance after Vietnam than after Korea and in a much higher state than after World War II. Operations could have started in a few months had a fourth call to arms sounded. Had the war lasted longer, had President

Bush followed the advice of the hard liners and chosen to fight in Iraq until Saddam Hussein was out of power, the powder lines on the Prairie might have run again. The plant was ready. It was just a matter of bringing the people back, training them and firing up the power plant, the acid works and the NC lines.

Jim Mattei was Olin's top manager at the plant. He recalls a phone call he received during the war. "The phone rings and a guy says to me, 'you don't remember me but I used to be the technical director.'

"I remembered. The guy was out in Utah, he must have been about eighty. 'I just want to tell you, if you need me, I can come back.'"[271]

So could Badger have come back, had it been called up once again. Instead, the brief war in the Gulf ended, so had the decades-long Cold War. Public sentiment and political pressure to close down inactive military installations grew. In 1998, the Army initiated decommissioning procedures at Badger, suspended all but the most basic maintenance and prepared to dispose of the powder plant on the Prairie. In a few years all that will be left of the plant is the legacy it leaves behind.

IV

Legacy

"Badger is a real treasure chest of history."

Mike Mossman, Badger History Group, 1998[1]

PAULA BUNCH

The Evil and the Good

During Badger's inactive days in the late 1970s, Dave Fordham, who became Commanding Officer's Representative in 1976, sent an unusual request up the chain to his Army superiors. Fordham wanted to pull a few acres of ground in the northwest corner of the plant out of cultivation by a farm leaseholder and replant it to prairie. Unlike tree planting, prairie restoration was still a relatively new conservation concept. It was a bit too new for the United States Army and Fordham had to overcome a "surprising amount of resistance." He persisted and succeeded in bringing back a piece of the ancient landscape of the Sauk Prairie. He also added another ironic curve to the course of the Badger story. The powder plant so guilty of assaulting the environment was, in the person of its manager, responsible for restoring a bit of the natural landscape that was erased by the settlers in the 1840s. For Fordham, who spent over thirty years of his working life at Badger, "It's one of the things that I've done here that I'm proudest of."[2]

Fordham's prairie was the result of deliberate human intervention, not unlike the firing of the grasslands by the Indians of pre-settlement times, and it set a precedent for other grassland restoration projects on the property. Without deliberate intervention, however, much of the powder plant had evolved into a refuge for wildlife threatened by the drastic changes in land use that had taken place outside the fence line since the end of World War II. Continuous row crop farming, early cutting of hay, the planting of pine trees in monotonous plantations, the dicing of forests into isolated woodlots and persistent suburban sprawl have severely reduced habitat for all grassland wildlife, especially birds. The Badger property has been less subject to these processes, and, as a result, supports approximately 100 bird species, eighteen of which are "considered critical in Wisconsin." The wide stretches of treeless space between the buildings and occasionally the buildings themselves, the magazine and reservoir areas, some of the farmed ground, as well as other relatively undisturbed tracts, have become important refuges for birds that have lost living space not just in Sauk County or Wisconsin, but throughout North America. For the meadowlark, bobolink, savannah sparrow and other grassland birds, the powder plant is a much more hospitable place than the modern farms across the fence. How odd it would seem to Leo Crowley, H.N. Wendle or Louis Reuter to know that when they set out to build a powder plant, they were creating a bird sanctuary too.[3]

Badger's accidental transformation into a wildlife refuge is not unique. Beneficiaries of benign neglect by the forces of progress in rural America for decades, many World War II military installations harbor an extraordinary abundance and diversity of plant and animal life compared to adjacent territory which has kept up with the times. Nor are these accidental refuges restricted to the United States. One of the most important havens for cranes in northeast Asia is the 2.5-mile-wide demilitarized zone between South and North Korea. Once a battleground where opposing armies exchanged artillery salvoes — with the South Koreans firing propellant made at Badger Ordnance — the DMZ became a tranquil no-man's-land where wildlife flourished. Recently, as tensions have somewhat eased between North and South Korea, the wildlife of the DMZ is threatened by development, just as the grassland birds that have taken refuge at Badger Ordnance may be threatened by future land use there.

Just as Badger has become a refuge from modern land use for wildlife, it may also be a refuge for humans. While ever more real estate in Sauk and neighboring counties is parceled into country lots, and villages expand into the towns, the plant remains in public ownership. It is a set-aside buffer zone spanning the geological divide defined by the terminal moraine of the Wisconsin glacier, where geographic zones — riverbank, wetland, prairie, savanna, hardwood forest — meet and where they may be preserved along with the history of the plant and that of the displaced farm and Indian communities that also occupied the site.

Badger may be a refuge for wildlife, and rich in geographic diversity, but it is not paradise — and far from it. The buildings, grounds and water beneath are contaminated with an encyclopedia full of toxic substances. Buildings — 900 of which are too contaminated to be sold for salvage—contain asbestos, lead and propellant residues. Process lines and sewer pipes are laden with deposits of nitrocellulose and rocket paste. The soil holds nitrates, sulfates, solvents and heavy metals. Groundwater is laced with carbon tetrachloride, trichloroethylene and chloroform. A total of thirty-three contaminated sites have been identified inside the plant and one outside. They range from the old acid and Oleum areas at the head of the production line, run through all the NC, Ball Powder, nitroglycerin and rocket paste areas, encompass the disposal pits, landfills and ditches, continue on to the settling ponds not constructed until 1972 and down to the drinking-water wells of homes located just south of the fence line.

As the route by which all the plant's wastewater has been discharged into the Wisconsin River, Gruber's Bay is a special case. Sediments there contain virtually every pollutant run into the ditches and drains, from organic compounds to heavy metals—at least those that have not washed down the Wisconsin River. In 1971, the Army forestalled a Wisconsin DNR lawsuit by pledging to clean up Gruber's Bay and was ready to begin dredging in 1977. Citing a need for additional studies on the impact of dredging and the disposal of the spoil removed, the DNR halted the work. The subject was not reopened until the late 1990s and dredging did not begin until 2001.

The listed sites cover only about two percent of the total land area at Badger, yet account for the bulk of the known contamination. By 1998, the Army had spent $66 million on cleanup work. It is estimated that another $167 million will be required to remove contamination, for a total of $233 million. More than one environmental activist has argued that more studies should be conducted because more contamination is present. Be that as it may, decades will pass before the entire site, inside and outside the fence, meets federal and state environmental standards. The genuine distress felt by Badger's immediate neighbors, who are living with contaminated groundwater, and neighbors farther removed, who feel threatened, is yet another unforeseen consequence of the plant's presence on the Prairie.[4]

The costs of cleaning up Badger represent only a portion of the investment made in the plant over the years. The bill to construct Badger's acid, smokeless and rocket areas, plus its infrastructure of power plant, water system, waste disposal, transportation, maintenance and other facilities, ran to $125 million in World War II-era dollars. Rehabilitation of these facilities for Korea in 1951-'53 added another $49 million to the tab. A few years later, the new Ball Powder plant required an expenditure of $21 million. By the time the plant reopened for Vietnam in early 1966, the Army estimated that replacing Badger would cost the taxpayers over $304 million. While rising costs were a factor in these estimates, inflation was not yet "galloping," as the pundits of the '70s were so fond of stating. By the end of operations for Vietnam, inflation was not merely galloping, but streaking. In a decade that saw no new major construction or remodeling except for the installation of pollution control equipment and the digging of settling ponds, the replacement cost of the plant rose to $700 million. When Badger marked its fiftieth anniversary in 1992, the figure had ballooned beyond $1.8 billion. In the oft-quoted words of Senator Everett Dirksen, "A billion here and a billion there, and soon you're talking about real money." Real or not, the cost of replacing Badger was not large enough to prevent the Defense Department from decommissioning the plant in 1998. Should a crisis ever come in which the nation needs the quantities of acids and propellants Badger is capable of churning out, the bill for new facilities will be much more real than $1.8 billion.

The charges for Badger, in millions or billions, real and unreal, are itemized on the sixty-year-old bill for national defense. They are part of the debt owed for a half-century of international conflict that generations unborn at the time are still paying and will be paying for many more years. As Wendell Phillips said, "Eternal vigilance is the price of liberty." So is eternal taxation. Those who enjoy the benefits of American life in the 21st Century will help pay for investments in defense made in the 1940s, '50s and '60s. The national debt is also the price of victory. World War II was won by brilliant strategists, valiant soldiers, sailors and aviators, and the mightiest industrial establishment on earth. As part of that establishment, Badger Ordnance played a key role in the victory. Badger also played a key role in the longer struggle that followed World War II and in which the United States was also the victor. The Cold War with the Soviet Union ended with the defeat of communism in the early 1990s. As part of the military-industrial system that ultimately wore down the Soviets without the use of nuclear weapons, Badger defeated the communists just as it defeated the Nazis. One Cold War battle, in Korea, may have ended in a stalemate, another with defeat in Vietnam, but in the end the United States won a war of forty-five years' duration.[5]

That Badger required and enabled citizens in Wisconsin to take part in these conflicts is also part of its legacy. Patriotic women who had family responsibilities and who could not serve in combat anyway, men too old for the military and the physically disabled but still willing men of draft age who wanted to serve their country could do so at the powder plant. Not only G.I. Joe and Rosie the Riveter but also Polly and Pete the Powdermakers volunteered to do their duty as citizens of the republic. And like the G.I.'s, they took casualties, dying in accidents at the plant or living but suffering from illnesses aggravated by exposure to hazardous substances. Veterans of Vietnam had to contend with Agent Orange. Veterans of Badger had "dynamite heart" from exposure to nitroglycerin, which left them as much as thirty times more susceptible to heart ailments than the rest of the population. They also encountered asbestos dust, ether fumes and exposure to lead, benzine, DNT, and so on down the list of workplace toxins whose hazards had yet to be discovered when the powder workers faced them minus protection and knowledge of their dangers.[6]

Also doing their duty for their country were the farmers who gave up their land for the plant. The pain and loss they endured created another class of war heroes. Their struggle to obtain just compensation by means of law, instead of by shotgun, and the government's response, also by means of law instead of by machine gun, explains and justifies their sacrifice and the war itself. Among the freedoms for which Americans waged World War II was the right of citizens to peacefully protest the actions of their government and for their grievances to be redressed under the rule of law. The embattled farmers of the Sauk Prairie and their adversaries in the federal prosecutor's office gave generations to come a lesson in how America is supposed to work. A generation later, when a stubborn government pursuing a failed policy was attacked by equally stubborn and inept terrorists, Badger became the classroom for a lesson in how America is not supposed to work.

As an economic entity, the powder plant had its greatest impact on the people of Sauk and its neighboring counties. The gradual shutdown that began in 1973 meant a loss of $78 million in gross regional output over the next few years. The earlier shutdowns brought about similar reversals followed by upswings when the plant was reactivated. With Badger the glass was either empty or full to overflowing. Overflow from the first filling of the glass swept the region out of the Great Depression and lifted the general economic status of everyone within forty miles of the plant. Paychecks from Badger improved tens of thousands of lives. The business activity those paychecks generated, as well as work the plant assigned to area contractors, gave birth to new businesses and bolstered established enterprises. When the Badger glass drained to empty, local business and government were forced to pursue economic development to replace the wealth created by the plant. In this sense, Badger was good for the region when it was open and good for the region when it closed.

One loss that would not be made up—except in the works of Erhart Mueller—was that of the farm community the powder plant displaced. In this matter, Badger presaged and accelerated the process of farm consolidation that took off with a vengeance as soon as World War II ended. In 1945, Sauk County had nearly 1,600 dairy farms; in 2000, fewer than 500. By the end of the century acreage equivalent to the land necessary for two or three powder plants the size of Badger was enrolled in federal set-aside programs and withdrawn from production. Farm productivity had so improved that the loss of thousands of acres of "the finest farmland in Wisconsin" to the powder plant made barely a blip on the radar screen of agricultural progress. It can be argued that the plant has paid its debt to farming by making available a thousand-plus acres to the United States Department of Agriculture for its Dairy Forage Research Center Field Station. When the station was sited along the Wisconsin River in 1979, in a corner of the property never used for powder production, Director Harlow Hodgson predicted that, after ten years of operation, the Forage Center would save dairy producers and consumers as much as $850 million a year. If so, then Badger has not only paid its debt to farming, but to the taxpayers as well.

The powder plant was on the front edge of American social history as it evolved in the last half of the 20th century. Some of the most significant changes in how all Americans, and especially rural Americans, have lived in the years since World War II occurred at Badger during the war. The increase in the number of women in the workforce often credited to the young adults of the baby boom generation in the 1970s occurred at Badger during World War II. By the end of the war, women comprised forty-eight percent of the workforce and were holding down jobs in every department of the plant—from the guard shack to the rocket presses. This was certainly up from the prewar industrial world where women made up about twenty-five percent of the workforce, and closer to the Wisconsin family farm average, where every women and man on the place worked the farm. At Badger, women were restricted to the lowest rungs on the occupational ladder. They were janitors, food servers, clerks, laundry workers, lab technicians and powder line laborers. All but a few supervisory slots and virtually all the skilled trade and managerial positions were held by men. The picture would change only slightly in the Korean and Vietnam periods, with more women in supervisory positions. Overall, however, less than one-third of the total workforce was female, a significant reduction from World War II. Nonetheless, a precedent was set in the 1940s and an example made for the daughters of the powdermakers who later entered the workforce, and who broke through the barriers blocking their mothers from supervisory and managerial positions.

That a workforce containing a large percentage of women who were also parents would place an increased demand for services to support families has been a much-discussed fact of life in America since the 1970s. It had already been well-covered at the Badger School in 1943. The Badger School provided day care on a 24/7 schedule for infants, toddlers, grade and high school students. It had after-school enrichment, recreation and sports programs, and monitored the health and family life of its young charges. As principal Clarence Krumm announced at the dedication ceremony, "Our schools are going to take care of the children of workers so that the workers do their best." He was voicing a commitment to the children of working families that would be music to the ears of harried parents decades later, only he did it in 1943. The passing of time would also prove him correct when he said that the "wartime child was also the peacetime child." It needed only a few decades for the social conditions of America at peace—with a majority of children raised in single parent families or in families with two parents working outside the home—to return to those of America at war.

Another feature of American life in the last half of the 20th century that was anticipated at Badger Ordnance was its reliance on the automobile. In fact, the auto enabled Badger to operate, because it was by auto that a majority of the plant's widely dispersed workforce travelled to and from the plant. Commuter rail services established to carry people to and from Madison in the spring of 1942 shut down after only a few months for lack of ridership. The extensive and economical Badger bus system, although well used, did not carry a majority of workers to the plant. The auto did carry them, despite unimproved country roads, fuel rationing, tire shortages and scarcity of parts for repairs. In this matter, the workers at Badger set a precedent for rural commuters driving their cars for ten, twenty, thirty or forty miles to and from work every day. Regular commutes of such length by such great numbers of rural people were unheard of in prewar Wisconsin. They would become commonplace after the war.

Nearly all of the workers commuting to Badger were farm men and women. They set the trend for a future in which the economically self-sufficient family farm was an endangered enterprise that required one or more family members to work off the farm to make ends meet. By the end of the century the farm family that did not derive a significant portion of its living from off-farm employment was in the minority. Going one step further, once Badger had demonstrated that it was possible, the countryside filled with residents who chose to live beyond the sidewalks not because they farmed, but because they liked country living and could commute to jobs many miles away. Just as workers can live just about anywhere and get to work, employers can locate in urban, suburban or rural areas knowing that in all but the most isolated locations the workforce will come to them by car, just as workers from Spring Green, New Lisbon and Madison drove to Badger during World War II. Not only concrete, but history, links the four-lane highway and bypass to the Sauk City bridge built for Badger in 1942 and the four-lane highway proposed to be built between Madison and Sauk City fifty years later.

Not all rural residents hold down jobs and Badger has also helped set the historical trend here. The conversion of Badger Village to senior citizen housing as Bluffview Courts condominiums in the early 1960s was an early example of what would become another commonplace of rural America. It would become the retirement location of choice for Americans who once — if they lived long enough to retire at all — preferred cities and villages.

The buildings, machinery and equipment still at Badger stand as mute witnesses to the labors of the powder workers and to the industrial technology they used to make war. The breadth and depth of the place, briefly glimpsed in the breath-catching panoramic view from the slope of Highway 12, demonstrates what was required to win World War II and defend the United States in the last half of the 20th century. There are numerous memorials to American military accomplishment and well there should be. There are also many museums of technology and American industrial enterprise in peace and war, including the railroad museum at North Freedom and a California facility dedicated to "Rosie the Riveter," but, except for a humble plaque or stone marker, there are no memorials or museums commemorating and depicting the labor and sacrifices of ordnance workers. Until funding dried up in 1998, most of the physical plant at Badger was well maintained. Much of the place remains as it was during World War II and, as such, is an ideal site for a memorial, museum and learning center where the record of citizens who did their duty at the ether still, the dipping floor or the angel walk, along with the hardening houses, boiling tubs, tramways, escape chutes and extrusion presses they used to make propellant, may be preserved. Shakespeare wrote that the evil men do lives after them, while the good is "oft interred with their bones." The evil consequences of wars fought sixty, fifty and thirty years ago are present on the Sauk Prairie, but so is the good. May each one be remembered.

Notes

I Prairie

[1] Butterfield, *The History of Sauk County*, p. 584.
[2] Lapham quoted in Cole, *Places List of Sauk County*, p. 41.
[3] Lange, *Postglacial Vegetational History*, p. 14.
[4] Butterfield, p. 490; Leopold, *A Sand County Almanac*, p. 26.
[5] Madsen, *Where The Sky Began*, p. 160.
[6] Dickcissel in Collins, *Complete Field Guide to American Wildlife*, p. 229, and Mossman, "Breeding Birds of the Badger Army Ammunition Plant," p. 13.
[7] Casper, "The Reptiles and Amphibians of the Badger Army Ammunition Plant," pp. 19–22 and *The Audubon Field Guide to North American Reptiles and Amphibians*, pp. 298, 387, 543, 596. Thompson and Walsh, "The Biological Inventory of the Badger Army Ammunition Plant."
[8] *Proceedings* of the Sauk County Board, 1895.
[9] Pyne, *Fire In America*, p. 87.
[10] Marryat, *Wisconsin Historical Collections*, XIV, pp. 138, 152.
[11] Leopold, p. 27.
[12] Lange, *Ancient Rocks and Vanished Glaciers*, p. 7.
[13] Clayton and Attig, *Geology of Sauk County*, p. 38.
[14] Clayton and Attig, pp. 46, 49.
[15] Clayton and Attig, p. 49; Canfield, *Outline Sketches*, 1861-1901, Fourth Sketch, p. 1.
[16] Butterfield, p. 677. Cochrane & Iltis, *Atlas of the Wisconsin Prairie and Savanna Flora*, p. 32.
[17] Lange, p. 11.
[18] Grassland acreage in Lange, p. 14; poetry in Longfellow, "Evangeline."
[19] Pyne, p. 76.
[20] Wittry, "The Raddatz Rockshelter," *The Wisconsin Archeologist*, v. 40, n. 2., pp. 58–61.
[21] Cronon, *Uncommon Ground*, p. 83.
[22] Wight quoted in Curtis, *Native Vegetation of Wisconsin*, p. 299.
[23] Pyne, p. 10.
[24] Dorney, *The Impact of Native Americans on Vegetation In Southeastern Wisconsin*, p. 32.
[25] Gartner, *Four Worlds Without an Eden*, p. 331.
[26] Inventory of mounds in *The Wisconsin Archeologist*, v. 1, n. 3, August, 1922.
[27] Russell, *Indian Set Fires in the Forests of the Northeastern United States*, p. 81.
[28] Parmalee, *Animal Remains from the Raddatz Rockshelter*, p. 90. Gartner, *Four Worlds Without an Eden*, p. 334.
[29] Wright and Bailey, *Fire Ecology*, pp. 65, 113.
[30] Gartner, *Late Woodland Landscapes of Wisconsin, Ridge Fields, Effigy Mounds and Territoriality*, p. 671.

[31] Gartner, p. 334.
[32] Carver, *Travels In North America*, quoted in *Wisconsin Archeologist*, v. 5, no. 2, p. 278.
[33] Lurie, *Winnebago Photohistory*, p. 796.
[34] Lurie, pp. 790–797.
[35] Mueller, *Only in Sumpter*, p. 30.
[36] Accounts of fires from Curtis, pp. 297-299.
[37] Owen, quoted in Curtis, p. 299.
[38] Rendtorff in Canfield, *Outline Sketches*, Second Sketch, p. 64. Muir, *My Boyhood and Life*, 1913, p. 63. "The Prairie" in *The Wisconsin Farmer*, v. 14, n. 11, November 1, 1862.
[39] Curtis, p. 276.
[40] Lueders, *Transactions,* p. 511.
[41] Rendtorff in Canfield, Second Sketch, Curtis, p. 269.
[42] Curtis, p. 279.
[43] Brewer, *Natural History*, p. 30. Mossman, *Breeding Birds*, p. 12–13.

II Settlement

[1] Canfield, *Outline Sketches*, 1861–1901, Second Sketch, p.64.
[2] William Johnson story in Canfield, Third Sketch, pp. 16–18, and Mueller, *Only in Sumpter*, pp. 29–31.
[3] Development of the breaking plow in Rogin, pp. 13–49.
[4] F.G.I. Lueders in Butterfield, p. 490; Muir in Rogin, p. 47.
[5] Ragatz, *Memoirs*, WMH, v. 19, pp. 213–214.
[6] Ragatz, *Memoirs*, WMH, v. 19, p. 215.
[7] Canfield, Third Sketch, p.17.
[8] Staines, *Transactions of Wisconsin State Agricultural Society*, v.1, 1851, p. 215.
[9] Greening in *Wisconsin Magazine of History*, v. 26 p. 214; Canfield, Third Sketch, p. 2.
[10] Mueller, *Return to Sumpter*, p.173.
[11] Knapp, Autobiography, p.13.
[12] Leopold, *A Sand County Almanac*, p. 29.
[13] Cole, *Stagecoach and Tavern Tales*, p. 212.
[14] Canfield, Second Sketch, 64.
[15] Mueller, *Forever in Sumpter*, p. 54.
[16] Farm statistics for 1869 in Canfield, Third Sketch, p. 9, and Fourth Sketch, p. 7; 1900 and 1901 numbers in Report of Supervisor of Assessments, *Proceedings* of Board of Supervisors, Sauk County, 1902–1903. Johnson's plow in Butterfield, p. 681.
[17] Robert Hatz, Reminiscences, unpublished manuscript, "Grain Production," p. 1.
[18] Wayne and Robert Hatz, Reminiscences, unpublished manuscript, p. 2.
[19] Mueller, *Forever In Sumpter*, p. 58.
[20] Mueller, *Also In Sumpter*, p. 137.
[21] Mueller, *Return To Sumpter*, p.73.
[22] Annual Report, State Highway Commission, 1925, p. 12
[23] *Baraboo News-Republic*, November 12, 1925.
[24] La Follette version of Seibecker story in La Follette, *Autobiography*, pp. 61–71. Proposal to close Lakeview school in Merritt, *Reorganization of the Elementary Schools of Sauk County*, 1941, p. 17.
[25] Clifford Schweppe's memories in Doris Litscher Gasser, *Sauk Prairie Star Stories*, 1998, no page number.

[26] *Baraboo News*, April 2, 1914.

[27] Sumpter power companies in Mueller, *Always in Sumpter*, p. 102. Mather Maples in Mueller, *Also in Sumpter*, p. 210. Farm electrification statistics in Goc, *Where the Waters Flow*, p. 107.

[28] Kirkpatrick and Boynton, pp. 138–140, and Sauk County Agricultural Statistics, 1942, p. 10.

[29] Luetscher barn in Mueller, *Only in Sumpter,* p. 213, and author's interview with Melvin Luetscher, January 16, 2001.

III Powder Plant

[1] Mayer, p. 2.

[2] Ecklund interview and Weiss, p. 113.

[3] Weiss, p. 113.

[4] Robert Ecklund interview with Leo T. Crowley, 1969, Tape 4, Side 1, Part 1.

[5] C&NW's public timetable, January 15, 1939, published in *Northwestern Line Magazine*, Winter, 1985, p. 32.

[6] The role of politics in siting of defense plants for World War II in Thomson and Mayo, p. 108, and Kane, p. 25. Resolution of Midwest Defense Council quoted in Dooley, p. 218.

[7] Poll cited in Kane, p. 26.

[8] Definitions in Kane, p. 7.

[9] *Milwaukee Journal*, October 30, 1941.

[10] *Milwaukee Journal*, October 29, 1941.

[11] Stevenson biography in *Wisconsin Blue Book*, 1944, p. 22. Siting list in Report of Badger Ordnance Works, Hercules Powder Co., v. 1., p. 34, hereafter cited as Hercules Report. Response to Quartermaster General in Hercules Report, p. 6.

[12] *Milwaukee Journal*, November 2, 1941.

[13] Hercules Report, p. 48.

[14] *Milwaukee Journal*, November 14, 1941.

[15] *Milwaukee Journal*, November 18, 1941. *Capital Times*, January 16, 1942.

[16] Hoar in *Milwaukee Journal*, November 18, 1941, and *Baraboo News-Republic*, November 24, 1941. Wendle in Hercules Report, v. 1, p. 36.

[17] Mayer, p. 7. *Baraboo News-Republic*, November 15, 1941. *Milwaukee Journal*, November 20, 27, 1941.

[18] *Proceedings* of Sauk County Board, 1941–42, p. 88.

[19] Derleth quoted in *Milwaukee Journal*, November 20, 1941.

[20] *Milwaukee Journal*, November 22, 1941.

[21] Copy of Evjue telegram in Erhart Mueller Papers, Archives of State Historical Society, Madison. "State Triple A Protest Powder Site" in *Capital Times*, November 24, 1941. Wickard visit in *Baraboo News-Republic*, November 29, 1941. Townsend letter in *Baraboo News-Republic*, November 26, 1941. "Now a Powder Plant" in *Milwaukee Journal*, November 22, 1941.

[22] Swanton to Mrs. Albert Mueller in Erhart Mueller Papers, Madison. Townsend in *Baraboo News-Republic*, November 26, 1941.

[23] Loomis and Engebretson in *Capital Times*, November 30, 1941. Gehrmann and Johnson in *Baraboo News-Republic*, November 13 and 28, 1941. "Now a Powder Plant" in *Milwaukee Journal*, November 22, 1941.

[24] *Milwaukee Journal*, November 24, 1941.

[25] Robert La Follette, Jr., to Albert Mueller, November 26, 1941, in Erhart Mueller Papers.

[26] Hercules Report, p. 40.

[27] Stevenson letter dated November 26, 1941, to Mr. and Mrs. Fred Wintermantel, Prairie du Sac, in Erhart Mueller Papers, State Historical Society of Wisconsin.

[28] *Milwaukee Journal*, November 24, 1941, Madison *Capital Times*, November 30, 1941, *Baraboo News-Republic,* November 24 and December 4, 1941.

[29] Nora Premo, *Capital Times*, November 23, 1941. Garth Premo in letter to Dora Mueller in Erhart Mueller Papers, Madison.

[30] Premo, Kindschi, Mather, Dahnke and Leutscher quoted in Madison *Capital Times*, November 23, 1941.

[31] *Baraboo News-Republic*, November 26, 1941. Iowa story in Fine and Remington, p. 181.

[32] Taped reminiscence of Walter Hoar, 1971, State Historical Society of Wisconsin, Madison.

[33] *Baraboo News-Republic*, December 8, 1941.

[34] Fine and Remington, p. 180. Hoar, Madison, 1971.

[35] *Baraboo News-Republic*, December 11, 1941.

[36] "To Whom It May Concern," memo in Hercules Report, v. 1., p.11. Silas Knudsen died in 1962. His feelings about the Badger land acquisition were related to the author by his son, George Knudsen, in the spring of 2001.

[37] Undated news article in Erhart Mueller papers and chart accompanying letter to Emma Kindschi from Attorney William Ryan, May 16, 1944.

[38] Wendle memo in Hercules Report, v. 1, p. 36.

[39] Oldfield memo to Hauck, January 30, 1942, in Reuter, p. 12.

[40] Gasser, *Sauk Prairie Stories*, p. 12.

[41] Barber in *Baraboo News-Republic*, February 12, 1942.

[42] Jessie in Gasser, p. 2.

[43] Gasser, pp. 6, 14.

[44] Landowners list in *Sauk County News*, March 5, 1942.

[45] Steuber story in Mueller, *Only in Sumpter*, p. 206. Hatz quote in conversation with author, February 21, 2001.

[46] Interview with Mel Luetscher, January 16, 2001.

[47] Gasser, p. 13.

[48] Knudsen to Commanding Officer, Badger Ordnance Works, January 21, 1942.

[49] *Milwaukee Journal*, February 26, 1942.

[50] Thomson and Mayo, *The Ordnance Department: Procurement and Supply*, p. 34.

[51] Bradford, pp. 92–109; Guthrie, *Roll On Columbia, Roll On*, 1940.

[52] Contract in Mayer, p. 7. TNT cancellation in Mayer, p. 25. E.C. plant in *Badger Ordnance News*, February 19, 1943.

[53] Hercules Report, v. 1, p. 103.

[54] Construction methods as described by department in Hercules Report, v. 1–5.

[55] Blow-out construction in Kane, p. 88.

[56] War Department Facilities Inventory of Badger Ordnance Works, October 1945.

[57] Contractors list in Hercules Report, pp. 168–192.

[58] Chronology assembled from Hercules Report, David P. Mayer, *Badger Ordnance Works News*, *Wisconsin State Journal*, *Baraboo News-Republic* and *Sauk Prairie News*. John Geoghegan recalled his memories to Lisa Hartmann, Badger History Group, in August 1999.

[59] *Wisconsin Dells Events*, September 8, 1942.

[60] Mayer, p. 22.

[61] *Badger Ordnance News*, February 28, 1944.

[62] Hercules Report, p. 226.

[63] Mason and Hangar, Monthly Progress Report, January 1945.
[64] French article excerpted in *Badger Ordnance News*, December 8, 1944.
[65] Employment figures in Reuter, Historical Report, p. 479, and in Hercules Report, pp. 365-367, 583-585, 766.
[66] Woman working in *Badger Ordnance News*, November 13, 1942, March 19, 1943, February 26, 1943. Latrines in Hercules Report, p. 545.
[67] Pay scales in "Field Instructions to United States Employment Service Managers and Hercules Powder Recruiters, p. 3.
[68] Hercules Report, p. 274. *Badger Ordnance News*, July 9, 1943; November 17, 1944.
[69] Penicillin in *Badger Ordnance News*, August 25, 1944; Union vote in *Badger Ordnance News*, November 3, 1944, and contract in Hercules Report, p. 286.
[70] Author's interview with World War II-era nurse, Carol Sorg; *Proceedings* of the Sauk County Board, 1942-43.
[71] Lola Huber interview with William Sisson, November 27, 2000; Hercules Report, p. 534; grocery prices in "OPA Community Ceiling Prices," May 20, 1944; sales flyer for Victory Heights, 1943.
[72] Reuter, Louis F. Historical Report, Badger Ordnance Works, Book 8, Period ending September 30, 1944, p.458. *Badger Ordnance News*, September 1, 1944.
[73] *Westfield Central Union*, September 21, 1944, *Badger Ordnance News*, June 6, 1944; July 7, 1944; January 19, 1945; March 23, 1945.
[74] *Badger Ordnance News*, June 16, 1944.
[75] *Badger Ordnance News*, January 22, 1945.
[76] Hercules Report, pp. 538-549, 579-583.
[77] Corps of Engineers, Great Lakes Division, Data on Buildings in War Department Facilities, Supplement, Badger Ordnance Works, V. 3.
[78] Hercules Report, p. 696; Badger History Group interviews with William Sisson, Norma Clavadatscher and Rita Slazek Lachmund. See Appendix for death of Jamaican worker Herman Beach.
[79] Schuette, Reedsburg Remembers, p. 108, Goc, *Many A Fine Harvest*, p. 122.
[80] *Badger Ordnance News*, September 8, 1944, February 9, 1945, March 30, 1945.
[81] Hercules Report, p. 696.
[82] Hercules Report, p. 697.
[83] Hercules Report, p. 375.
[84] Davies to Evjue, October 21, 1942, Evjue to Davies, October 26, 1942 in Evjue Papers at Wisconsin Historical Society Archives.
[85] Hercules Report, p. 376.
[86] Hercules Report, p. 288.
[87] Hercules Report, p. 397.
[88] Hercules Report, p.402.
[89] Hercules Report, p. 406.
[90] Hercules Report, p. 404.
[91] Hercules Report, p. 416.
[92] Lola Huber interview with Rita Slezak Lachmund, August 28, 1998.
[93] VanWhy in *Badger Ordnance News*, March 2, 1945; sidewalk in *Badger Ordnance News*, Friday, May 25, 1945.
[94] Hercules Report, p. 312.
[95] Hercules Report, pp. 314-315.
[96] Lola Huber interview with Rita Slezak Lachmund, August 28, 1998.
[97] *Badger Ordnance News*, August 8, 1944. Lola Huber interview with Mardetta O'Brien, July 18, 1998, and William Sisson, November 2000.
[98] Hercules Report, p. 348.

[99] Deaths in explosions are covered in text. See Appendix for other fatalities at Badger Ordnance. Safety policy in *Badger Ordnance News*, June 3, 1943. DNT exposure policy in Hercules Report, p. 344-345, 780-781.
[100] Hercules Report, p. 282.
[101] Ambrose, pp. 65-66.
[102] Hercules Report, p. 721.
[103] Hercules Report, p. 365, 681.
[104] Hercules Powder Monthly Report, May 1944. Lola Huber interview with William Sisson, November 2000.
[105] Sisson interview.
[106] Lola Huber, "Badger Worker's Daughter," *Wisconsin Academy Review*, Fall 2000, p. 36. Mardetta O'Brien in Lola Huber interview.
[107] Hercules Powder Monthly Operating Report, October 1943.
[108] Hercules Report, pp. 882-888. Hercules Powder Monthly Operating Reports, November-December, 1943; January-October, 1944.
[109] Hercules Report, pp. 390-396, 503,
[110] Hercules Report, pp. 390, 855. Badger Ordnance Works Plot Plan, December 1, 1945.
[111] Rocket powder production as described in Kane, p. 132-135.
[112] *Badger Ordnance News*, July 27, 1945. Hercules Powder Monthly Operating Reports, July, August, 1945, Sauk County Register of Deeds, Death Certificates, v. 27, pp. 516, 521, 532, 536.
[113] Hercules Report, p. 905. Hercules Powder Monthly Operating Report, April 1945.
[114] Hercules Report, p. 912.
[115] Hercules Report, p. 638.
[116] Henry Ringling to Julius Heil, December 29, 1941 in Heil Papers, Archives, State Historical Society of Wisconsin.
[117] Madison *Capital Times*, February 26, 1942.
[118] Madison *Capital Times*, April 16, 1942.
[119] Madison *Capital Times*, February 6, 1942.
[120] William Blount letter to Carol Sorg, Badger History Group, September 30, 1999.
[121] *Badger Ordnance News*, October 9, 1942.
[122] Madison *Capital Times*, February 25 and 27, 1942.
[123] Madison *Capital Times*, February 27, 1942, and June 30, 1942.
[124] Clifford Schweppe in Gasser, p. 10.
[125] Biennial Report, Wisconsin Beverage and Cigarette Tax Division, 1940-42, 1944-45.
[126] Lola Huber interview with Rita Slezak Lachmund.
[127] *Badger Ordnance News*, Friday, April 6, 1945.
[128] St. Mary's Ringling Hospital, 1922-1997, p. 7.
[129] School enrollment figures were found in several sources: *Baraboo News-Republic*, September 8, 1941, September 5, 1942, September 9, 1942; *Sauk City Pioneer Press*, September 10, 1942; Wisconsin Department of Public Instruction, Thirty-First, Thirty-Second and Thirty-Third Biennial Report, 1940-42, 1942-44, 1944-45; Report of the Sauk County Superintendent of Schools, 1943-44.
[130] *Sauk City Pioneer Press*, May 21, 1942.
[131] *Sauk County News*, August 20, 1942.
[132] Krumm Report, p. 129.
[133] *Wisconsin State Journal*, undated clip in Erhart Mueller scrapbook, State Historical Society of Wisconsin.
[134] Hercules Report, p. 533-538.
[135] Lola Huber, interview with William Sisson.
[136] *Sauk County News*, June 25, 1942.

[137] *Sauk County News*, May 6 and September 2, 1942.
[138] Monthly Reports, Mason and Hanger, July 1944, February, 1945.
[139] *Milwaukee Journal*, April 1, 1945.
[140] Shooting in *Wisconsin State Journal*, March 23, 1945; Geigerich in *Wisconsin State Journal*, April 1, 1945.
[141] Undated newsclip, spring of 1942, in Erhart Mueller Scrapbook.
[142] *Sauk County News*, February 15, 1946.
[143] *Badger Ordnance News*, October 9, 1942.
[144] Lisa Hartmann, Badger History Group interview with Mabel Gillingham and Loretta McCarthy, August 1999.
[145] Krumm, *A School-Centered Community*, p. 14.
[146] Krumm, p. 12.
[147] Krumm, p. 14.
[148] *Baraboo News-Republic*, October 7, 1943.
[149] Carol Sorg interview with Krumm, April 7, 1993.
[150] Sorg interview with Carol Bohn Hulterstrom, April 7, 1993.
[151] Sorg interview with Krumm, April 7, 1993.
[152] Sorg interview with Krumm, April 7, 1993, and Krumm, p. 26.
[153] Krumm, p. 25.
[154] Krumm, pp. 29–30.
[155] Krumm, pp. 30, 34.
[156] Krumm, p. 69.
[157] Hartmann interview with Loretta McCarthy, August 1999.
[158] Sorg interview with Mary Edward Weigand, April 7, 1993.
[159] Krumm, pp. 82, 90–91.
[160] Krumm, p. 92.
[161] Sorg interview with Krumm, 1993.
[162] Hartmann interview with Gillingham, August 1999.
[163] Krumm, p. 79.
[164] Krumm, p. 115.
[165] Final Decree, January 24, 1945. United States vs 10,565 acres of land situated in Sauk County, Civil Case No. 307, United States District Court for the Western District of Wisconsin, January 10, 1942. Hereafter identified as Case No. 307.
[166] State Supreme Court Justice Horace Wilkie quoted in Madison *Capital Times*, January 14, 1963.
[167] Ryan obituary in Madison *Capital Times*, March 27, 1958; Doll, Historical Sketches, p. 19.
[168] United States vs 10,565 acres of land situated in Sauk County, Civil Case No. 307, United States District Court for the Western District of Wisconsin, January 10, 1942. Hereafter identified as Case No. 307. As recorded by the court, Ryan represented Anton Pulvermacher, Roy C. Kindschi, Frank Steuber, Louisa Kirner, Virgil H. Roick as Trustee of the Ida Roick Estate, Mary Jane Roick, Paul Shimniok, Ralph Kindschi, Raymond Wieglow, William E. Wieglow, Edward Shimniok, Anna Knipschield, George Jenewein, Robert Waffenschmidt, Richard Heimlich, Walter L. Kindschi, Edward W. Mockler, William H. Simon, Charles Kunz, Arthur Weirich, Louise Frese, Jacob W. Gasser, Fred Henry, Julius. P Steidtman, Gust Kinzler, Fred W. Kurtz, Anna Magli, William Reusch, Alex W. Steuber, Frank Wiatrok, John Shimniok, Wm. Eschenbach, Theo. Wiegand, August Wittmer, Lyman Kindschi, Ray Ederer, E.M. Waffenschmidt, C.C. Waffenschmidt, L.G. Eschenbach, Paul Kirner, Raymond Jessie, John Litscher, Wm. Burkhardt, Walter Schlag, Garth Premo, Chris Huber, Harold Thoelke, Elmer C. Keitel, Fred Steuber, Henry Weirich, Wm. P. Ballweg, Bernard Oelke, Swain H. Mather,

Emma Dahnke, Bertha Roick, Fred Waffenschmidt, Frank Roznos, Alvin W. Hoppe, R.E. Thomas, Dr. F.M. Asma, R.E. Organ, Matilda Weigand, Kate Knipschield, Lyman Hatz, J.M. Woods, Charles Herr, John Herr, Irvin Luetscher, W.H. Stoneman, Leroy Welch for the Walter Welch Estate, Herbert Alt, Charles Eschenbach, Frank Eschenbach, Charles Weirich, Metha Bauer Henry. Hill, Miller and Hill represented Raymond Wieglow, Wm. E. Wieglow, Ed Schweppe, Wm H. Simon, Theo Wiegand, Matilda Wiegand, Swain H. Mather, Alvin W. Hoppe, R.E. Thomas, Dr. F.M. Asma, R.E. Organ. Hill and Dithmar represented Lakeview School District 5, Joe Licari as Creditor of William Ballweg, and the Trustees of the First German Independent Christian Association. Ryan quoted in *Wisconsin State Journal*, February 8, 1942.

[169] Holmes Obituary in *Capital Times*, July 28, 1946. Case No. 307, Instructions to Commissioners, June 8, 1942.

[170] Ryan report to Kindschi, May 16, 1944. Letter from Thomas Blount to Carol Sorg, Badger History Group, September 30, 1999. Blount was a Lieutenant in the Army Ordnance Corps in January 1942 when he encountered the armed farmer.

[171] Case No. 307, Agreements, March 6, 1942, May 1, 1942. Declaration of Taking, May 6, 1942, Petition, June 19, 1942.

[172] Valuations in report from William Ryan to Emma Kindschi, May 16, 1944. Ryan quote in *Sauk County News*, October 8, 1942.

[173] Civil Case 307, Affidavit, February 22, 1943.

[174] Ryan to Kindschi, May 16, 1944, Commission valuations in Ryan Report to Kindschi, May 16, 1944 and *Sauk County News*, October 8, 1942. The favorable 1944 verdicts might also have been affected by the lingering illness and death of John Boyle, who contracted pneumonia in November 1943 and died in March 1944.

[175] U.S. Army Corps of Engineers, War Department Facilities Inventory of the Badger Ordnance Works, v. 3.

[176] Ryan Report and Ryan letter to Kindschi, May 16, 1944.

[177] *Badger Ordnance News*, July 6, 1945.

[178] Reuter, *History of Badger Ordnance Works*, 2/9/45–1/7/51, p. 22.

[179] Hart to William Evjue, in Evjue Papers, Archives, State Historical Society of Wisconsin.

[180] Reuter, p. 22.

[181] Reuter, p. 23.

[182] Report No. 3 of Board to Recommend Destruction of Buildings and Improvements Because of Hazards Incident to Contamination with Powder and Explosives, April 8, 1946. Exhibits 1, 2 & 3, Facilities Historical Report (Bi-Annual) 1/1/46–30/6/46.

[183] Reuter, p. 27.

[184] Telegrams from Goodland to Surplus Property Administrator and Senators La Follette and Wiley, January 23, 1946. Letter from Goodland to Robert Whittet, Chief, Institutional Division, Real Property Disposal, War Assets Administration, May 9, 1946, in Goodland Papers, Archives, State Historical Society of Wisconsin.

[185] Memo from Colonel R.R. Klanderman to Ordnance Commanding Officer, July 22, 1946. Exhibit Q, History of Badger Ordnance Works, 2/9/45–1/7/51.

[186] Klandermann memo.

[187] Reuter, p. 32.

[188] Reuter, p. 32.

[189] Memo to Commanding Officer, Badger Ordnance Works, from the Field Director of Ammunition Plants, Joliet, Illinois, December 9, 1947. Reuter, p. 39.

[190] *Milwaukee Sentinel*, May 6, 1952, *Portage Daily Register*, December 12, 1952, Madison *Capital Times*, March 10, 1953, *Baraboo News-Republic*, May 12, 14, 15, 16, 17, 18, 1953.

[191] Madison *Capital Times* and *Wisconsin State Journal*, July 19, 1944.

[192] Moulton letter in Madison *Capital Times,* January 29, 1946.

[193] Lorenz to Goodland, April 4, 1946, in Goodland Papers.

[194] Wisconsin Surplus Property Procurement Agent to War Assets Administration, June 15, 1946, in National Archives Records Group 270, Records of War Assets Administration, Chicago Region Real Property Liquidation File.

[195] Cronon & Jenkins, The University of Wisconsin, v. 4, pp. 40–45, and Lawrence Halle, Interview No. 360, UW-Madison Oral History Project, 1990.

[196] Phyllis Young, Badger Village Interview No. 390, UW-Madison Oral History Project, 1990. Enrollment figures in Mueller, Badger Village, Bluffview Courts, p. 90

[197] On the list of well-known Badger Village alumni are Wisconsin Governors Lee Sherman Dreyfus and John Reynolds, Supreme Court Justice Nathan Heffernan, University of Wisconsin System President Edwin Young and many others who distinguished themselves in professional, scientific, academic, government and business careers.

[198] August Lemberger, Interview No. 390, UW-Madison Oral History Project, 1990.

[199] Halle and Schreck interviews.

[200] Phyllis Young, UW-Madison Oral History Project, 1990.

[201] Marjorie Johnson, UW-Madison Oral History Project, 1990.

[202] Madison *Capital Times*, February 22, 1946.

[203] Peggy Baine interview, UW-Madison Oral History Project, 1990.

[204] Baime Interview, 1990.

[205] Halle interview, 1990. Mueller, pp. 11, 13.

[206] Young interview, 1990.

[207] Young and Schereck interviews, 1990.

[208] Johnson interview, 1990.

[209] Baime interview, 1990.

[210] Lemberger interview, 1990.

[211] Liberty Powder Defense Corporation, Monthly Operating Reports, January 1952–December 1957.

[212] *Badger Ordnance World*, April 2, 1954, and April 29, 1955.

[213 Frozen NG notice in Production and Safety Training Program, 1972–73, BAAP Historical Files. Reuter Historical Summary, July–December, 1952, pp. 33–34; Ruth Corrao quoted in Badger Army Ammunition Plant 50th Anniversary booklet, 1992, unnumbered.

[214] *Wisconsin State Journal*, October 7 and October 29, 1953.

[215] Reuter, Semi-Annual History, July–December, 1951, p. 19.

[216] *Baraboo News-Republic*, May 22, 1952.

[217] *Baraboo News-Republic*, March 3, 1953

[218] Reuter, p. 57.

[219] Reuter, p. 57.

[220] Reuter, p. 58 and Semi-Annual History, July-December 1952, p. 52.

[221] *Baraboo News-Republic*, July 31, 1953.

[222] BAAP 50th Anniversary booklet.

[223] Reuter, Semi-Annual History, July–December, 1952, p. 17.

[224] Step-by-step description of Ball Powder production in BAAP Safety and Training Program, 1972–73.

[225] Maroney in *Badger Ordnance World*, January 31, 1958.

[226] Historical Summary of Badger Ordnance Works Covering the period 1959-1966.

[227] Basic Unit History, Badger Army Ammunition Plant, January 1, 1965–December 31, 1967; Historical Monograph, Army History Program in Unit History, Fourth Annual Supplement, July 1, 1970–June 30, 1971, p. 4.

[228] Page to Reuss, September 30, 1957, in Reuss Papers.

[229] Reuss to Nichols, February 27, 1958; Hamilton to U.S. Army Engineer District, Chicago, March 18, 1958; Nichols to Reuss, March 25, 1958.

[230] "For Lease Facilities At Badger Ordnance Works, Baraboo, Wisconsin," fact sheet, April 1958.

[231] Proxmire and Reuss to Secretary of the Army Wilber M. Brucker, March 1, 1960, in Reuss Papers.

[232] Brig. General H.A. Gerhardt to Henry Reuss, April 11, 1960, in Reuss Papers.

[233] *Wisconsin State Journal*, December 30, 1965.

[234] Author's interview with Jim Mattei, December 18, 2001.

[235] *Milwaukee Sentinel*, September 9, 1965.

[236] Number of rounds fired quoted from House Armed Services Committee Report as reported in *Army Times*, November 8, 1967.

[237] Badger History Group interview with Walter Schuck, August 1999.

[238] Unit History, July 1, 1969–June 30, 1970; July 1, 1970–June 30, 1971, p. 101, and *Baraboo News-Republic*, April 20, 1970. Author's interview with Dave Fordham, November 1, 2001.

[239] Schommer in Monthly Operating Reports, July–December, 1970; Hardy in Monthly Operating Reports, July 1970–June, 1971; Turner in *Badger World*, February, 1971.

[240] *Wisconsin State Journal*, April 13 and April 14, 1971.

[241] Monthly Operating Reports, July 1, 1971–June 30, 1972. *Wisconsin State Journal*, December 9, 1971.

[242] *Wisconsin State Journal*, *Milwaukee Journal*, December 31, 1965.

[243] Madison *Capital Times*, December 30, 1965.

[244] Author's interview with Jim Mattei, December 18, 2001.

[245] Schaefer, "The Vietnam-Era Draft" in *The Legacy*, p. 69.

[246] Quick and Alberts in *Badger World*, October 1967; Eno in *Badger World*, April 1969; Cormican in *Badger World*, March 1968; Hellenbrand and Carpenter in *Badger World*, March 1970.

[247] *Badger World*, June 1968.

[248] Badger History Group interview with Susan Sweeney, August 1999. Harmon and McCullough in *Wisconsin State Journal*, May 4, 1975.

[249] Lenerz in *Badger World*, February, 1969.

[250] Malone, Saunders and Denman in *Badger Ordnance World*, October 1967. Turdot in *Badger World*, August 1970.

[251] Wages in "BAAP Work Force and Imprint on Southern Central Wisconsin Communities, Employee Record Review," March 1971, BAAP Historical Files. Labrenz in *Badger World*, September 1967.

[252] Zantow, Badger Village–Bluffview manuscript, p. 1.

[253] Zantow, Badger Village–Bluffview manuscript, p. 1.

[254] *Wisconsin State Journal*, August 4, 1970.

[255] *Wisconsin State Journal*, October 29, 1953.

[256] *Badger World*, April 1968.

[257] CNRA suit in *Wisconsin Magazine of History*, Autumn 1978. Suits against Olin and United States in BAAP Unit History, July 1, 1969–June 30, 1970.

[258] Supplement to Basic Unit History, July 1, 1969–June 30, 1970, p. 58.

[259] Madison *Capital Times*, June 10, 1966.

[260] *Madison Capital Times*, June 10, 1966.
[261] *Wisconsin State Journal* and *Milwaukee Journal*, June 19, 1966.
[262] This account of Karl Armstrong's exploits relies heavily on *Rads* by Tom Bates.
[263] Supplement to Basic Unit History, July 1, 1969-June 30,1970, p. 58.
[264] Author's interview with Frank Wolf, November 21, 2001, and with Tom Robertson, December 17, 2001. Enoch in *Badger World* article in Supplement to Basic Unit History, July 1, 1969–June 30, 1970, p. 70.
[265] *Wisconsin State Journal*, March 15, 1975 and *Baraboo News-Republic*, March 19, 1975.
[266] *Baraboo News-Republic*, May 30, 1975.
[267] Author's interview with Dennis Thurow, December 13, 2001, and Economic Adjustment Program, Sauk County, Wisconsin, p. 59.
[268] Economic Adjustment Program, Sauk County, p. 59, Janzen in *Baraboo News-Republic*, June 23, 1975.
[269] Economic Adjustment Program, Sauk County, p. 41 and *Baraboo News-Republic*, June 23, 1975.
[270] Author's interview with Dennis Thurow, December 13, 2001.
[271] Author's interview with Jim Mattei, December 18, 2001.

IV Legacy

[1] *Sauk Prairie Star*, July 23, 1998. Mossman was one of the founders of the Badger History Group.
[2] Author's interview with Dave Fordham, November 1, 2001.
[3] Mossman, *Breeding Birds of the Badger Army Ammunition Plant*, p. 37.
[4] Adams, *The Influence of Environmental Contamination on the Reuse of Closing Military Bases: A Case Study of the BAAP*, pp. 79–87.
[5] Often attributed to Thomas Jefferson, the price of liberty quote was pronounced by the anti-slavery activist Wendell Phillips, who paraphrased Jefferson's contemporary, John Philpot Curran.
[6] "Dynamite Heart," a study of Badger workers in the 1960s, conducted by Dr. Ramon Lange and reported in the Madison *Capital Times*, April 9, 1975.

Appendices

Family Names/Properties Purchased for BOW

Name	FLB Appraisal	CDM Comm	Final Settle	Type*
Alt, Herbert B. & Viola	n/a	$6,000	$6,000	1
Andriese, Catherine	n/a	875	875	1
Asma, F.H.	n/a	1,200	1,200	1
Asma, Susan (2 parcels)	n/a	400	400	1
Baker, William F. & Irene	n/a	100	100	1
Ballweg, William P. & Gertrude	12,000	14,263	13,700	2
Bass, Floyd K.	n/a	12,000	12,000	1
Bass, Frank & Edna	n/a	125	125	1
Block, Patrick & Katherine	n/a	12,000	12,000	1
Burkhardt, William & Amelia	19,000	26,916	21,500	3
Cemetery Society at Kingston	n/a	1,340	1,340	1
Christ, Frank G., William F. & George C.	n/a	1,350	1,350	1
Dahnke, Fred & Emma	7,000	9,920	9,000	2
Ederer, Raymond & Caroline	7,600	9,271	8,700	2
Ehret, John & Ida	n/a	33,600	33,600	1
Erickson, Otto G. & Lottie	n/a	7,200	7,200	1
Eschenbach, Charles & Alvina	n/a	9,530	9,530	1
Eschenbach, Frank & Will (2 parcels)	n/a	423	516	1
Eschenbach, Louis & Amy	11,000	15,100	13,600	2
Eschenbach, Louis	7,500	10,069	9,400	2
Eschenbach, William A. & Caroline	240	207	410	2
Eschenbach, William W.	8,500	10,457	9,500	2
Frese, Louis & Eunice	14,500	20,700	19,400	2
Gasser, Benjamin C. & Barbara	n/a	18,000	18,000	1
Gasser, Jacob W. & Emma	11,800	14,179	12,500	3

continued next page

Name	FLB Appraisal	CDM Comm	Final Settle	Type*
Gattwinkel, Charles (2 parcels)	n/a	8,000	8,000	1
German Lutheran Church	n/a	400	400	1
Gishwiler, Elmer	n/a	925	925	1
Hamlin, Edward A. & Estella E.	n/a	1,600	1,600	1
Hartman, Louis & Margaret	n/a	975	975	1
Hatz, Lyman & Ruth	7,200	10,413	9,000	2
Heimlich, Max	n/a	491	491	1
Heimlich, Richard & Laura	9,100	11,105	10,800	2
Henry, Metha B.	6,500	8,975	8,200	2
Herr, Charles & Lucille	11,550	14,425	12,600	3
Herr, John & Ellen	850	1,410	1,410	2
Hoppe, Alvin W. & Evelyn	16,200	20,505	19,800	2
Huber, Christ & Anna	32,200	43,192	36,800	2
Immanuel Evangelical Association (Thoelke Church)	n/a	1,200	1,200	2
Jenewein, George	240	516	500	2
Jessie, Raymond & George Carpenter	19,500	22,931	22,300	2
Keitel, Elmer, Victor & Norene	6,000	8,246	7,000	2
Kern's Corners School	n/a	2,015	2,015	1
Kessler, Fred	n/a	900	900	1
Kindschi, Lyman & Essie	16,500	24,801	22,500	3
Kindschi, Ralph M. & Celia	14,000	21,309	20,500	2
Kindschi, Roy C. & Lillian M.	13,600	20,105	18,000	2
Kindschi, Walter L. & Edith	18,600	23,907	20,500	3
Kinzler, Gustav A. & Clara	7,200	8,679	8,100	2
Kirner, Louisa, Fred & Mildred	9,600	12,808	12,400	2
Kirner, Paul & Mildred	8,800	11,853	11,200	2
Knipschield, Anna M. & John	12,200	15,608	14,700	2
Knipschield, Kate	11,200	16,685	14,600	2
Kunz, Carl A. & Margaret	7,200	10,239	10,100	2
Kurtz, Fred W. & Lena	9,400	13,019	13,000	2
Lakeside School	n/a	1,200	1,200	1
Lakeview School	n/a	1,700	1,700	1
Litscher, John J. Jr. & Mae	20,000	24,851	23,200	2
Luetscher, Irvin & Margaret (2 parcels)	31,600	40,186	39,900	2
Magli, Anna	200	258	250	2
Magli, Anna	15,000	19,285	17,600	2
Martin, R.H.	n/a	675	675	1
Mather, Roy H. & Clara	n/a	19,000	19,000	1

Name	FLB Appraisal	CDM Comm	Final Settle	Type*
Mather, Swain H. & Cora	18,700	24,735	23,800	2
Methodist Church	n/a	3,550	3,550	1
Miller, Erma C.	n/a	1,340	1,340	1
Miller, W.G.	n/a	580	580	1
Mockler, Edward G. & Maud	9,300	11,343	10,000	3
Oelke, Bernard L. & Eleanor	5,200	6,882	6,600	2
Organ, R.E.	n/a	1,300	1,300	2
Osborne, Lottie	n/a	1,300	1,300	1
Pioneer Cemetery Association	n/a	925	925	1
Pobjoy, Henry & Eva	n/a	150	150	1
Premo, Garth R. & Nora (2 parcels)	21,900	26,942	26,942	2
Pulvermacher, Anton J. & Helen	6,000	8,880	8,200	2
Reusch, Frank & Ida (2 parcels)	n/a	15,250	15,250	1
Reusch, William B & Shirley K.	9,000	12,842	11,500	2
Roick, Bertha	17,600	20,234	19,500	2
Roick, Mary J. (2 parcels)	20,900	24,490	23,200	2
Roick, Virgil & Ida (2 parcels)	24,100	27,550	27,100	2
Roznos, Frank	7,200	9,815	8,700	2
Scheufler, Fred D. & Lena	n/a	8,000	8,000	1
Schlag, Clara & Rudolph	n/a	22,975	22,975	1
Schlag, Walter	7,000	8,975	8,000	2
Schroder, Edward & Bertha (2 parcels)	n/a	7,815	7,815	1
Schweppe, Edwin & Iva	n/a	16,818	16,818	1
Shimniok, Edward F. & Frieda	13,000	15,705	15,100	2
Shimniok, John G. & Elsie	12,800	15,255	14,700	2
Shimniok, Paul & Anna	14,300	17,504	17,000	2
Simon, William H. & Anna	6,200	8,743	8,300	2
Spoors, Jennie Johnson	n/a	13,750	13,750	1
Steidtman, Julius C. & Agnes, Lewis & Jessie	13,450	19,071	16,500	2
Stempke, John, George & Bertha	n/a	500	500	1
Steuber, Alex W. & Alma	22,000	26,809	25,500	2
Steuber, Frank W. & Ella	8,500	11,854	11,400	2
Steuber, Fred H. & Cora	9,600	12,676	11,700	2
Stoneman, William H. & Rachel L.	n/a	2,500	2,500	1
Sumpter Town Hall	n/a	800	800	2
Thoelke Cemetery Association	n/a	415	415	1
Thoelke, Harold & Alma (3 parcels)	19,825	26,192	23,570	3
Thomas, Ralph E. & Bertha F.	1,800	3,302	2,900	3

continued next page

Name	FLB Appraisal	CDM Comm	Final Settle	Type*
Van Matre, Dean	n/a	775	775	1
Waffenschmidt, E.M. & C.C. & Hazel	19,350	29,179	26,500	3
Waffenschmidt, Fred & Amy (2 parcels)	9,000	16,094	14,400	2
Waffenschmidt, Robert & Irma	11,500	14,618	13,500	2
Walter, Rudolph	n/a	600	600	1
Walter, Rudolph & Carl	n/a	900	900	1
Weigand, Matilda	1,600	3,777	3,400	2
Weigand, Theodore	10,000	17,229	16,900	2
Weigand, Theodore	700	1,552	1,550	2
Weirich, Arthur & Selma (2 parcels)	9,750	14,539	14,000	2
Weirich, Henry & Charles (4 parcels)	22,815	31,114	30,500	3
Welch, Walter J.	8,400	10,330	10,000	2
Wiatrok, Frank	n/a	600	600	1
Wieglow, William E. & Mollie (2 parcels)	9,200	14,168	12,900	2
Wisconsin Power & Light Co.	n/a	2,500	2,500	1
Wittmer, August & Caroline	n/a	1,275	1,275	1
Woods, Emily	n/a	1,575	1,575	1

*TYPE

1 Accepted government offer of January-May, 1942

2 In condemnation proceedings, accepted government offer of January 1942-May 1944

3 In condemnation proceedings-Jury Verdict

Periods of Operation of Badger's Production Areas*

World War II

Acid	January 1943–August 1945
Oleum	January 1943–August 1945
Smokeless Propellant	March 1943–July 1945
Rocket Propellant	March 1945–September 1945
Ethylcentralite Powder	July 1943–October 1945

Korea

Acid	July 1951–November 1957
Oleum	October 1952–October 1956
Smokeless Propellant	October 1951–December 1957
Rocket Propellant	November 1951–December 1954 & March 1955–September 1955
Ball Powder	September 1955–September 1956 & 1957 part time

Vietnam

Acid	August 1966–June 1975
Oleum	September 1966–June 1974
Smokeless Propellant	August 1967–August 1973
Rocket Propellant	June 1967–June 1975
Ball Powder	May 1966–May 1975

*Dates indicate when a production area was fully operational and when it halted completely. Portions of an area often began work months ahead of the entire area and continued after most of the area shut down. For example, only the C-Line Smokeless ran until July 1945, the other lines stopped in April and May.

Production

World War II

Small Arms Powder	109,267,500 pounds
Cannon Powder	148,701,400 pounds
Rocket Grains	13,394,700 pounds

Korea

Ball Powder Propellant for 30 Caliber Rifle	3,902,500 pounds
Ball Powder Propellant for 20 mm Gun	1,856,100 pounds
Smokeless Powder for 75 mm Rifle	3,087,500 pounds
Smokeless Powder for 75 mm Gun	24,753,500 pounds
Smokeless Powder for 76 mm Gun	3,151,300 pounds
Smokeless Powder for 90 mm Gun	26,433,000 pounds
Smokeless Powder for 105 mm Rifle	450,000 pounds
Smokeless Powder for 105 mm Howitzer	26,685,300 pounds
Smokeless Powder for 155 mm Howitzer	95,666,900 pounds
Smokeless Powder for 8 inch Howitzer	43,605,600 pounds
Smokeless Powder for 240 mm Howitzer	11,999,800 pounds
MK-18 Rocket Mortar	22,343,900 pounds
MK-22 Rocket Mortar	421,500 pounds
MK-21 Rocket Mortar	438,300 pounds
M-16 Rocket Mortar	12,641,500 pounds
115 Mortar	17,400 pounds

Vietnam

Ball Powder Propellant for 7.62 mm Rifle	28,014,600 pounds
Ball Powder Propellant for 5.56 mm Rifle	37,455,200 pounds
Ball Powder Propellant for 30 Caliber Rifle	9,934,300 pounds
Ball Powder Propellant for 20 mm Gun	19,410,700 pounds
Ball Powder Propellant for 50 Caliber Guns	5,170,800 pounds
Smokeless Powder for 105 mm Howitzer	1,513,200 pounds
Smokeless Powder for 155 mm Howitzer	160,620,000 pounds
Smokeless Powder for 175 mm Howitzer	86,587,000 pounds
Smokeless Powder for 8 inch Howitzer	53,430,900 pounds
5" Navy Gun	12,869,500 pounds
MK-43 Rocket Mortar	71,718,600 pounds

Commanding Officers

Lt. Col. George F. Griffith	January 1942–October 1942
Major Richard P. Staeffler	October 1942–June 1943
Captain F.I. Richards	June 1943–July 1943
Major Victor L. Keldsen	July 1943–November 1944
Lt. Col. Charles H. Pruden	November 1944–November 1945
Captain R.C. Phillips	November 1945–December 1945
Captain J.W. LeMaistre	December 1945–March 1946
Captain E.E. Wells	March 1946–August 1946
Lt. Col. Charles E. Kelley	August 1946–November 1946
Captain E.E. Wells	November 1946–November 1948
1st Lt. Raymond S. Stein	November 1948–January 1949
Major J. King Wallace	January 1949–July 1949
1st Lt. Raymond S. Stein	July 1949–July 1949
Major Charles E. Branson	July 1949–November 1950
Lt. Col. Oliver Williams	November 1950–June 1952
Major Ulysses G. Stewart, Jr.	June 1952–August 1952
Lt. Col. William A. Evans	August 1952–July 1954
Major Ulysses G. Stewart, Jr.	July 1954–January 1955
Lt. Col. Jewell G. Haynie	January 1955–May 1958
Major Charles E. Ramsey	May 1958–December 1959
John Troyer (Commander's Rep.)	December 1959–February 1966
Major Sidney I. Davis	February 1966–April 1968
Lt. Col. Russell O. Enoch	April 1968–June 1970
Lt. Col. Eugene R. Belcher	July 1970–July 1972
Lt. Col. Phillip D. Thorne	July 1972–February 1974
Lt. Col. Robert G. Hill	February 1974–July 1975
Captain James H. Konicki	July 1975–September 1975
Elmer G. Johnson (Commander's Rep)	September 1975–July 1976
David C. Fordham (Commander's Rep)	July 1976–2002

Resident Managers

Henry Chase	1942-1943
Charles H. Grant	1943-1945
F.L. Boddicker	1945-1946
George Ripley	1946
N.A. Hamilton	1951-1952
J.C. Stewart	1952-1957
H.B. Maroney	1957-1958
M.F. Lindsley, Jr.	1959
C.E. Voegeli	1959-1965
John DalSassa	1965-1969
Howard Shehorn	1969-1970
J.R. Katic	1970-1975
L.M. Unverzagt	1975-1979
James Mattei	1979-2000
Dennis Thurow	2000-2001
Richard Hess	2001-

Casualties

— World War II —

August 20, 1942
Fitzhugh Davidson Charlton, 21, a Mason & Hanger construction laborer from Virginia, was the first worker to die as a result of an accident at Badger Ordnance. He was struck in the chest and fatally injured by an unloading hook when the rig broke while offloading pipe. (Corps of Engineers Historical Record; *Baraboo News-Republic*)

September 6, 1942
William Joseph Cinealis, 34, a Mason & Hanger construction laborer from Sheboygan, died of a fractured skull after being hit on the head when the truck he was riding on drove beneath a low-hanging pipe. (Corps of Engineers Historical Record; *Baraboo News-Republic*)

October 23, 1942
Joseph Clarencc Fjelstad, 66, an Armco Railroad worker, died of pneumonia after he was hospitalized in Madison with nine cracked ribs resulting from his collision with a heavy timber protruding from the side of a construction truck. (Corps of Engineers Historical Record)

December 30, 1942
Frank O. Smith, a steamfitter from Chicago, died in a Chicago hospital of internal injuries suffered after the wall of a trench he was working in collapsed and drove a steel pipe into his abdomen. The accident occurred at Badger Ordnance on September 1, 1942. (Corps of Engineers Historical Record)

January 22, 1943
Walter P. Ring, 27, Plain, maintenance mechanic for Hercules, died when sections of an unloading platform at the Oleum Plant inadvertently closed, pinning down his body and "strangling" him. (Hercules Report; Death Register, Sauk County Register of Deeds)

March 13, 1943
About five miles south of the plant, bus driver Paul Leach was killed and four passengers injured when Leach's bus slid into a semi-trailer on an icy stretch of Hwy 12. (Hercules Report)

April 9, 1943
Harry O. Borland, Madison, C-Line wringerman, died after suffering second-degree acid and fire burns when his wringer exploded while he was peering into the basket. He was the first person to die while making powder at Badger. (Hercules Report; *Baraboo News-Republic*)

October 23, 1943
Arnold Swansby, 30, Baraboo, and Alexander Libke, 42, Columbia County Town of Scott, died instantly in an explosion of undetermined cause which completely destroyed the E.C. Blending House in which they were working. The blast was powerful enough to rattle windows over the bluff in Baraboo. (Hercules Report; Death Register, Sauk County Register of Deeds)

November 28, 1943
Charles G. Bailey, 40, Baraboo, a plant guard, was fatally injured while riding his motorcycle on routine patrol at about 3:00 AM. He failed to negotiate a turn on the two-wheeler and, with practically no embankment along the side of the road, was thrown from the motorcycle and died almost immediately. (Hercules Report; Death Register, Sauk County Register of Deeds)

June 20, 1944
Byron James Mortimer, 49, Endeavor, died after falling beneath the wheels of a loaded flatcar on the narrow gauge railroad (Hercules Report; *Baraboo News-Republic*)

February 20, 1945
Cyrenus S. VanWhy, 37, East Stroudsburg, Pennsylvania, one of the furloughed soldiers working on construction of the rocket plant, died after being hit by a truck while walking in a snowstorm on Highway 12. (*Badger Ordnance News, Baraboo News-Republic*)

April 4, 1945
Herman Beach, 21, St. Catherine, Jamaica, died at St. Mary's-Ringling Hospital, of a peritonitis infection caused by a ruptured appendix. According to the terms of the work contract he signed, his body did not have to be shipped home. He was buried in an unmarked grave at St. Joseph's cemetery, Baraboo. (Hercules Report; Death Register, Sauk County Register of Deeds)

July 19, 1945
Erwin Ellis Pugh, 52, Baraboo; Mark Edward Shearer, 27, Duluth, Minnesota; Edwin Ellsworth Goff, 24, North Freedom; and William M. Denny, 45, Oneida, Wisconsin, were killed instantly when Nitroglycerin Storehouse B exploded. (*Badger Ordnance News*; Death Register, Sauk County Register of Deeds)

— Korea —

July 29, 1952

Rehab worker Roy Henry Elmer, 30, Hillsboro, fell through the opening left in the second floor of a boiling tub house where a tub had been removed and died of a skull fracture. (Liberty Powder Defense Company Report; Death Register, Sauk County Register of Deeds)

August 6, 1954

Henry A. Grunke, 41, Endeavor, an electrician with the Commonwealth Electric Company, died beneath a load of utility poles which rolled off a rail flatcar. (Liberty Powder Defense Company Report; Death Register, Sauk County Register of Deeds)

November 30, 1955

Leslie Barkley, 45, Lake Delton, died as a result of an explosion and flash fire in the ball powder hardening house. Barkley was a chief operator in the hardening house and also the barber at Badger Village. (Liberty Powder Defense Company Report; Sauk County Register of Deeds)

— Vietnam —

October 7, 1970

Victor R. Schommer, Jr., 31, while performing his duties as a carpenter repairing shingles on Activated Carbon Recovery Building 2555, came into contact with live steam and suffered severe thermal burns on the upper half of his body and respiratory tract. He developed pulmonary edema and died of respiratory failure at St. Mary's Hospital, Madison. (BAAP Monthly Operating Report)

January 13, 1971

Richard D. Hardy, 31, Reedsburg, died when several "low order" explosions destroyed the salt coat and glaze house in which he was working. He was found about three hours after the explosions in one of the bays, either immediately killed or incapacitated by the first blast. (BAAP Monthly Operating Report; *Baraboo News-Republic*)

February 11, 1971

Leo G. Turner, 58, Richland Center, died while working in a coal hopper car at the powerhouse. The load of frozen coal suddenly broke up, slid down on Turner and suffocated him. (*Badger World*; Death Register, Sauk County Register of Deeds)

December 8, 1971

John F. Mitchell, 58, Juneau County Town of Lyndon, died in a low order explosion and fire in the rocket paste breaker and blender building. Three other workers were injured but escaped with their lives. (BAAP Monthly Operating Reports; Death Register, Sauk County Register of Deeds)

Bibliography

— Books —

- Ambrose, Stephen E., *Citizen Soldiers, The U.S. Army From the Normandy Beaches to the Bulge to the Surrender of Germany June 7, 1944-May 7, 1945*, Simon & Schuster, New York, New York, 1997.
- Bates, Tom, *RADS, The 1970 Bombing of the Army Math Research Center at the University of Wisconsin and Its Aftermath*, HarperCollins, Publishers, Inc., New York, 1992.
- Beck, Bill, *Transforming the Heartland, The History of Wisconsin Power & Light*, Wisconsin Power & Light Company, Inc., Madison, 1990.
- Behler, John L., King, F. Wayne, *The Audubon Society Field Guide to North American Reptiles and Amphibians*, Alfred A. Knopf, Inc., New York, 1979.
- Blair, Emma Helen, *The Indian Tribes of the Upper Mississippi Valley and Region of the Great Lakes, Volume II*, Arthur H. Clark Co. Cleveland, Ohio, 1912.
- Brown, Lauren, *The Audubon Society Nature Guides, Grasslands*, Chanticleer Press, Inc., New York, New York, 1985.
- Butterfield, C.W., *History of Sauk County, Wisconsin*, Western Publishing Company, Chicago, 1880.
- Canfield, William H., *Outline Sketches of Sauk County; Including its History, From the First Marks of Man's Hand to 1861, and its Topography*, A.N. Kellogg, Printer, Baraboo, Wisconsin, 1861.
- Carlisle, Rodney P., *Powder and Propellants: Energetic Materials at Indian Head, Maryland, 1890-1990*, United States Navy, 1990.
- Carver, Jonathan, *Travels through the interior parts of North America in the years 1766, 1767, 1768*, London, 3rd edition, 1781.
- Collins, Henry Hill, Jr., *Complete Field Guide to American Wildlife, East, Central and North*, Harper & Row, New York, New York.
- Costello, David F., *The Prairie World*, Thomas Y. Crowell Company, New York, 1969.
- Cronon, E. David & Jenkins, John W., *The University of Wisconsin A History 1945-1971, v. 4, Renewal to Revolution,* University of Wisconsin Press, Madison, 1999.
- Cronon, William, *Nature's Metropolis, Chicago and the Great West*, W.W. Norton & Company, 1991.
- Cronon, William, editor, *Uncommon Ground, Toward Reinventing Nature,* Maple-Vail Book Manufacturing Group, 1995.
- Curtis, John T., *Vegetation of Wisconsin, An Ordination of Plant Communities,* University of Wisconsin Press, Madison, Wisconsin, 1971.
- Denevan, William M., editor, *Native Population of the Americas in 1492,* The University of Wisconsin Press, second edition, 1992.
- Diamond, Stanley, editor, *Culture in History, Essays in Honor of Paul Radin*, Columbia University Press, New York, 1960.

- Doll, Walter, *Historical Sketches of Prairie du Sac*, Tripp Memorial Library, Sauk Prairie, Wisconsin, undated.
- Dutton, William S. Du Pont, *One Hundred Forty Years*, Scribner's Sons, New York, 1942.
- Ebner, James A., *The Butterflies of Wisconsin*, Milwaukee Public Museum, Milwaukee, 1970.
- Engelbrecht, H.C., and Hanighen, F.C., *Merchants of Death, A Study of the International Armament Industry,* Dodd, Mead & Company, Inc., New York, 1934.
- Fairchild, Byron, Grossman, Jonathan, *The Army and Industrial Manpower*, Department of the Army, Washington D.C., 1959.
- Featherstonhaugh, George W., *A Canoe Voyage up the Minnay Sotor, Vol. 1*, Minnesota Historical Society, St. Paul, Minnesota, 1970.
- Fine, Lenore and Remington, Jesse A., *The Corps of Engineers: Construction in the United States*, Office of the Chief of Military History, U.S. Government Printing Office, Washington, D.C., 1972.
- Goc, Michael J., *Others Before You, The History of Wisconsin Dells Country*, New Past Press Inc., Friendship, Wisconsin, 1995.

  ——, *Many A Fine Harvest, Sauk County, 1840–1990*, New Past Press, Inc., Friendship, Wisconsin, 1990.

  ——, *Stewards of the Wisconsin, Wisconsin Valley Improvement Company*, New Past Press Inc., Friendship, Wisconsin, 1993.

  ——, *Where the Waters Flow, A Half Century of Regional Development, 1941–1991*, New Past Press Inc., Friendship, Wisconsin, 1991.
- Gallup Company, *The Gallup Poll*, New York, 1964–1971.
- Green, Constance McLaughlin, et al., *The Ordnance Department, Planning Munitions for War*, Office of the Chief of Military History, Washington D.C., 1955.
- Gregory, Chester, W., *Women in Defense Work During World War II, An Analysis of the Labor Problem and Women's Rights*, Exposition Press, Inc., New York, 1974.
- *History of Northern Wisconsin*, The Western Historical Company, Chicago, Illinois, 1881.
- Hole, Francis, *Soils of Wisconsin*, University of Wisconsin Press, Madison, Wisconsin, 1976.
- Hughes, David R., *The History and Development of the M16 Rifle and its Cartridge*, Armory Publications, Oceanside, California, 1990.
- Huston, James A., *The Sinews of War: Army Logistics 1775–1953*, Center of Military History, United States Army, Washington D.C. 1988.
- Kessler-Harris, Alice, *Out to Work, A History of Wage-Earning Women in the United States*, Oxford, University Press, New York, 1982.
- Kimball, Jeffrey, *Nixon's Vietnam War*, University Press of Kansas, Lawrence, 1998.
- Kindscher, Kelly, *Medicinal Wild Plants of the Prairie, An Ethnobotanical Guide,* University Press of Kansas, Lawrence, Kansas, 1992.
- Lambert, David, *The Field Guide to Early Man*, Facts On File, Inc., New York, 1987.
- Lange, Kenneth I., *A County Called Sauk: a Human History of Sauk County, Wisconsin*, Sauk County Historical Society, 1976.
- Lemert, Ann Arnold, *First You Take A Pick & Shovel, The Story of The Mason Companies*, The John Bradford Press, Kentucky, 1979.
- Madson, John, *Where the Sky Began, Land of the Tallgrass Prairie*, Iowa State University Press, Ames, 1995.
- Mann, Robert, *A Grand Delusion, America's Descent into Vietnam*, Basic Books, New York, 2001.

• Marshall, George C., *The Papers of George Catlett Marshall, Vol. 2, "We Cannot Delay," July 1, 1939-December 6, 1941*, Johns Hopkins University Press, Baltimore, Maryland & London, 1986.
• Mayo, Lida, *The Ordnance Department: on Beachhead and Battlefront*, Center for Military History, U.S. Army, Washington D.C., 1991.
• McDonald, Forrest, *Let There Be Light, The Electric Utility Industry in Wisconsin*, 1881-1955, The American History Research Center, Inc., Madison, 1957.
• *Memorial & Biographical Record and Illustrated Compendium of Biography of Columbia, Sauk and Adams Counties, Wisconsin*, George Ogle & Co., Chicago, 1901.
• Merk, Frederick, *Economic History of Wisconsin During the Civil War Decade*, State Historical Society of Wisconsin, 1916 and 1971.
• Milne, Lorus, Milne, Margery, *The Audubon Society Field Guide to North American Insects and Spiders,* Alfred A. Knopf, New York, 1980.
• Mueller, Erhart A., *Also In Sumpter*, Worzalla Publishing Co., Stevens Point, Wisconsin, 1980.
——, *Always in Sumpter*, Worzalla Publishing Co., Stevens Point, Wisconsin, 1983.
——, *Badger Village and Bluffview Courts*, Thomas J. Luitink, editor, Bluffivew Acres, Inc., Prairie du Sac, Wisconsin, 1982.
——, *Forever in Sumpter*, Worzalla Publishing Co., Stevens Point, Wisconsin, 1987.
——, *Only In Sumpter*, Worzalla Publishing Co., Stevens Point, Wisconsin, 1977.
——, *Return to Sumpter*, Amherst Press, Amherst, Wisconsin, 1994.
• Ostergren, Robert C. and Vale, Thomas R., editors, *Wisconsin Land and Life*, University of Wisconsin Press, Madison, Wisconsin, 1997.
• Pielou, E.C., *After the Ice Age, the Return of Life to Glaciated North America*, University of Chicago Press, 1991.
• Pyne, Stephen J., *Fire In America, A Cultural History of Wildland and Rural Fire*, Princeton University Press, 1982.
• Rogin, Leo, *The Introduction of Farm Machinery in its Relation to the Productivity of Labor in the Agriculture of the United States during the Nineteenth Century*, University of California Press, Berkeley, California, 1931.
• Runkel, Sylvan T., and Roosa Dean M., *Wildflowers of the Tallgrass Prairie: The Upper Midwest*, Iowa State University Press, Ames, Iowa, 1989.
• Sauer, Carl O., *Seventeenth Century North America*, Turtle Island, Berkeley, 1980.
• Schorger, A.W., *Wildlife in Early Wisconsin*, University of Wisconsin, Stevens Point, 1982.
• Schultz, Gwen M., *Wisconsin's Foundations, A Review of the State's Geology and Its Influence on Geography and Human Activity*, Cooperative Extension Service, University of Wisconsin, Madison, Wisconsin, 1986.
• Sesquicentennial History Committee, *Reedsburg Remembers 150 Years, A History of Reedsburg, Wisconsin 1848-1998*, BookCrafters, Chelsea, Michigan, 1997.
• Shafer, D. Michael, editor, *The Legacy, The Vietnam War in the American Imagination*, Beacon Press, Boston, 1990.
• Smith, David Lee, *Folklore of the Winnebago Tribe*, University of Oklahoma Press, Norman, Oklahoma, 1997.
• Smith, R. Elberton, *The Army and Economic Mobilization*, Center of Military History, U.S. Army, Washington, D.C., 1991.
• Sterling, Keir B., *Serving the Line with Excellence: the Development of the U.S. Army Ordnance Corps, as Expressed Through the Lives of its Chiefs of Ordnance,* Aberdeen Proving Grounds, U.S. Army Ordnance Center and School, Washington, D.C., 1987.
• Thibaudeau, May Murphy, *Depression, Draft, Dance, Democrats or World War II For Us at Home*, Ramur Publishing, South Milwaukee, Wisconsin, 2000.

• Thomson, Harry C. and Mayo, Lida, *The Ordnance Department: Procurement and Supply*, Office of the Chief of Military History, Department of the Army, Washington, D.C., 1960.
• Thornton, Russell, *American Indian Holocaust and Survival, A Population History Since 1492*, University of Oklahoma Press, Norman, 1987.
• Weaver, J.E., *Prairie Plants and Their Environment, A Fifty-Year Study in the Midwest*, University of Nebraska Press, Lincoln, Nebraska, 1968.
• Weiss, Stuart L., *The President's Man, Leo Crowley and Franklin Roosevelt in Peace and War*, Southern Illinois University Press, Carbondale, Illinois, 1996.
• Wendel, C.H., *Encyclopedia of American Farm Implements & Antiques*, Krause Publications, Iola, Wisconsin, 1995.
• Wright, Henry A. and Bailer, Arthur W., *Fire Ecology*, United States and Southern Canada, John Wiley & Sons, Inc., New York, New York, 1982.
• Zilg, Gerard Colby, DuPont, *Behind the Nylon Curtain*, Prentice-Hall, Inc., New Jersey, 1974.

Manuscripts, Theses

• Adams, Diane L., *The Influence of Environmental Contamination on the Reuse of Closing Military Bases: A Case Study of the Badger Army Ammunition Plant*, UW-Madison, 1999.
• *Chronology of Dates Leading to the Construction of Badger Ordnance Works*, Baraboo, Wisconsin, 27 January, 1971.
• Hatz, Robert and Wayne, *Family Stories*, 1996.
——*Farming Past and Present*, 1998.
• Krumm, Clarence J., *A School-Centered Community, The Story of Badger Schools*.
• Nuzzo, Victoria A., *The Extent and Status of Midwest Oak Savanna at the Time of Settlement and in the mid-1980s and the Effect of Soil Scarification on Seedling Establishment in an Oak Savanna Restoration*, Wisconsin, University of Wisconsin, Masters Thesis, 1985.
• Pritchard, Bernetta, American Economic History, Hs. 411, *The Economic Effects of the Badger Ordnance Works,* December 15, 1970.
• Zantow, Forrest, *Badger Village, Bluffview*, January, 1998.

Newspapers

• *Baraboo News*
• *Baraboo News-Republic*
• *Baraboo Times-Republic*
• Madison *Capital Times*
• *Milwaukee Journal*
• *Milwaukee Sentinel*
• *Portage Daily Register*
• *Sauk City Pioneer Press*
• *Sauk County News*
• *Sauk Prairie News*
• *Sauk Prairie Star*
• *Westfield Central Union*
• *Wisconsin Dells Events*
• *Wisconsin State Journal*

Videos

• Dave Erickson, *Powder on the Prairie, Remembering Erhart*, 1991.
• Dave Erickson and Mike Mossman, *Powder to the People*, 2000.
• Bob France, *Rusty*, 2000.

Articles

- Bailey, Kenneth, Editor/Translator, Journal of Joseph Marin, French Colonial Explorer and Military Commander in the Wisconsin Country, August 7, 1753-June 20, 1754, Henry E. Huntington Library and Art Gallery, 1975.
- Bibliography of Fire Effects and Related Literature Applicable to the Ecosystems and Species of Wisconsin, Technical Bulletin No. 187, Department of Natural Resources, Madison, Wisconsin, 1995.
- Birmingham, Robert A., "The Last Millennium: Wisconsin's First Farmers." *Wisconsin Academy Review*, Winter 1999-2000.
- *Bluffviews*, January 1967-December 1971, Bluffview, Sauk County, Wisconsin.
- Bray, J. Roger, "The Composition of Savanna Vegetation in Wisconsin." *Ecology,* Vol. 41, No. 4, October, 1960.
- Brewer, Richard, "Death by the Plow, Natural History." *The Journal of the American Museum of Natural History*, Vol. LXXIX, No. 7, August-September, 1970.
- Cleland, C.E., "Re-analysis of Faunal Remains from Raddatz Rockshelter, Sauk County, Wisconsin." The Prehistoric Animal Econogy and Ethnozoology of the Upper Great Lakes Region, pp. 98-108, Anthropol. Pap. Mus. Anthropol., University of Michigan, No.29, 1966.
- Croft, Josie Greening, "A Mazomanie Pioneer of 1847" (excerpt of diary by John Greening, 1847) *Wisconsin Magazine of History*, 26:208-18, 1942.
- Crudon, Ed, "Killing Three Indians In Bear Creek Valley," *Baraboo Times-Republic*, May 22, 1914.
- Day, Gordon M., "The Indian as an Ecological Factor in the Northeastern Forest." *Ecology*, Vol. 34. No.2, April, 1953.
- Dooley, Patricia L., "Gopher Ordnance Works, Condemnation, Construction, and Community Response," *Minnesota History*, Summer, 1985.
- Dorney, John R., "The Impact of Native Americans on Presettlement Vegetation in Southeastern Wisconsin." *Transactions of the Wisconsin Academy of Sciences, Arts and Letters*, 69:26-36, 1981.
- Dunlap, Thomas R., "DDT on Trial: The Wisconsin Hearings 1968-1990." *Wisconsin Magazine of History*, Autumn 1978.
- Follmar, Joe, "The C&NW Joins the Race for Passengers." *North Western Lines*, Winter, 1985.
- Gartner, William Gustav, "Four Worlds Without an Eden, Pre-Columbian Peoples and the Wisconsin Landscape." *Wisconsin Land and Life*, 1997.

  ——, "Late Woodland Landscapes of Wisconsin: Ridged Fields, Effigy Mounds and Territoriality." *Antiquity* 73:671-83, 1999.
- Gasser, Doris Litscher, "Sauk Prairie Star Stories," 1998.
- Hulbert, Lloyd C., "Causes of Fire Effects in Tallgrass Prairie." *Ecology*, 69(1), 1988, pp. 46-58.
- Inama, A., "Letters of the Reverend Adelbert Inama, O. Praem." *Wisconsin Magazine of History*, 11:437-58.
- Kaplan, Fred, "JFK'S First Strike Plan." *The Atlantic Monthly,* October 2001.
- Lueders, Herman F., "The Vegetation of the Town of Prairie du Sac." *Transactions of the Wisconsin Academy of Sciences, Arts and Letters*, v. 10, 1894-1895.
- Olson, Keith, "World War II Veterans at the University of Wisconsin." *Wisconsin Magazine of History*, v. 53, n. 2, winter 1969-1970.
- Parmalee, Paul W. "Animal Remains From the Durst Rockshelter, Sauk County, Wisconsin." *The Wisconsin Archeologist*, Vol. 41, No. 1, March, 1960.
- "Powder Pains." *Time Magazine*, September 8, 1967.

- "The Prairie." *The Wisconsin Farmer*, J. W. Hoyt, Editor, Vol. XIV, Hoyt & Campbell, Madison, Wisconsin, 1862.
- Ragatz, Lowell J., "A Swiss Family in the New World: Letters of Jakob and Ulrich Buhler, 1947-1877." *Wisconsin Magazine of History*, 6:317-33, 1922.
- Ragatz, O., "Memoirs of a Sauk Swiss." *Wisconsin Magazine of History*, 19:182-227, 1935.
- Russell, Emily W.B., "Indian-set Fires in the Forests of the Northeastern United States." *Ecology* 64(1), 1983, pp. 78-88.
- Staines, Henry B., "Agriculture of Sauk County," *Transactions of the Wisconsin State Agricultural Society*, with an Abstract of the Correspondence of the Secretary, Vol. I, 1851, Madison, Wisconsin, 1852.
- Thomson, Olive S., "Wild Flowers of Our Prairies," *Audubon Nature Bulletin*, Series 34. No.1, National Audubon Society, New York, New York, 1967.
- Tordella, Stephen J., "Net Migration for Wisconsin Counties, 1960-1980." Demographic Change in Wisconsin Trends and Outlook, Applied Population Laboratory, Department of Rural Sociology, University of Wisconsin, Madison, January, 1984.
- *Transactions of the Wisconsin Academy of Sciences, Arts and Letters*, v. 24, Madison, Wisconsin, 1935.
- *Transactions of the Wisconsin Academy of Sciences, Arts and Letters*, v. 30, Madison, Wisconsin, 1937.
- *Transactions of the Wisconsin Academy of Sciences, Arts and Letters*, v. 86, Madison, Wisconsin, 1998.
- "Wisconsin Winnebagoes," *Wisconsin Historical Collections*, v. 12.

Pamphlets

- Badger Army Ammunition Plant, *50th Anniversary Booklet*, 1992.
- Clayton, Lee, and Attig, John W., *Geology of Sauk County, Wisconsin*, Information Circular 67, 1990.
- Casper, Gary S., *The Reptiles and Amphibians of the Badger Army Ammunition Plant, Sauk County, Wisconsin,* The Nature Conservancy, Wisconsin Chapter, Madison, 1993.
- Cochrane, Theodore S. and Hugh H. Iltis, *Atlas of the Wisconsin Prairie and Savanna Flora*, Wisconsin Department of Natural Resources Technical Bulletin No. 191, 2000.
- Hercules Powder Company, *Badger Ordnance Works, Wisconsin's No.1 War Plant*, July, 1944.

  ——, *Nitrocellulose, Properties and Uses*, 1944.
- Henderson, Richard, *Plant Species Composition of Wisconsin Prairies*, Wisconsin Department of Natural Resources Technical Bulletin No. 188, 1998.
- Holman, Ernest H., *The First Fifty Years of Barksdale Works, Wisconsin Pioneer Explosives Plant, 1904-1954*, E.I. du Pont de Nemours & Co. Washburn, Wisconsin, 1954.
- Lange, Kenneth I., *A Postglacial Vegetational History of Sauk County and Caledonia Township, Columbia County*, South Central Wisconsin. Technical Bulletin No. 168, Department of Natural Resources, Madison, Wisconsin, 1990.
- Mossman, M.J., *Breeding Birds of the Badger Army Ammunition Plant, Sauk County, Wisconsin*, Bureau of Integrated Science Services, Wisconsin Department of Natural Resources, 1999.
- Mossman, M.J. and K.I. Lange, *Breeding Birds of the Baraboo Hills, Wisconsin, Their History, Distribution, and Ecology,* Wisconsin Department of Natural Resources and Wisconsin Society of Ornithology, 1982.
- Plat Book of Sauk County, Sauk County Publishing Company, Baraboo, Wisconsin, 1936.

• Risch, Erna, *The Quartermaster Corps, Organization, Supply and Services*, Center of Military History, U.S.Army, Washington D.C. 1995.

Documents

• Carey, Bill, Sauk County Resource Agent, Sauk County Overall Economic Development Program, 1977.
• Department of Public Instruction,Thirtieth Biennial Report,Wisconsin, 1940-1942.
——,Thirty-First Biennial Report,Wisconsin 1942-44.
——,Thirty-Second Report,Wisconsin 1944-45.
• Economic Adjustment Program, Sauk County,Wisconsin, Office of Economic Adjustment (Installations and Logistics) The Pentagon,Washington D.C., October, 1975.
• GAO, Military Bases, Cost to Maintain Inactive Ammunition Plants and Closed Bases could be Reduced, CAO/NSIAD• 97-56, February, 1997.
• Kane, Kimberly L., Historic Context for the World War II Ordnance Department's Government-Owned Contractor-Operated (GOCO) Industrial Facilities, 1939-1945, Report of Investigations, Number 1, October, 1995.
• Kimbrell, K. Diane, Hiatt, Kathleen E., Badger Army Ammunition Plant Supplemental Photographic Documentation of Archetypal Buildings, Structures, and Equipment for U.S. Army Materiel Command National Historic Context for World War II Ordnance Facilities, Report of Investigations, Number 2B,April 1995.
• Kirkpatrick, E.L. and Boynton, Agnes M., Wisconsin Human and Physical Resources, A Graphic Presentation of Conditions Affecting Rural Rehabilitation, Madison, July, 1936.
• New Industries and Plant Expansions Reported in Wisconsin, 1958, Prepared by Division of Industrial and Port Development, March 1959.
• Proceedings, Sauk County Board of Supervisors, 1895, 1941-47.
• Salkin, Philip H., An Archaeological Survey of Proposed Timber Harvest Sites at the Badger Army Ammunition Plant in Sauk County,Wisconsin, Report of Investigations, July, 1999.
• Sauk County Agriculture, County Agricultural Statistics 2d. Series, Wisconsin Crop and Livestock Reporting Service, USDA, Wisconsin State Department of Agriculture, Madison,Wisconsin, 1046
• Schoenoff, Kurt, R. Sauk County Annual, 1941-42.
• Shaffer, Scott, C., Crown, Deborah L., Bear Creek Archeology, Inc. Badger Ordnance Works, Baraboo, Wisconsin 1941-1945 Detailed Historical Investigation, Report of Investigations, Number 2A, July, 1995.
• Soil Survey of Sauk County,Wisconsin, USDA Bureau of Chemistry and Soils, Number 29, Series 1925.
• Soil Survey of Sauk County, Wisconsin, USDA Soil Conservation Service and College of Agricultural and Life Sciences University of Wisconsin, 1980.
• State Highway Commission of Wisconsin, A History of Wisconsin Highway Development, 1835-1945, Madison,Wisconsin, 1947.
• Wisconsin Academy Review, Wisconsin Academy of Sciences,Arts and Letters, Fall 2000, Volume 46, Number 4.
• Wisconsin Agriculture, Bulletin 140, Cooperative Crop and Livestock Reporting Service, 1932.
• Wisconsin Agricultural Statistics, Wisconsin Statistical Reporting Service, 1968.
• Wisconsin Archeologist, The, Vol. 40, No.1, March, 1959, The Wisconsin Archeological Society, Milwaukee,Wisconsin, 1959.
——,Vol. 74, Numbers 1-4, March-December, 1993.
• Wisconsin Dairying, Bulletin 200, Cooperative Crop and Livestock Reporting Service, 1939.

- Wisconsin Department of Agriculture, Bulletin No. 202, pp. 43-48, Sauk County, Series No. 46, Madison, Wisconsin, 1942.
- Wisconsin Rural Resources, Adams County, Wisconsin State Department of Agriculture, 1958.
- Wisconsin's Biodiversity as a Management Issue, A Report to Department of Natural Resources Managers, Department of Natural Resources, May, 1995.

Archival Material

- Civil Case No. 307, United States of America vs. 10,565 acres of land, more or less, in Sauk County, Wisconsin, Henry Weirich, et al., in Records of the District Court of the United States for the Western District of Wisconsin, February 22, 1943, National Archives and Records Administration Great Lakes Region, Chicago, Illinois.
- Crowley, Leo T., taped interview, 1969, Wisconsin Historical Society.
- Evjue, William Theodore, Papers, Wisconsin Historical Society.
- Goodland, Walter S., Gubernatorial Papers, Wisconsin Historical Society.
- Heil, Julius, Gubernatorial Papers, Wisconsin Historical Society.
- Hoar, Walter G, taped interview, 1971, Wisconsin Historical Society.
- Kaeser, William V., Papers, Wisconsin Historical Society.
- Knapp, George Nelson, Writings, 1940-1947, Wisconsin Historical Society.
- Mueller, Erhart, Papers, Scrapbooks, Memorabilia, Wisconsin Historical Society, Sauk County Historical Society, Sauk Prairie Historical Society, Sauk City Library.
- Records of the Baraboo Industrial Development Committee, Sauk County Historical Society.
- Records of the Office of the Chief of Ordnance, Chicago Ordnance District, 1940-51, Records Group 156, Badger Ordnance Works, National Archives and Records Administration, Great Lakes Region, Chicago, Illinois.
- Records of the War Assets Administration, Chicago Region, Real Property Liquidation Files, Records Group 270, Badger Ordnance Works, National Archives and Records Administration, Great Lakes Region, Chicago, Illinois.
- Records, Sauk County, Wisconsin Council of Defense, 1917-18, Wisconsin Historical Society.
- Reuss, Henry S., Papers, Wisconsin Historical Society.
- Richards, Roberts, Reminiscences, 1843-1855, Wisconsin Historical Society.

Interviews

- Michael Goc, Badger History Group: O.J. Befera, September 24, 2001; David Fordham, November 1, 2001; Al Gallagher, September 25, 2001; Elmer Johnson, September 24, 2001; Melvin Luetscher, January 16, 2001, James Mattei, December 11, 2001: Dennis Thurow, December 13, 2001;Tom Robertson, December 17, 2001; Frank Wolf, November 21, 2001.
- Lisa Hartmann, Robert France video, Badger History Group: John Geoghegan, Loretta McCarthy, Mabel Gillingham, Mary Grignano, Susan Sweeney, Clayton Weber, Clarence Zech, August 7, 1999.
- Lola Huber, Robert France video, Badger History Group: Norma Clavadatscher, November 2, 2000; Rita Slezak Lachmund, August 28, 1999; Mardetta O'Brien, July 18, 1998; William Sisson, November 27, 2000; Walter Zechel, January 15, 1999; Harold Thoelke, Carl Graf, Clifford Schweppe, Reven Marshall, Beatrice Dailey, undated.
- Carol Sorg, Sauk County Historical Society: Carol Bohn Hulterstrom, Clarence Krumm, Mary Edwards Weigand, April 7, 1993.

Documents–Badger Ordnance Works

United States Army Corps of Engineers

- Mason and Hanger Company, Monthly Progress Report, April 1944–July 1945.
- Mayer, David P., Corps of Engineers Historical Record, Badger Ordnance Works Sauk County, Baraboo, Wisconsin, December 31, 1942.
- War Department Facilities Inventory of the Badger Ordnance Works, Supplements, vols. 1-5, 1946.

United States Army Ordnance Department

- Historical Summary, July 1951–December 1957.
- Historical Summary of Badger Ordnance Works Covering the Period Calendar Year 1959–June, 1967.
- Installation Historical Files, History of Badger Ordnance Works, 2 September 1945–1 July 1951.
- Semi-Annual Historical Summary, January 1953–June 1958.

Hercules Powder Company

- *Badger Ordnance News*, newspaper, April 1942–August 1943; May 1944–September 1945.
- Historical Report of Badger Ordnance Works, 1942–1945, v. 1,2,3, Badger Ordnance Works, October, 1945.
- Historical Record, Badger Ordnance Works, 1942–1945, v. 1,2,3,4,5.
- Monthly Operating Reports, January 1943–August 1945.
- Photo inventory of properties taken for Badger Ordnance Works, 1942.
- Quarterly Operating Reports, January 1944–December 1945.
- Scrapbooks of Wisconsin newspaper clippings re Badger Ordnance Works, January 1942–September 1945, v. 1-4.

Liberty Powder Defense Corporation

- *Badger Ordnance World*, newspaper, October 1952–January 1958.
- Monthly Operating Reports, July 1951–December 1957.

Documents–Badger Army Ammunition Plant

Department of the Army

- Basic Unit History, January 1965–December 1978.
- Basic Unit History, Annual Supplements, January 1965–December 1978.
- Basic Unit History, Supplements, Fiscal Years, 1969–1975.
- Historical Summary, Basic Unit History, 1942–1964; January 1965– December 1967.
- Installation Assessment of Badger Army Ammunition Plant, Records Evaluation Report, No. 111, v. 1., May 1977.
- 2.75 Rocket and Ball Powder Propellant Production Process, Gov't File 783.

Olin-Matheson Chemical Corporation/Olin Corporation

- *Badger World*, newspaper, August 1967–December 1975.
- Briefing by Olin-Badger for Project Manager-Modernization, January 30, 1976.
- Current Outlook and Plant Potential, A Briefing for Congressman Robert W. Kastenmeier, November 14, 1975.
- GAO Briefing, Modernization Projects, FY 71–74, January 1973.
- Monthly Operation Reports, May 1968–June 1972.
- Production and Safety Training Program, 1972–73.
- Vertical Integration, Badger Army Ammunition Plant, September 1974.

Index

A
A&P grocery, 280
Abbott Construction Company, 283
Abbott, Charles, 52
Accola, John, 57; Lorene, 57; Margaret, 57;Valentine, 57
Adams, 77, 151-152, 180
Adams County, 17, 20, 29, 65-66, 76, 88-90, 95, 178, 289, 346
Adams, John, 312
Administration Barracks, 127, 130, 134, 141
African-American, 138
Agent Orange, 356
Agriculture Department, 86, 357
Agricultural Adjustment Administration, 87
Air Force, 342, 248, 338
Alabama Ordnance Works, 137
Albert, Phyllis, 289
Alberts, Ruth, 321
American Federation of Labor (AFL), 123, 140, 292
American Legion, 60, 142, 178
Ammunition Procurement and Supply Agency (APSA), 307, 310, 316
Anderson, Alfred, 246
Archibald, George, 305
Argonne National Laboratory, 306
Arlington, 107, 109
Armistice Day, 60
Armstrong, Donald, 337; Dwight, 316, 337-341; Karleton, 316, 337-343; Ruth, 337
Army; Chief of Real Estate, 99; Combat Develop ment Command, 314; Corps of Engineers, 46, 71, 100, 106, 112-113, 122-123, 125-127, 129, 131-132, 134, 191, 256, 259, 268-269, 286, 288-289; Material Command (AMC), 307; Math Research Center, 339, 341, 343; National Defense Research Committee, 168; Ordnance Ammunition Production Office, 77; Ordnance Department, 78, 113, 269, 271, 175, 294, 296; Quartermaster Corps, 84, 112; Springfield Armory, 298; United States, 177
Arsenal of Democracy, 128
Arthur Overgaard, co. name, 122
Ashland, city, 88
Atlas Powder Corporation Wilmington, 78
Avery, Frank, 83
Avoca, 319
B
Badger Army Ammunition Plant, 321, 331, 333, 338, 346
Badger Ordnance City, 243
Badger Ordnance News, 129, 133, 140, 143, 146, 169, 180, 262
Badger Ordnance Works, appraisals, 101-102, 256; Ball Powder process, 300-301; busing, 144, 151-152; construction, 112-136; design safety, 116-119; investigations, 147-148, 290; M-2, 132-136; racial segregation, 144-145, 186; rocket powder process, 169-172; shutdown, 172, 254, 315, 344-345; siting, 75-76, 84, 102; Smokeless Powder process, 160-165; TNT plant, 115,128; wages, 123, 139, 293, 324; women at, 138-140, 146-147, 154, 158, 321, 323; workforce, 138-139, 289-290, 320-323, 345
Badger School, 183, 246-247, 251-252, 275, 280, 283, 325, 358
Badger Village, 52, 131, 135, 169, 191, 243, 252, 263, 274-275, 277-283, 285, 287, 290, 293, 319, 325, 328, 337, 359
Badger World (news), 317, 320, 323-324
Baime, Peggy, 279-280, 282
Ball Powder, 286-290, 298-301, 303, 307, 310-315, 317, 337, 340, 343, 348, 354, 355
Bar B Que Restaurant, 180
Baraboo city, 52-53, 60, 63, 68, 81, 83-85, 88, 90-92, 96-97, 105, 109, 122, 137, 142-143, 147, 151, 155, 166-167, 169, 173-186, 191-194, 255, 269, 272, 278, 280, 285, 294, 303-306, 312, 316, 319, 322, 325-326, 335, 340
Baraboo American Legion Post, 128; Bluffs, 44, 59, 112, 127, 342; Chamber of Commerce; 90, 91; City Council, 91; Federal Savings and Loan, 180; High School, 182, 321; Hills, 15, 17-18, 20, 44-46, 55, 66, 272, 290; Industrial Expansion Corporation, 305; News-Republic, 89-92, 294, 316; Range, 18, 20; Range Transportation Company, 150; River, 17, 20, 29, 54
Baraboo town, 107
Barber, J.D. , 106
Barksdale, 78, 88, 297
Barkley, Leslie, 291
Bascom Hill, 337
Bass, Floyd K., 107, 110
Baxter, 49
Baxter's Hollow, 15
Bayfield, city, 88
Bayles, Donald C., 180; Kathleen, 180; Michael, 180
Bear Creek town, 81
Befera, O. J., 289
Beggs, John, 64

Belcher, Ed, 112; Eugene, 331
Bell, Ferguson 289
Belmont, 39
Belsky, Ed, 92, 103
Bentley Construction, 121
Berkley Chevrolet Garage, 122
Berlin Wall, 306
Bickford-Lawson Creamery, 53
Big Bear, 60
Bills, Alexander, 46
Black Earth, 151
Blackhawk, 43
Black Hawk, lake, 21
Black River Falls, 100
Black River, 29
Blaine, James, 60
Bliss, Della, 146; Elmer, 146; Melvin, 146
Blount, Thomas H., 175-76
Blue Mounds, 47
Bluff, church, 47; settlement, 47
Bluffs, 46, 325
Bluffview, 242, 312, 329-331, 341, 359
Boddicker, F.L. , 137
Bogholt, Carl, 274
Bohn, Henry, 190
Book of the Month Club, 256
Borden's, 141
Borland, R., 162
Bowen & McLaughlin-Birch, 121, 126
Boy Scouts, 252
Boyle, John J., 104, 253, 258-260
Briggsville, 151
Brigham, Hannah, 47; Prescott, 55, 46-47
Brittingham & Hixon, 90
Broadway, 179
Brown, 49; Addison, 53; Mary, 139
Bryant, Charles, 43
Buck, Horace M., 112, 122-123
Buell, Robert, 289
Building Destruction Board, 264-266
Building Trades Council, 123
Burgy, Anna (Hill), 50; Cecelia, 50; Jacob, 50
Burckhardt, Traugott, 50
Burkhardt, Amelia, 257-258, 260; William, 102, 107, 109, 259-260
Burlington, Iowa, 96
Burma Road, 47
Bush, Maybelle, 246
Bush, President, 349
Butterfield, Consul W,. 13, 21
Byers, Bernard, 143; Frances, 143; John, 143; Virgil, 143; Virginia, 143; William, 143
C
Cahoon Iron Mine, 272
Caledonia, 82, 107, 109
California Institute of Technology, 168
Callahan, John C., 246
Cambria, 269
Camouflage Department, 248
Camp Douglas, 60
Camp McCoy, 177, 320
Camp Randall Stadium, 275
Camp Williams, 177
Canfield, William, 21, 44
Capital Times, 87, 91, 97, 148, 177, 263
Capitol Square, 259
Carley, David, 306
Carpenter, 49; George, 127; Isabell, 322
Carpenters and Joiners Union, 123
Carrier Corporation, 121
Case, Russell, 192
Cassell Prairie, 22
Castle Rock, 19
Catlin, George, 30
CBS Evening News, 309
Cenex, 304
Central Colony, 337
Central States Construction, 122
Central Wisconsin Conservation Area, 100
Century 21, 15
Charleston, Indiana, 92
Charlton, Fitzhugh Davidson, 129
Chase, Henry V,. 122, 137
Cheek, Arthur, 85
Chemical Workers Local, 292
Chicago and North Western Railroad, 54, 75-77, 80, 82, 89, 103, 123, 125, 149, 167, 174, 290, 291, 343
Chicago River, 114
Chicago Tribune, 326
Chippewa County, 88
Christenson, Ann, 144
Church, Leroy, 105
Churchill Winston, 247
Cinealis, William Joseph, 129
Civil War, 43, 45, 50, 180, 250, 180, 298
Civilian Conservation Corps, 124, 127, 144
Clark County, 88
Clavadatscher, Norma, 145
Clean Air Act, 348
Clean Water Act, 348
Clement, Bill, 289
Clock Alley, 156-157
Coast Guard, 173, 248
Colby, 92; Earl, 103; Lorin, 107
Cold War, 286, 302, 306, 308, 356
Cole, Bessie, 58; H.E. 45
Colt firearms company, 308
Columbia County, 20, 82-84, 107, 124, 144, 166, 169, 174, 177, 193, 258, 269, 346
Columbia County Board, 90
Columbia River, 114
Columbia University, 140
Columbus, 107, 109
Commodity Credit Corporation (CCC), 269-270
Condemnation Commission, 104, 258-259
Congress, U.S., 91
Congreve, William, 116
Conservation Natural Resources Association, 333
Conway, Byron B., 254
Conway's Acres, 182
Copps, Michael, 289
Cormican, Elmer, 321; Marjorie, 321; Roger A., 321
Cornell University, 298
Cornhusker Ordnance, 265

Corrao, Ruth, 289, 291
Council of Defense, 83
County Z, 47, 53
County ZZ, 102, 112
Cramer, Earl, 85
Cravens, Guy, 269-270
Cronce, Otto, 124, 154
Cronkite, Walter, 309
Cronon, William, 23
Cross Plains, 339
Crowley, Ken, 93
Crowley, Leo T., 73-77, 81, 92-93, 262, 290, 354
Curtis, John, 33, 34
Cyclone Fence Company, 123
D
Dahnke, Emma, 96, 107, 258
Dairy Forage Research Center Field Station, 357
Dalton, 166
Dane County, 44, 61, 63, 65, 83, 107, 124, 145, 174-175, 177, 193, 254-255, 278, 335
Daniels, Luther, 48
Davies, Paul, 148
Davis, Sidney I., 320-321, 335
Day, Roland, 329
De Forest, 107
De Weese, Addison, 107
Defense Department, 344, 348, 355
DeLaney, James J., 326
Dell Creek, 60
Dell Prairie, 89-90
Dellona town, 65
Delton State Graded, 183
Delton town, 17, 29, 65, 94
Denman, Jane, 323
Dennett, John, 52
Denny, William, 169
Denzer, 21, 178-179
Deppe, Wilbur, 329
Derleth, August, 86, 154
Detroit Tank Arsenal, 112
Devils Lake State Park, 20, 47, 80, 124, 127, 144-146, 152, 177, 182, 186, 193, 332, 336, 340, 343
Diamond Transportation System, 150-151
Diem, Ngo Dinh, 307
Dipperette, 161
Dipperman, 161
Dirksen, Everett, 355
Dodgeville, 150
Doll, Walter, 83
Douglas Aviation, 173
Douglas County, 169
Dow Chemical Company, 337
Doyle, James, Sr., 343
Dresen Brothers, The, 193
Dreyfus, Lee, 281
Drumb, William, 131
Du Drop, 179
Du Pont (E.I. Du Pont de Nemours & Company), 77-79, 88, 114, 116, 155, 298, 309, 314
Duerr, Carl, 51; Emilie, 51
Dugan, Anthony, 315
Durand, 78
Durst Rockshelter, 23, 26
Durward's Glen, 58
E
East Alton, 298, 309-310, 314
East High School, 337
East Sauk Road, 46-50, 102
Eau Claire County, 88
Eau Claire River, 88
Eau Claire, Wisconsin Ordnance Plant, 113
Eberhardt, Jack, 109
Ecklund, Laurence, 75-76, 83
Ederer, Ray, 107
Education Department, 253
Edward Kraemer & Sons, 122, 188
Edwards, Mary Wiegand, 247
Effinger, 281
Ehret, 52, John, 107, 257; Ida, 257
Elder, David, 319
Elite Club, 58
Elmer, Roy, 291
Elperin, Nell, 279
Elroy, 76, 78, 151
Elwood Ordnance Plant, 91
Endeavor, 151
Engebretson, George, 88
Eno, LaCretta, 321
Enoch, Russell, 342
Environmental Defense fund, 333
Equitable Powder Company, 298
Esch, John, 67
Eschenbach, 110; Adam, 50; Barbara, 50; Charles, 106-107, 257; Louis, 107; Lowell, 107; William, 62, 107, 254
Evangelical church, 255
Evans, 258; Isaac, 255; William A., 289, 291, 296, 332
Evjue, William, 87, 148
F
Fairfield town, 81-82
Farm Bureau, 58
Farm Holiday, 67
Farm Security Administration, 127
Farmer's Institutes, 56
Farmers and Citizens Bank of Sauk City, 257
Farmers Union Central Exchange, 304
Farmers union, 254
Fassnacht, Robert, 341, 343
Federal Bureau of Investigation, 148, 177-178, 342
Federal Deposit Insurance Corp. (FDIC), 74-75
Federal Employment Service, 123
Federal Housing Administration, 189, 282
Federal Land Bank, 84, 98-99, 101-102, 254-257, 268
Federal Public Housing Admin., 244, 275, 284
Federal Works Administration, 268
Fegles Construction Company, 286
Feight, Mrs. Kenneth, 146
Ferry Bluff, 19
First Methodist Church, 105, 122, 181
First National Bank of Baraboo, 255
Fish and Wildlife Service, 307
Fjelstad, Joseph Clarence, 129
Flambeau Plastics, 305

Fleming, Robert, 177
Flying Tigers, 248
Fond du Lac, 17
Fordham, Dave, 353
Fort McCoy, 60, 84-85, 137
Fort McHenry, 116
Fort Sheridan, 320
Fort Sumter, 51
Fort Winnebago, 30
Fosdahl, Susanna, 143
Fox Indians, 27, 29
Fox River, 17, 27
Freedom town, 50, 107, 194
Freie Gemeinde, 50-51
French, Lewis, 135
French-Canadian, 16
Frenz, Ruthann, 122
Frese, George, 50; Louis, 107, 258, 260; Eunice, 258
Friendship, 64, 88, 285
Froker, Fred, 272
Fuchs, Bill, 326-327, 330
Fullerton, Thomas, 47
G
G.I. Bill, 275
G.I. Joe, 356
Gaber, John, 50
Gallagher, Al, 322, 334
Gant, C.H., 250
Gasser, 52; Ben, 107, 257; Emma, 101; Jacob, 101, 259-260; Jake, 107; Roy, 85
Gate 10 barracks, 134, 135
Gateway Transportation, 290
Gavol, George, 85
Gehrmann, B.J., 88
Geier, Joseph, 285
Geigerich, Bert, 191
Gem City, 305
General Accounting Office, 344
General Motors, 173
General Paper Company, 73
General Services Administration, 268, 319, 325
Geoghegan, John, 122
Gerhardt, H.A., 306
Ghost Line, 134
Gibbs, Carl, 257; Emma, 257; Isaac, 48-49
Gibson, Shirley, 122
Giertsen, A.E., 293
Giese, Kenyon, 343-344
Gillette Tire factory, 275
Gillingham, Mabel, 244
Gillingham village, 244
Gin's Pizza, 339
Gisholt Machinery, 337
Glacial Lake Wisconsin, 20
Gleason, H.A., 31
Glenville, 181-182
GOCO (government-owned, contractor-operated ordnance facility), 79
God Bless America, 95
Goerk's Tavern, 179
Goff, Ellsworth, 169, 337
Goldwater, Barry, 307
Gollmar, Robert H. 83, 184
Goodland, Walter, 266-267, 272
Gopher Ordnance Works, 98
Gotham, 151
Gottfried, Johann, 50
Governmental Services Administration, 306
Graf, 52; Carl, 50; Graf Siebecker, Christina, 50; William, 50
Grand Coulee Dam, 114-114, 152, 185, 189
Grange, 58
Grant, Charles H., 137
Graves, Leroy, 318
Great Depression, The, 67, 69, 74, 177, 180, 298, 657,
Great Lakes, 49
Great Plains, 61
Green Bay Police Department, 124
Green Bay, 19-20, 88, 169
Greenfield, 107
Greening, John, 44
Greenwood, 144
Greyhound bus, 152-153
Griffith, George F., 122, 137, 150, 154, 176, 186
Gross, Andreas, 50
Gruber, 92; Chris, 103; Christ, 96
Gruber's Bay, 51, 54, 66, 101, 102, 103, 125, 334, 355
Gruber's Grove, 58
Grunke, Henry, 291
GSA, 326
Guard Force, 153-154
Guernsey cattle, 56, 108
Gulf of Tonkin Resolution, 307
Guthrie, Woody, 115
H
H.K. Ferguson contractor, 286
Haase, George M., 85
Hackett, 258; Charles, 255
Halbach, Richard, 306
Haley, Harry, 255
Halle, Lawrence, 276, 278, 282
Halsey, Admiral "Bull," 173
Hamilton, Norl, 289, 304
Hanighen, F.C., 78
Hanko, Ronald, 318
Haraszthy, Agoston, 48
Harding, Ralph, 289
Hardy, Richard, 317
Harlan Electric, 121
Harmon, Howard, 323
Harper, Charles A., 176-178
Harrison, Roy, 107
Hart, William O., 67
Hartley, Avery, 318
Hatz, Jacob, 56; Lyman, 107, 258; Robert, 56-57; Wayne, 57, 108, 224, 319, 325-329, 330
Hauck, Mrs., 173-175; Wayne, 92, 101, 103, 109, 112, 115, 122-123, 126, 137, 173-175
Hayes, William, 91, 303
Haynie, Jewell G. , 289
Hazeltine, Ruth, 263
Heckman, C.H., 289

Hedman, Amelia, 144
Hegerfeld, Richard, 326
Heil, Julius, 77, 82, 17-175
Heimlich, 55
Heimlich, Richard, 107
Heimlich, William, 107
Hein, Dora, 143-144; Frank, 318
Heintz, Raymond, 127, 188
Heinz baby food, 280
Hellenbrand, Elaine, 322
Hercules Chemical Company, 76
Hercules Powder Company, 15, 78, 80-82, 84, 98, 113-117, 122-124, 127-128, 155, 131-170, 182, 185-191, 245, 262, 264, 266-267, 271, 276, 284, 286-289, 295, 309, 344,
Herr, Charles, 102, 107, 259-260; Lucille, 102
Hewitt, 143
Hickham Field, Hawaii, 105
Highway 12, 46-49, 53, 59, 64, 103, 109, 120, 123-124, 127-129, 136, 146, 151-153, 174-178, 184-187, 191, 194, 243-244, 278, 281-282, 284-285, 312, 316, 335, 340-341, 359,
Highway 12 bridge, 60
Highway 14, 151, 174, 278,
Highway 16, 317
Highway 23, 151
Highway 33, 151, 174
Highway 60, 174
Highway 78, 59, 62, 102, 125-126, 151, 174, 224, 278
Highway 113, 47, 174, 176
Highway 136, 151
Hill and Dithmar, 255
Hill, Harold K., 53, 87; Henry 50
Hill, Miller and Hill, 255, 258
Hilliard, S.G., 176
Hillsboro, 146, 151-152, 320
Hines, E.M., 282
Hitler, Adolf, 78, 336
Ho Chunk, 25, 27-30, 32, 36, 39, 94, 138, 271
Hoar, Walter G., 84-85, 92, 97, 99-100
Hoard, William, 66
Hod Carriers and Common Laborers, 123
Hodgett, Andrew, 46
Hodgson, Harlow, 357
Hofsteter, Stephan, 49
Holmes, Fred, 255, 256
Holstein cattle, 56, 108
Home Circle Club, 57
Honey Creek, 15, 18, 21, 43
Honey Creek town, 42, 44, 55, 68, 85, 93, 107, 109-110, 194
Honey Creek Valley, 39
Hoover, John, 47
Hoppe, Alvin, 105, 107, 110, 257; Evelyn, 105, 107, 110, 257
Horicon, 27
Hornung, Carlton, 91-92
Hotel Raulf, 250
Huber, 52; Anna, 257; Christ, 103, 107-108, 257
Huber, Lola O'Brien, 163, 181
Hudson River, 114
Hudson's Bay, 18, 20
Hulburt, 23; Creek, 27; Dell, 20
Hulterstrom, Carol Bohn, 247, 249
Huntley-Brinkley Report, 309
Hurwitz, Ken, 281
Hussein, Saddam, 349
Hutchens, David, 312

I

Ice Age, 20
Ichord Committee, 314
Ichord, Richard, 314
Idelewild, 106
Illinois Indians, 28
Illinois River, 114
Illinois, 17, 18, 22, 28-29
Indiana Ordnance Works, 92, 133, 295
Indiana, 16-17
Industrial Expansion Corporation, 319
Industrial Food Systems, 127
Insull, Samuel, 64
International Crane Foundation, 305
Investigations Department of BOW, 147, 148
Iowa County, 39
Iowa State University, 275
Isenberg, George, 90
Italy, 78

J

J.H. Findorff & Son, 121
J.J. Duffy, 121
Jackson County, 29, 100
Jackson, Oliver, 186; Robert, 253
Jacobsen, Hubert L., 289
Jameson, Albert, 46
Jameson's Hollow, 46, 59
Jantz, James, 336
Janzen, Durlin, 345
Jefferson, Thomas, 41
Jenewein, George, 256, 258, 260; William, 62
Jenkins, Leroy, 150
Jersey cattle, 56
Jessie, Frantje, 108; Raymond, 106-107, 127
Jim Crow, 144
John Deere, 41
Johnson, Elmer, 289; Grace, 82; Gus, 88; Harry, 82, 176; Herman, 289; Lyndon, 307, 310, 311, 335; Mahalia, 40; Marjorie, 278, 281; Norvel, 145; Roswell, 39-40, 46; William "Uncle Billy," 39-43, 46, 52-54, 111
Johnson's Corners, 60
Joliet Army Ammunition Plant, 311
Jones, Hugh A., 177
Jorgenson, Emil, 86, 88, 96
Juneau County, 17, 20, 29, 60, 65, 76, 87-88, 100, 143, 178, 346
Justice Department, Lands Division, 259

K

Kading, John, 144
Kaiser-Frazer Company, 272
Kaleidoscope, 342
Kanevsky, Ann Ruth, 254
Kankakee Ordnance Works, 155
Karasik, John, 312
Karch, Charles, 67

Kastenmeier, Robert, 344
Kehl, 125
Keitel, Elmer, 257
Keldsen, Colonel, 182; Victor L., 137, 142
Keller, Caspar, 49
Keller, Eugene, 49; Frank, 59; Maria, 49
Kellogg, Grace, 93
Kennedy, John, 307, Joseph, 75; Robert, 335
Kenosha County, 30, 65
Kentucky Ordnance Works, 265
Kern, Charles, 48; Charlotte, 48
Kern's Corner School, 58
Kern's Corners, 46-48, 50, 54, 60, 102, 107
Key, Francis Scott, 116
Keystone Ordnance Works, 137, 151
Kilbourn, 64
Kindschi, 92, 110; Bernice, 254; Donald, 330; Emma, 87, 102, 184, 254-255, 258, 261; Essie, 102, 258; Glen, 96, 102-103, 184, 254-255; Harry, 96; John Peter, 52, 54, 254; Lillian, 95; Lyman, 102, 109, 123, 187, 243, 258-260; Ralph, 102, 256, 259; Roy, 107, 124; Walter, 107, 258-260
King, Martin Luther, 335
King, Thomas, 48, 84-85, 91-92, 96, 114
King's Corners, 47-48, 82, 108, 182-183, 326
Kingston Cemetery, 50-51, 111, 127, 169, 265
Kingston, Jamaica, 145
Kingston/Sumpter town, 49
Kinzler, John, 322
Kirner, 110; Carl, 107; Fred, 107; Paul, 107, 256
Klanderman, R.R., 267, 274
Knapp, George, 45
Knights of Pythias Hall, 122
Knipschield, 110
Knowles, Warren, 337
Knudsen, 111; Silas J, 100, 101, 106, 111
Knutson, Ed, 328
Korean war, 172, 193, 283, 286; cease-fire, 288; conflict, 286
Kraemer, Herbert, 190
Krumm, Clarence, 179, 182, 243, 244-253, 280, 358; Guy, 143
Ku Klux Klan, 73
Kunz, Carl, 258; Charles, 107; Fred, 107; Herman, 50
Kurtz, Lena, 110
L
Labor Department, 123
La Crosse County, 81
La Crosse, 81
La Follette, Josephine, 61
La Follette, Phillip, 74
La Follette, Robert M. Sr., 61, 256
La Follette, Robert M. Jr., 75, 89-90, 97, 254, 266
La Follette's Magazine, 256
La prairie, 16
La Valle, 76
Lachmund Lumber Company, 325-326
Lachmund, Rita Slezak, 145, 153, 156, 179-181
Lafayette County, 253
Lake Black Hawk, 21
Lake Butte des Morts, 27
Lake Delton, 60, 141, 177, 186-187, 282
Lake Koshkonong, 27
Lake Merrimac, 20-21
Lake Michigan, 19, 27, 114
Lake Pepin, 30
Lake Street, 338
Lake Superior, 18-19, 88
Lake Winnebago, 17, 28, 124
Lake Wisconsin, 20, 25, 51, 62, 64, 89, 155, 169, 174, 253, 333
Lakeside School, 51, 61-62, 102
Lakeview School, 51-52, 62
Lange, John, 91, 98, 305, 319; Laurel, 322
Langer, Harold M., 175
Lapham, Increase, 15
Lathrop, Nelson, 46
Law, James R., 83, 94, 182
Le Due, Steve, 184
Leach, Paul, 152
League of Wisconsin Municipalities, 83
Lebrenz, Ronald, 224
Lee, Abner, 272; Charles, 169
Leigel, Lorraine, 181
Leigh, Delzie, 291-292
Leiser, Roy, 103
Leland, 21, 43
Lemberger, August, 278, 281
Lenerz, Howard, 323; Joan, 323
Leopold, Aldo, 15, 18, 45
Leroy, Melvin, Wilbur, 68, 69
Leslie, Jack, 335
Lewiston Basin, 20
Leykauf's Bluff, 19
Liberty cargo ships, 272
Liberty Powder Defense Company, 283-289, 292-294, 302, 304, 307, 310
Libke, Alexander, 166; Anice, 166
Lincoln Tunnel, 114
Lindgren, Bruce, 318; Roy, 305
Litscher, John, 87, 107, 254, 258; Pete, 106
Littell, Norman M., 259-260
Lodi, 52, 91, 107
Loganville, 43, 145
Lone Rock, 346
Loomis, Orland "Spike," 87, 175; Percy, 312, 342
Lorenz, William, 272
Lovecraft, H.P., 154
Lowell, 144
Loyal Americans of German Descent, 148
Lueders, Frederick Herman, 33
Luetscher, Arthur, 110; Irv, 68-69, 87, 96, 102, 107, 110, 256, 260; Margaret, 68, 69, 87, 96, 102, 107, 110, 256, 260; Mel, 109
Lyndon Station, 152
M
MacArthur, Arthur, 149
MacFarland, George, 85
Machinists Union Local, 272
Madison, 73, 84, 91, 104, 147-152, 174, 178-179, 182, 224, 255, 259, 259, 261, 275-277, 280, 282 284-285, 317, 319, 322, 330, 333, 335-338, 341-342, 346, 358-359
Madison and Wisconsin Foundation, 91, 271

Madison Central High, 73, 81
Magli, Anna, 113, 123, 126
Mahlkuch, Sam, 255
Malone, Arline, 323
Mandt, Ted, 305, 319
Marachowsky, Jule, 269
Marathon County, 17
Marines, 248, 343
Maroney, H.B, 289, 302
Marquette County, 20, 31, 76, 289, 346
Marryat, Frederick, 17
Marshall, George C., 77-76
Marshfield, 89
Martiny, George, 192, 305; Rachael, 192
Mason and Hanger Construction Company, 112-114, 121-123, 127-128, 131-132, 134, 142, 150-151, 159, 176, 286
Mason City, 115
Mather, 49, 64; Cora, 102; Roy, 107, 257; Swain, 87, 96, 99, 102, 107, 124, 256; Urban, 58
Mattei, Jim, 312, 320, 349
Mauston, 151, 269
Mayer, David P,. 71, 132
Mazomanie, 44, 146, 151, 175, 320
Mazzacano, Vito, 292-293
McCarthy Improvement, 121
McCarthy, Joseph, 281; Lorna, 244, 249, 251
McCarthyite, 321
McCormick-Deering, 108
McCullough, 323
McGlashan, John, O., 123
McKillip, Kathleen, 140
McNamara, Robert, 306
McQueen, Steve, 308
McVeigh, Timothy, 334
Medicare, 327
Menasha, 27
Menominees, 29, 138
Merchants of Death, 78
Merrimac Defense Housing Project, 188
Merrimac, Massachusetts, 82
Merrimac State Graded School, 183
Merrimac, 48-49, 54, 62, 64, 76, 80-82, 84-85, 123, 174-177
Merrimac town, 15, 21, 42, 46, 49-55, 57, 61-62, 66, 84-85, 89, 91, 93, 103, 106-107, 110
Meyers, Majorie, 144
Middle Sauk Road, 47
Middleton, 107, 151, 179, 278, 339,
Midland Co-op, 280
Milk Strike, 67
Milton Junction, 73
Milwaukee Journal, 75, 80, 82, 88, 90, 135
Milwaukee Road, 54, 73, 76-77, 83, 89, 92, 103, 123, 149, 167, 290, 317
Miner, Bernadine, 144
Mineral Point, 40, 285, 320
Mississippi River, 16-23, 29, 54
Missouri Ordnance, 295
Mitchell, John F., 318
Mobilization Plan, 302-303, 307
Mockler, Edward, 101, 107-108, 259-260; Maude, 101
Modernization Plan, 346-348
Moeller, Henry, 305
Moely, 92; Ella, 246; John, 103
Moersch, Mae, 122
Mohr, William, 83
Monroe County, 60, 84, 100
Montello, 151-152
Moran, Leo, 143; Michael, 143
Morey Airport, 338-339
Morgenthau, Hans, 75
Morrisonville, 107
Mossman, Mike, 351
Moulton, August, 272, 292
Mount Horeb, 150
Mueller, Albert, 86, 89; Dora Accola, 86-87; Erhart, 30, 57-59, 61, 67, 86, 93, 99, 152, 191, 279, 281, 283, 330, 357
Muha, Ann, 144
Muir, John, 31-32, 42
Mullen, Joseph E., 190
Murphy, Catherine, 143; D.F., 85
Murrah federal building, 334
Muscoda, 152
Myron Park, 192

N

Naffz, Charles, 50-51; Rosalie, 50-51
Narrows Creek, 20
National Farm Bureau, 86
National Farmers Union, 86
National Grange, 86
National Guard Armory, 176
National Guard, 177, 306, 317, 321, 336-337, 342
National Housing Authority, 243
National Institute of Concrete Construction, 328
Natural Bridge, 23
Naval Ammunition Depot, 171
Navy, 165, 190
Neale, Kermit, 124
Necedah National Wildlife Refuge, 100
Necedah, 143, 152
Neenah, 27
Neff, Grover, 64
Nelson, Gaylord, 303, 307, 344
New Lisbon, 144, 151-152, 320, 359
New River Ordnance Plant, 114, 154
New York Times, 337
Newman, Ralph, 60
Nichols, 327
Nichols, Don, 330; Ron, 325; Thomas S., 304
Nicolet, Jean, 28
Nietz, John, 90
Nike missile, 302
Nikolay, Frank, 270, 291
Nixon, Richard, 224, 326, 333, 337,
Nobel, Alfred, 116
Nolden, Adolf, 179; Conrad, 85; Natalie, 179
Nolden's Café and Bar, 179
North Badger, 274, 276, 278, 281, 283, 325, 328
North Freedom, 183, 322, 359
Norton Company, 126-127, 130
Norton, James A., 121

O
O'Brien, 260; Clara, 182-183; John, 182, 259; Mardetta, 156, 163-164
O'Keefe, John R. , 315
O'Konski, Alvin, 321
Oak Villa Resort, 100
Oelke, Bernard, 101; Eleanor, 101
Office of Agricultural Defense Relations, 97
Office of Price Administration, 193
Ohio River, 19, 92
Ohio, 17, 39, 41
Ojibwas, 138
Old Red Gym, 338, 341
Oldfield, Alfred A. "Barney," 103, 112, 185
Olin, Franklin W., 298
Olin's Winchester div., 289
Olin Corporation, 286, 333-334, 344-345, 349
Olin-Mathison Chemical Corporation (O-M), 304, 309-310, 314, 320
Olson, Arthur, 194; Robert, 194
Oneida County, 62
Oneidas, 138
Operation Rolling Thunder, 307
Ordnance Ammunition Command, 307
Ordnance Ammunition Production Office, 81
Ordnance Department, 77, 80, 118, 122, 131-137, 191, 264, 267, 286, 289-290, 299
Organ, Robert, 62
Osborne, R.J., 176
Osterhus Publishing Company, 326
Otter Creek, 20-21, 39-40, 45-46, 48, 103, 129, 191
Owen David, Dale, 31
Oxford, 151
P
Packard, Harlan, 325, 327, 330
Page, Curt, 294, 303, 305, 309-310
Page, Harwood, 107
Panama Canal, 63-64
Panmunjon, 286
Pantex Ordnance, 265
Parent-Teacher Associations, 250, 252
Parfrey's Glen, 52, 62
Park View school, 182-183, 193
Patrick, Katherine, 107; Paul, 107
Patterson, 91; Robert F., 89
Pawlisch, Otto, 190
Peace Day, 60
Pearl Harbor, 77, 94, 96-97, 104
Pechan, Dennis, 318
Peck, George, 67, 74
Peck's Prairie, 22
Pennsylvania Ordnance Works, 135
Persian Gulf War, 348
Phi Kappa Sigma, 338
Philipp's Boulevard, 175, 340
Philipp, Emmanuel, 39
Phillips, Wendell, 356
Picatinny Army Arsenal, 175
Pine Hollow, 46-47
Pioneer Cemetery, 47, 111, 127, 155
Pitt Construction Co., 121
Pittsville, 317
Plain, 21, 43, 151
Plover River, 29
Plover-to-Portage "P" line, 89
Point Sauk, 17
Pontarelli & Sons, 121
Pope Pius XI, 74
Portage, 28, 30, 66, 83, 91, 130, 150-151, 174, 179, 186, 250, 269, 282, 317, 323
Portage County, 17, 20, 29, 66
Portage Warehousing Company, 269-270
Potratz, Russ, 316
Poynette Game Farm, 258
Prairie du Sac, 15, 20, 40, 42, 48, 51-52, 54, 60, 63-67, 76, 83, 86, 89, 92-93, 103, 107, 109, 120, 123-125, 127, 130, 134, 136, 142, 150, 152, 155, 169, 173-178, 182-185, 188-191, 193, 244, 255, 269, 275, 287, 312, 322, 326, 330, 338, 341
Prairie du Sac town, 17, 33, 44, 47, 53, 85, 110, 177, 194, 341
Prairie Plumbing and Heating, 326
Premo, Charles, 52; Elizabeth, 52; Garth, 62, 83, 85-86, 93-97, 102, 107, 254, 256-258; Grace, 93, 94; Ken, 194; Nellie Shaw, 57; Nora, 93, 95, 102, 110, 256-258; Roscoe, 109; Steve, 93, 94
Pretsch, 52
Priebe, Robert, 289
Primate Research Center, 338, 341
Primrose town, 61
Probert, John, 342
Progressive Party, 254
Progressive Republicans, 61, 74
Proxmire, William, 303, 306-307, 309, 344
Pruden, Charles, 137
Public Service Commission, 150
Pugh, Erwin, , 169
Pulvermacher, 112
Pyne, Stephen, 23
Q
Quartermaster Corps, 80, 92, 248
Quartzite rock, 272
Quick, Robert, 320-321
Quinn, Edwin, 291-292
R
Raddatz Rockshelter, 23-26
Radford Army Ammunition Plant, 344
Radford Ordnance Works, 82, 84, 114, 122, 124, 133, 140, 142, 168, 170, 288, 290, 297
Ragatz, 52; Barthomelew, 68; Henry, 42;
Randall Park, 275
Randolph, 317
Ray-O-Vac, 338
Reader's Digest, 328
Reconstruction Finance Corp., 263, 268-270
Reed, Dr. Howard L., 126-127
Reedsburg, 43, 76, 145, 186, 304, 321,
Reedsburg-LaValle, 150
Reedsburg school, 183
Reedsburg town, 65
Rendtorff, Edmund, 31, 33, 37, 46
Reno, Milo, 67
Republican Party, 61, 66, 74
Resettlement Administration, 95, 100

Reusch, 110; Frank, 107, William, 107
Reuss, Henry, 303-307, 333
Reuter, Louis, 263-264, 288-289, 295, 354
Rib Mountain, 18
Richland Center, 152, 285, 319-320
Richland Center schools, 183
Richland County, 53, 81, 244, 346
Richmond Virginia, 154
Riley, Roderick, 304
Ring, Walter, 326
Ringling Mansion/Elks Club, 122
Ringling, Henry, 173-175; Jean, 173
Ripley, George, 137
Ritz Corners Development Corporation, 282
River Street, 179
Rivers and Harbors Act of 1899, 333
Roach, John W., 177-178
Roberts, George, 112
Roche, E.M. 121
Rock Island, 81
Rock River, 28
Rock Springs, 43, 319
Rockies, 16, 17
Roetter, Dr. Friedrich H., 148
Roicks, 49, 52, 110
Romell, Louis, 289
Roosevelt, Franklin, D., 66-67, 74-75, 78, 84, 86-89, 247, 253-254
Roosevelt, Theodore, 331
Rosemount, Minnesota, 98
Rosie the Riveter, 139, 356
Rossing, Geraldine, 345
ROTC, 338, 341
Rotzoll, Ben, 319
Rowlands, Mervyn, 269-270
Roznos, Frank, 103, 112
Ruesch, 257
Ruesch, Howard, 107
Rural Electrification Administration, 64
Ryan, Katherine, 73; William, 255, 257-261

S

Sabby, Gerald, 318
Sachtjen, Robert, 343
Sansun, Elizabeth, 140
Sauco, 63
Saucomac, 63
Sauey, Norman, 305; Ron, 305; William, 305
Sauk Indians, 29
Sauk City, 21, 27, 39-41, 44, 47, 54, 60, 76, 83, 85, 91, 103, 124, 127, 146, 150, 152, 155, 167, 173, 175, 178-179, 184, 282, 285, 312, 325, 340, 343, 359
Sauk City Association, 51; bridge, 174; Creamery, 54; school, 183;
Sauk County, 19-22, 27-29, 35, 43, 47, 49, 51, 53, 56, 59, 60, 63, 65, 67, 77, 79, 81, 82-86, 87, 89-91, 94, 97, 107, 123-124, 134-135, 141, 145-146, 148, 173-185, 191, 193, 245, 250, 259, 272, 287, 294, 302, 306-312, 318-319, 325, 327, 329-330, 336-337, 345-346, 354, 357
Sauk County Airport, 339; Board of Supervisors, 85, 97, 104; Federation of Labor, 272; Highway Commission, 124; News, 91-92, 186; Normal School, 141; Nurse, 141; Ration Board, 149; Sheriff's Department, 148
Sauk Hill, 131, 272
Sauk Prairie, 17-27, 30-36, 39-40, 42-49, 59-71, 77, 82, 84, 86-90, 93, 95-99, 100, 102, 105, 110-112, 128, 149, 152, 179, 181, 194, 252, 255, 261, 267, 272, 274, 280, 303, 325, 330, 332, 340, 349, 353, 357, 359
Sauk Prairie Airport, 340-341; Farmer's Club, 58; Farmers Committee, 94; Gazetteer, 15; Landowners Committee, 94
Saunders, Rose, 323
Savanna Illinois Army Depot, 342
Sawyer, Philetus, 61
Schedeman, Albert, 74
Schereck, William J., 278, 280
Schlag, 52; Clara, 107, 257; Walter 126
Schneider, John, 83
Schoenoff, Kurt, 183, 245-246
Scholl, Martin, 190
Schollenberger, Emma, 254
Schommer, Victor R., 316-317
Schroeder, Ed, 107
Schubring Dorothy, 143
Schubring, Ryan, Petersen and Sutherland, 255
Schuck, Walter, 316
Schultz, Lynn, 338, 340
Schulz, Frank, 83
Schwarz, R.J. 83
Schweppe, 179; Clifford, 63, 178; Edwin, 107, 178; Iva, 178
Schwichtenberg, Thomas J., 313
Scott, Lillian, 127
Scrap Board, 264
Seabees, 248
Selective Service, 309, 341
Shanks, Nean, 85
Shaw, Elhanon, 47
Shearer, Mark, 169
Shimniok, Anna, 256, 107, 109, 256; Carl, 50; Ed, 107; John, 107; Paul, 102, 107, 109, 256
Siebecker, 62; Christina Graf, 61; Siebecker, Robert, 61; William, 50, 61
Shimniok farm, 155
Sigma Kappa, 339
Simon, William, 107
Sisson, William, 141, 144-145, 156, 163-164, 188
Skelgas LP, 325
Skillet Creek, 44
Smith, Al, 74; Elaine, 304; Frank O., 130; Kate, 95; Lewis, 47; R.R., 99
Soils Research, Inc., 304
Solie, Bruce L., 278
Somervell, Brehon, 81
Sommers, Forrest, 289
South Badger, 328, 244, 274-276, 278, 283-284
South Boulevard, 142
Southern Chemical Company, 160
Southern Wisconsin Power Company, 63-64
Spencer, Dorothy "Dolly," 335-336; Merle "Mike," 335-336
Spiegel, Pauline Euker, 103

Spoors, Jennie Johnson, 127, 129, 191-192, 257
Sprague, William, 50
Spring Green, 22, 54, 66, 85, 150, 174, 255, 359
Spring Green town, 81-82
Springfield, 179; Hill, 278
Sputnik, 302
St. Joseph's Church, 181
St. Louis Ordnance Plant, 298-299
St. Mary's Hospital, 73
St. Mary's-Ringling Hospital, 84, 141, 182
St. Raphael's Cathedral, 285
Staeffler, Richard P., 137
Staff Village, 39, 129, 149, 191, 276, 290
Staines, Henry, 43
Standard Gas Company, 74-75
Stanley, city, 88
Starrett Brothers, 121
State Board of Health, 140
State Emergency Board, 276
State Highway Commission, 109
Steidtman, 52, 95; Agnes, 110, 257; August, 50; Christian, 50; Frederika Gattwinkel, 50; J.P., 107; John, 50; Julius, 64, 109-110, 257; Louis, 107; Nora, 93; Wilhelmina Schlag, 50
Steidtmann lots, 76
Steinke, Gladys, 143
Stekl, Isabelle, 250
Stempke, John, 106
Sterling Hall, 341
Steuber, 52, 110; Aaron, 107; Alex, 102-103, 107-108, 124; Holmer, 107; John Frederick, 50
Stevens Point, 19, 89
Stevenson, William H., 81, 92, 124, 174-175
Stewart, Ulysses G., 289
Stimson, Henry, 104
Stoffel, John, 68
Stoltz, Albert E., 137
Stone, Nellie Hatz, 57; Nellie Luetscher, 58; Patrick T., 104, 253, 255-259, 270, 253, 270, 291-292
Stone's Pocket school, 21, 54
Stoughton, 107
Stout State University, 322
Strong's Prairie, 90
Struciform Company, 244
Suez Canal, 105
Sumpter town, 15, 21, 39, 44, 46-59, 61-64, 66, 76, 83-84, 90, 97, 92-96, 104, 106, 110, 136, 141, 173, 177, 194, 246, 252, 254, 281, 284, 330
Sumpter Evangelical church, 50; German Lutheran church, 49-50; Methodist Church, 48, 58, 93
Sun Prairie, 304, 320
Sunflower Ordnance Works, 135, 137, 168, 170, 289-290, 296, 343
Suppy, Willard, 291-292
Swalheim, Carl, 108
Swansby, Arnold, 166; Bessie, 166
Swanton, Milo K., 87
Sweeney, Susan Mossman, 322
Swenson, Magnus, 63-64
T
Table of Interplant Distances for High Explosives, 117
Tammaro, Alfonso, 137
Teel, Charles, 46-47; Charlotte, 46-47
Thayer, A.H., 128
Thiede, Ralph, 322
Thies, H.H., 141; John, 50; Mary, 50
Thoelke Cemetery, 102, 111, 127
Thoelke, 52, 110; Harold, 102, 107, 289, 259-260; Henry, 64; John Henry, 50; Maria, 50
Thomas, 323; Bertha, 257, 259-260; Norman, 281; Ralph, 257, 259-260
Thomson, Vernon, 307, 309, 326, 344
Thorp Agency, 108
Thurow, Dennis, 345, 348
Tomah, 64
Tomahawk, 253
Toman, Jean, 143; June, 144
Townsend, Clifford, 87
Tripp, 63
Troy town, 55, 107
Troyer, John, 289, 308-310, 313, 318, 326, 332
Truax Field, 85, 276
Truman, Harry, 75, 173, 295
Turdot, Nancy, 323
Turner, Leo, 317
Twenty-First Century Community Learning Center, 253
Twin Arches tavern, 179, 190
Twin Cities Ordnance Plant, 132, 317
259th Ordnance Detachment, 342
U
Ulrich, Betty, 346
Union Center, 76, 146
United Nations, 286
University of Wisconsin, 81, 143, 260, 254-255, 265-266, 272, 274, 276, 279, 284, 288, 304, 322, 335-338
University of Wisconsin Extension, 56, 86, 246
Uphatel, Frank, 154
Upper Fox River Valley, 29
Up-To-Date Auto, 179
USDA Office of Agric. Defense Relations, 87
USO, 248
V
Valley of Our Lady Monastery, 39
VanWhy, Cyrenus, 153
Verona, 107, 109, 330
Victory Heights, 141, 192-193
Vietnam War, 172
Vilas County, 23, 62
Viroqua, 275, 285
Vogts, Harry F. ,282
Voll, Violet, 143
Volunteer Ordnance Works, 295
W
Wabash, 16
Wabash River Ordnance Works, 299
Waffenschmidt, 52, 56, 110, 112; Chris, 107; E.M., 259-260; C.C., 259-260; Emilia, 102; Emma, 107; Fred, 64, 107, 256; Robert, 107
Wage Stabilization Board, 294
Wall Street Journal, 75
Wallace, Henry, 75

Walsh & Steenburg, 121
Wampler, William, 344
War Assets Administration, 266, 268-269, 272
War Department, 78-81, 84, 88, 90-91, 98, 103, 132-133, 139, 142, 152, 243-244, 262-263, 269-270, 275; Facilities Board, 80
War Food Administration, 144
War Manpower Commission, 142
War Production Board, 192
Warren Hotel, 91, 179
Warren, Robert, 333-334
Warrington, E.W., 289
Wartner, Herman, 82
Washburn, Wisconsin, 88
Washington D.C., 74-75, 79, 82, 97, 173, 254, 265-266, 303, 337, 344
Washington, George, 39, 41
Wasielewski, Theodore, 84-85, 114
Water Street, 190, 191, 340
Waterbury School, 93, 246
Waterbury, James Isaac, 42, 44-46
Watertown, 78, 107
Waukesha County, 65
Waupaca County, 66
Wausau, 253
Waushara County, 20, 66
Wayne, John, 308
Weidenkopf, Harold, 194
Weigand, Mary Edwards, 249; Theodore, 62, 107, 258
Weigand's Bay, 66, 87, 93, 102, 106, 108, 125, 155, 192, 257
Weirich, 52, 55, 110, 112; Arthur, 107; Charles, 259-260; Henry, 50, 259-260; Harry, 108; John, 50; Michael, 50
Welch, Roy, 258; Lila, 107, 258; Walter, 58
Wendle, H.N., 84, 90, 102-103, 134, 187, 267, 354
West Badger, 244
West Point town, 107, 137, 255
West Sauk Road, 47, 59
Western Cartridge Company, 298
Western Union, 87
Westfield, 151, 224
Westfield schools, 183
Westlund, Elsie, 146
Westmoreland, 309; William C., 308-309
Weyerhaeuser, 160
Weynand, 92; George, 103
WHA radio, 95
White Eagle, Mrs. Winslow, 152
White, Aimee, 140; John, 335; Wesley, 106
WIBA, 143
Wickard, Claude, & Mrs. Claude, 87-88
Wight, Thomas, 24
Wigwam, 179
Wilcox, C.H., 80-81
Wiley, Alexander, 84-85, 266
Wilkinson, Clark F., 91, 98, 291-292
Williams, David, 67-68, 97; Oliver D., 271, 289; Roland, "Bud," 75, 77, 167
Wilmington, 78, 82, 84, 97, 103, 114, 145
Wilson, Charles, 319; Louis, 154, 156
Wilson, Woodrow, 67, 74
Wimmer, Virgil, 289
Winch, Thomas, 320
Winchester Repeating Arms Company, 298
Wirtz, Willard, 310
Wisconsin Agricultural Conservation Com., 87
Wisconsin Central Railroad, 89
Wisconsin Conservation Dept., 265, 285, 308
Wisconsin Council of Agriculture, 87
Wisconsin Dells, 20, 23, 54, 66, 88-89, 100, 131, 143, 150-151, 177, 179, 186, 193, 304, 316-317, 328
Wisc. Dept. of Natural Resources, 328, 333, 355
Wisconsin Dept. of Public Welfare, 266, 268, 272
Wisconsin Dept. of Resource Development, 306
Wisconsin Desert, 22
Wisconsin Farm Bureau, 87
Wisconsin Farmer, The, 31
Wisconsin Power and Light Company, 64, 122, 253, 260, 338, 342-343
Wisconsin Psychiatric Institute, 272
Wisconsin Rapids, 89, 254
Wisconsin River, 15, 20-21, 27, 30, 43, 46, 51, 62, 66, 81, 89, 91, 125, 174, 333, 343, 355, 357
Wisconsin State Fair, 44
Wisconsin State Journal, 186
Wisconsin State Police, 127
Wisconsin Territory, 30
Wisconsin Valley, 29
Withrow, Gardner, 303
Witwen, 21
Woerth, George F., 175
Wolf, Frank, 342
Wolfes-Jenson, 121
Women Ordnance Workers, "WOWS," 139
Women's Clubs, 252
Wonewoc, 151, 246
Wood County, 17, 20, 29, 100
Wood, George, 88; J.W., 44
Woodland Indians, 26, 30
World War I, 60, 65, 67, 78, 116, 152, 158, 193, 296, 298, 310
World War II, 24-25, 57, 64, 66-67, 75-79, 90, 95, 104, 112-120, 136-140, 148-151, 157-159, 162-164, 168, 172-175, 178-180, 184, 243, 262, 265, 268, 271, 277, 280-290, 294-300, 302, 305-306, 310, 315, 318-319, 325, 330, 341, 345-348, 353-359
Wyocena, 151-152
Wyttenbach, Fred, 103
Y
Yanke, Herbert, 246
Yankee Street, 175
Yellow River, 29
Yellow Thunder, 29, 94
Young, Edwin, 278, 337; John, 280; Peter, 53; Phyllis, 278, 280
Z
Zantow, Forrest "Woody," 319, 324-330
Zech Rockshelter, 23
Zech, Milton, 341
Zechel, Walter, 289
Zierdt, John, 271

About the Author

Michael J. Goc is the author or editor of over sixty books on Wisconsin history, six of which have received awards of merit from the State Historical Society of Wisconsin. The author of numerous magazine articles and radio commentaries on historical subjects, he served as principal scriptwriter of the first comprehensive history of Wisconsin produced on compact disc. He is also the founder and president of the New Past Press, Inc. Dedicated to producing fine, limited-edition history books, New Past Press is the largest publishing house in Friendship, Wisconsin.